Semantic Technologies for Intelligent Industry 4.0 Applications

RIVER PUBLISHERS SERIES IN COMPUTING AND INFORMATION SCIENCE AND TECHNOLOGY

The "River Publishers Series in Computing and Information Science and Technology" covers research which ushers the 21st Century into an Internet and multimedia era. Networking suggests transportation of such multimedia contents among nodes in communication and/or computer networks, to facilitate the ultimate Internet.

Theory, technologies, protocols and standards, applications/services, practice and implementation of wired/wireless networking are all within the scope of this series. Based on network and communication science, we further extend the scope for 21st Century life through the knowledge in machine learning, embedded systems, cognitive science, pattern recognition, quantum/biological/molecular computation and information processing, user behaviors and interface, and applications across healthcare and society.

Books published in the series include research monographs, edited volumes, handbooks and textbooks. The books provide professionals, researchers, educators, and advanced students in the field with an invaluable insight into the latest research and developments.

Topics included in the series are as follows:-

- Artificial intelligence
- Cognitive Science and Brian Science
- Communication/Computer Networking Technologies and Applications
- Computation and Information Processing
- Computer Architectures
- Computer networks
- Computer Science
- Embedded Systems
- Evolutionary computation
- Information Modelling
- Information Theory
- Machine Intelligence
- Neural computing and machine learning
- Parallel and Distributed Systems
- Programming Languages
- Reconfigurable Computing
- Research Informatics
- Soft computing techniques
- Software Development
- Software Engineering
- Software Maintenance

For a list of other books in this series, visit www.riverpublishers.com

Semantic Technologies for Intelligent Industry 4.0 Applications

Editors

Dr. Archana Patel

School of Law, Forensic Justice, and Policy Studies,
National Forensic Sciences University, Gandhinagar, Gujarat, India

Prof. Dr. Narayan C. Debnath

School of Computing and Information Technology,
Eastern International University, Vietnam

NEW YORK AND LONDON

Published 2023 by River Publishers
River Publishers
Alsbjergvej 10, 9260 Gistrup, Denmark
www.riverpublishers.com

Distributed exclusively by Routledge
605 Third Avenue, New York, NY 10017, USA
4 Park Square, Milton Park, Abingdon, Oxon OX14 4RN

Semantic Technologies for Intelligent Industry 4.0 Applications / by Dr. Archana Patel, Prof. Dr. Narayan C. Debnath.

Routledge is an imprint of the Taylor & Francis Group, an informa business

ISBN 978-87-7022-782-7 (print)
ISBN 978-87-7022-996-8 (paperback)
ISBN 978-10-0096-410-3 (online)
ISBN 978-1-003-44113-7 (ebook master)

Contents

Preface

Semantic technologies work with the concepts and relations that are very close to the working of the human brain. By using these technologies, data becomes a real-world entity rather than a string of characters. Semantic technologies are highly valuable tools to simplify the existing problems of the industry, leading to new opportunities. However, there are some challenges that need to be addressed to make industrial applications and machines smarter. The proposed book aims to provide a roadmap for semantic technologies and highlights the role of these technologies in industry. The book covers semantic search engine, semantic web services, semantic web of things, AI compatible key hardware design, knowledge-graph-based integration, ontology development methodologies, knowledge extraction, prediction of semantic relations of biocontrol agents, integration of biomedical knowledge, semantic checking of information support, a tool for automatic anomaly identification in OWL ontologies, challenges and prospects of industry 4.0, and ontological modeling. This book will potentially serve as an important guide toward the latest industrial applications of semantic technologies for the upcoming generation, and it, thus, becomes a unique resource for scholars, researchers, professionals, and practitioners in the field.

Editors:

Dr. Archana Patel
School of Law, Forensic Justice, and Policy Studies, National Forensic Sciences University, Gandhinagar, Gujarat, India

Prof. Dr. Narayan C. Debnath
School of Computing and Information Technology,
Eastern International University, Vietnam

List of Contributors

Abdullahi, Muhammad Bashir, *Department of Computer Engineering, Padre Conceicao College of Engineering, India*

Agarwal, Trisha, *Department of Computer Science, Pondicherry University, India*

Akbar, Zaenal, *Research Center for Computing, Research Organization for Electronics and Informatics, National Research and Innovation Agency, Indonesia*

Aliyu, Hamzat Olanrewaju, *Department of Information & Media Technology, Federal University of Technology, Nigeria*

Aminu, Enesi Femi, *Department of Computer Science, Federal University of Technology, Nigeria*

Avdeenko, Tatiana, *Novosibirsk State Technical University, Russia*

Bala, P. Shanthi, *Department of Computer Science, Pondicherry University, India*

Cimino, James J., *Informatics Institute, School of Medicine, the University of Alabama at Birmingham, USA*

Debnath, Narayan C., *School of Computing and Information Technology, Eastern International University, Vietnam*

Ebietomere, Esingbemi Princewill, *Department of Computer Science, University of Benin, Nigeria*

Ekuobase, Godspower Osaretin, *Department of Computer Science, University of Benin, Nigeria*

Gorrepati, Rajani Reddy, *Department of Computer Science and Engineering, Koneru Lakshmaiah Education Foundation, Vaddeswaram, Guntur, India*

Guntur, Sitaramanjaneya Reddy, *Department of Electronics and Communication Engineering, Vignan's Foundation for Science, Technology, and Research, India*

Indrawati, Ariani, Research Center for Computing, Research Organization for Electronics and Informatics, National Research and Innovation Agency, Indonesia

Islam, Noman, *PAF KIET, Karachi*

Jeyakodi, G., *Department of Computer Science, Pondicherry University, India*

Jing, Xia, *Department of Public Health Sciences, College of Behavioural, Social and Health Sciences, Clemson University, USA*

Jonnala, Prathiba, *Department of Electronics and Communication Engineering, Vignan's Foundation for Science, Technology, and Research, India*

Kartika, Yulia Aris, *Research Center for Computing, Research Organization for Electronics and Informatics, National Research and Innovation Agency, Indonesia*

Kieu, Manh-Kha, *Becamex Business School, Eastern International University, Vietnam*; *School of Business and Management, RMIT University, Vietnam*

Kim, Do-Hyeun, *Department of Computer Science and Engineering, Jeju National University, Republic of Korea*

Le, Ngoc-Bich, *School of Biomedical Engineering, International University, Vietnam; Vietnam National University Ho Chi Minh City, Vietnam*

Le, Ngoc-Huan, *Mechanical and Mechatronics Department, Eastern International University, Vietnam*

Li, Shijuan, *Department of Information Management, Peking University, China*

Moreira, Dilvan A., *University of São Paulo, ICMC, Brazil*

Murtazina, Marina, *Novosibirsk State Technical University, Russia*

Nguyen, Duc-Canh, *Mechanical and Mechatronics Department, Eastern International University, Vietnam*

Nguyen, Vu-Anh-Tram, *Becamex Business School, Eastern International University, Vietnam*

Nguyen, Xuan-Hung, *Mechanical and Mechatronics Department, Eastern International University, Vietnam*

Ninh, Tran-Thuy-Duong, *Becamex Business School, Eastern International University, Vietnam*

Orlando, João Paulo, *Federal Institute of São Paulo, IFSP, Brazil*

Oyefolahan, Ishaq Oyebisi, *Department of Information & Media Technology, Federal University of Technology, Nigeria*

Pan, Xuequn, *Changsha Medical University, China*

Pustovalova, Natalia, *Novosibirsk State Technical University, Russia*

Saleh, Dadan Ridwan, *Research Center for Computing, Research Organization for Electronics and Informatics, National Research and Innovation Agency, Indonesia*

Shaikh, Zubair A., *Muhammad Ali Jinnah University, Karachi*

Shynkarenko, V., *Ukrainian State University of Science and Technologies, Ukraine*

Syed, Darakhshan, *Iqra University, Karachi*

Waskita, Arya Adhyaksa, *Research Center for Computing, Research Organization for Electronics and Informatics, National Research and Innovation Agency, Indonesia*

Zhuchyi, L., *Ukrainian State University of Science and Technologies, Ukraine*

List of Figures

List of Tables

List of Abbreviations

AeDNV	Aedes densonucleosis virus
AGV	Automated guide vehicle
AI	Artificial intelligence
AIC	AGV inlet conveyor
AM	Additive manufacturing
AMQP	Advanced message queuing protocol
AOC	AGV outlet conveyor
API	Application programming interface
AR	Augmented reality
AS	Automated storage
BAS	Behavioral activation system
BCGO	Breast Cancer Grading Ontology
BFO	Basic formal ontology
BIS	Behavioral inhibition system
BPL	Broadband over power lines
CDS	Clinical decision support
CDSS	Clinical decision support system
CLIPMERGE PGx	Clinical implementation of personalized medicine through electronic health records and genomics – pharmacogenomics
CMRTR	Current maintenance of railway track rules
CMTS	Connectivity management semantics ontology
CNTRO	Clinical narrative temporal relation ontology
CoAP	Constrained application protocol
COBASEN	Context-based search engine
CPOE	Computerized physician order entry
CQ	Competency question
CSIRO	Common wealth Scientific & Industrial Research Organization
DAML	DARPA agent markup language
DB	Database
DCM	Data checking model

DCS	Distributed control system
DDCD	Data-driven change discovery
DDT	Dichlorodiphenyltrichloroethane
DENV	Dengue virus
DHFSRW	Driver handout form of speed restriction warnings
DIM	Data integration model
DOM	Document objective model
DSL	Digital subscriber line
DT	Digital twin
DWM	Data wrangling rules
EER	Electronic educational resource
EHR	Electronic health record
EKR	Electronic knowledge record
EMR	Electronic medical record
EpSO	Epilepsy and seizure ontology
FAIR	Findable, accessible, interoperable, reusable
FBO	Functional basis ontology
FLOWS	First-order logic ontology for web services
FTP	File transfer protocol
GAICW	Guidelines for automated issuance and cancellation of warnings
GATE	General Architecture for Text Engineering
GIS	Geographic information system
GO	Gene ontology
GPA	General academic performance
GWPQ	Gray–Wilson personality questionnaire
H3-IoT	Home health hub IoT
HER	Electronic health record
HL7	Health level 7
HMI	Human–machine interface
TML	HyperText Markup Language
HTTP	Hypertext transfer protocol
ICE	Information connection engine
ICS	Industrial control system
IDDAP	Infectious disease diagnosis and antibiotic prescription
IDF	Inverse document frequency
IDO	Infectious disease ontology

IDOMAL	Infectious Disease Ontology for Malaria disease
IDP	
IDR	Infectious disease risk
IEL	Infrastructure element lists
IIoT	Industrial Internet of Things
IoMT	Internet of medical things
IoS	Internet of service
IoST	Internet of semantic things
IoT	Internet of Things
IP	Internet protocol
IRAC	Insecticide Resistance Action Committee
IRB	Inspection record books
IRB	Institutional review board
IRI	Internationalized resource identifier
IRM	Insecticide resistance management
ISA	Intelligent speed adaptation
ISO	International Standardization Organization
IST	Intellect structure test
IT	Information technology
ITF	Inverse term frequency
JSON	JavaScript object notation
KA	Knowledge agent
KB	Knowledge base
LHS	Learning health system
LM	Language Model
LRS	Learning record store
LSM	Linked sensor middleware
M2M	Machine-to-machine
MES	Manufacturing execution system
MeSH	Medical subject headings
MetaFOR	Metadata description for ontologies/rules
MIRC	Minimal inconsistent resolve candidate
MIRO	Mosquito insecticide resistance ontology
ML	Machine learning
MoA	Mode of action
MPMO	Manufacturing predictive maintenance ontology
MQTT	Message queue telemetry transport
MR	Matching rules
MVC	Model-view-controller

NCBO	National Center for Biomedical Ontology
NER	Named entity recognition
NIAID	National Institute of Allergy and Infectious Diseases
NLP	Natural language preprocessing
OIL	Ontology interchange language
OntoKBCF	Ontology-based knowledge base on cystic fibrosis
OT	Operational technology
OWL	Web ontology language
OWL-DL	Web ontology language-description logics
OWL-S	Ontology web language for services
PLC	Programmable Logic Controllers
PLE	Personal learning environment
POM	Probabilistic ontology model
POS	Part of speech
PRISMA-ScR	Preferred reporting items for systematic reviews and meta-analysis extension for scoping reviews
R2ML	Rewerse rule markup language
RAD	Rapid Application Development
RBRSI	Records book of railway switches inspections
RBRTI	Records book of railway track inspections
RBSRW	Records book of speed restriction warnings
RDAO	Research digital artifacts ontology
RDF	Resource description framework
RDFS	Resource description framework schema
REST	Representational state transfer
RFID	Radio frequency identification
RIDL	Release of insects carrying a dominant lethal
RIF	Rule interchange format
RIV	Railway infrastructure vocabulary
RNN	Recurrent neural network
ROWS	Rule-based ontology for web services
RPC	Remote procedure call
RRR	Railway regulation rules
RS	Resting state
RS	Retrieval systems
RSL	Railway switch list
RTL	Railway track list

RV	Resources vocabulary
SAREF	Smart applications reference ontology
SCADA	Supervisory control and data acquisition
SCO	Sensor core ontology
SHOE	Simple HTML ontology extensions
SIC	Semantic interoperability conflict
SIT	Sterile insect technique
SKOS	Simple knowledge organization system
SLE	Systemic lupus erythematosus
SNOMED	Systematized nomenclature of medicine
SOA	Service-oriented architecture
SOAP	Simple object access protocol
SOS	Smart onto sensor
SPARQL	SPARQL protocol and RDF query language
SPELTA	Speech and language therapy assistant
SRWBF	Speed restrictions warnings books and forms
SSE	Semantic search engine
SSL	Secured socket layer
SSN	Semantic sensor network
SVM	Structure validation rules
SVR	Structure validation rules
SWM	Structure wrangling rules
SWoT	Semantic Web of Things
SWRL	Semantic web rule language
SWS	Semantic web services
SWSE	Semantic web search engine
SWSF	Semantic web services framework
TAP	The Semantic Web Application Framework
TCP	Transmission control protocol
TERR	Technical operation rules of railways
TF	Term frequency
TIMER	Temporal information modeling, extraction, and reasoning
TLS	Transport layer security
TMR	Train management rules
TOVE	Toronto Virtual Enterprise
TS	Table structure
TSCR	Table software classification rules
TSTCR	Table station classification rules

TTCR	Table type classification rules
TWSAR	Track works safety arrangements rules
UDDI	Universal description discovery and integration
UDP	User datagram protocol
UI	User interface
UID	Unique identifier
UML	Unified modeling language
URI	Uniform resource identifier
UUID	Universally Unique Identifier
UX	User experience
VBD	Vector-borne disease
VSMO	Vector surveillance and management ontology
W3C	World Wide Web Consortium
WaWO	Waste water treatment christened
WMS	Warehouse management software
WoT	Web of things
WSDL	Web services description language
WSMO	Web services modeling ontology
WSMX	Web service execution environment
WSSN	Wireless semantic sensor network
XML	Extensible markup language
XTM	XML Topic Map
ZIKV	Zika virus

1

Semantic Search Engine in Industry 4.0

Esingbemi Princewill Ebietomere and Godspower Osaretin Ekuobase

Department of Computer Science, University of Benin, Nigeria
princewill.ebietomere@uniben.edu; godspower.ekuobase@uniben.edu

Abstract

The core of Industry 4.0 is knowledge operationally encapsulated as data, which must be of peak veracity for efficient, effective, and sustainable control and coordination of its cyber–physical components. However, the volume, velocity, and variety of these data demand a search and retrieval mechanism that guarantees peak veracity in time, context, and situation of the data consumed from an available pool of data. Specifically, the machine sovereignty of Industry 4.0 demands that this mechanism be semantically deepened. This chapter explicates this mechanism called semantic search engine (SSE) and its roles and challenges in Industry 4.0. The chapter begins with the exploration of information retrieval, its processes, models, and evaluations, and then the concept of search engine, generally, its evolution, components, development techniques, and categorization along selected dimensions. Thereafter, focus is shifted to the concepts and operations of Industry 4.0 and how SSE contributes to its seamless operations. At the end of this chapter, readers will be exposed to the concepts, technologies, features, operations, and challenges of SSE in Industry 4.0 and also get acquainted with the state-of-the-art standards, methods, and tools for the design, development, and deployment of SSE in Industry 4.0.

Keywords: Industry 4.0, knowledge representation, search engine, semantic search, semantic retrieval.

1.1 Introduction

The advent of Industry 4.0 – an advanced manufacturing model that, at provenance, is used to refer to modifications linked to automation fields fused with information technology, with its emergence driven by three fundamental players: knowledge, experimentation, and innovation [1] – has revolutionized the way goods and services are handled from the point of design, through production, supply, and distribution to consumption. This digitalization, on the one hand, has led to increased investment in solutions that help integrate all participating entities in a network (processes, machines, people, and products) [2] and, on the other hand, increased the return on investment by lowering the overall cost burden on organizations and also increased output. At the heart of Industry 4.0 are cyber–physical systems [3] that generate knowledge operationally encapsulated as data that must be necessarily gathered, analyzed, and evaluated for efficient and effective decision making. Interestingly, these data are usually big data – characterized by vast volume, high velocity, and variety [4]. However, since discovery, access, and retrieval of such data are central for efficient, effective, and sustainable monitoring, control and coordination of these cyber–physical systems, a search and retrieval mechanism that can guarantee peak veracity of data consumed from its pool (repositories), is of utmost importance.

Information retrieval is a concept that is used in describing any mechanism that involves the activities of searching, accessing, and fetching of data/information from a repository for consumption by man and machines alike. A typical information retrieval system is formally defined by the quadruple $[D, Q, F, \text{sim}]$, where D is a set of documents, Q is a set of queries, and F is a framework for modeling documents, queries, and their relationship, and sim: $Q\ X\ D \rightarrow U$ is a ranking function that defines an association between queries and documents with U denoting an ordered set of query results [5–7]. This concept of information retrieval predates the digital age, but the advent of the internet and the web being its major service with massive amount of data/information to be discovered and mined both by man and machine for consumption made information retrieval more popular and desirable, with some of its implementations being recommender engines, question answering engines, and search engines.

The search engine is the most popular technology for finding and harvesting the vast data/information on the web; thus, emphasis will be on this type of information retrieval system. The importance of the search engine has contributed to its proliferation and diffusion across several fields and industries that support digitalization, which has seen the search engine industry growing into a multi-billion dollar industry; and just like every other technology, it came with challenges, which has seen it being metamorphosed over the years across several dimensions including its underlining algorithms, models, and strategies.

The goal of this chapter is to explicate the search mechanism with special focus on the semantic search engine and its place in Industry 4.0. This chapter begins with the exploration of information retrieval, the concept of search engine, its evolution, components, development techniques, and categorization along selected dimensions. Subsequently, attention shifts to the concepts and operations of Industry 4.0 and how semantic search engine contributes to its seamless operations. The chapter is finalized with a brief description of some of the challenges militating against the full realization of the benefits of semantic search engines in Industry 4.0 and conclusion drawn.

1.2 Information Retrieval

The multiplicity of concepts and terminologies in information retrieval has led to pedagogical fuzziness in the field. This confusion stems from several perspectives of the view of the concepts and terminologies used by authors and creators/curators hugely influenced by the field of applications. Our focus is not to argue on whose view of these concepts is right or wrong but to clarify these concepts and terminologies via definitions/descriptions so that they can be properly positioned for pedagogical reasons. Popular and important among these are search and retrieval system, retrieval systems, search systems, and search engines that, overtime, have been used to mean the same by many authors.

A search and retrieval system is a set of interacting components encumbered with the capability of finding, accessing, and fetching basically data/knowledge from web (repository). The data/knowledge could be from a single central store or disparate sources with disparate contents – e.g., text, images, videos, and audios. Search and retrieval systems typically consist of different entities including people, machines, software, procedures, and policies. However, search and retrieval systems can simply be referred to as either retrieval systems or search systems. Nonetheless, a search engine is simply an implementation of retrieval techniques [8]. A typical retrieval system consists of several processes as depicted in Figure 1.1.

From Figure 1.1, three fundamental processes are evident. These processes are query processing, indexing, and matching (searching and ranking) and are described in the following.

i. **Query processing**: This entails analyzing and transmuting the user need to an internal form. This could involve several activities such as part of speech (POS) tagging, lemmatization/stemming, and filtering.

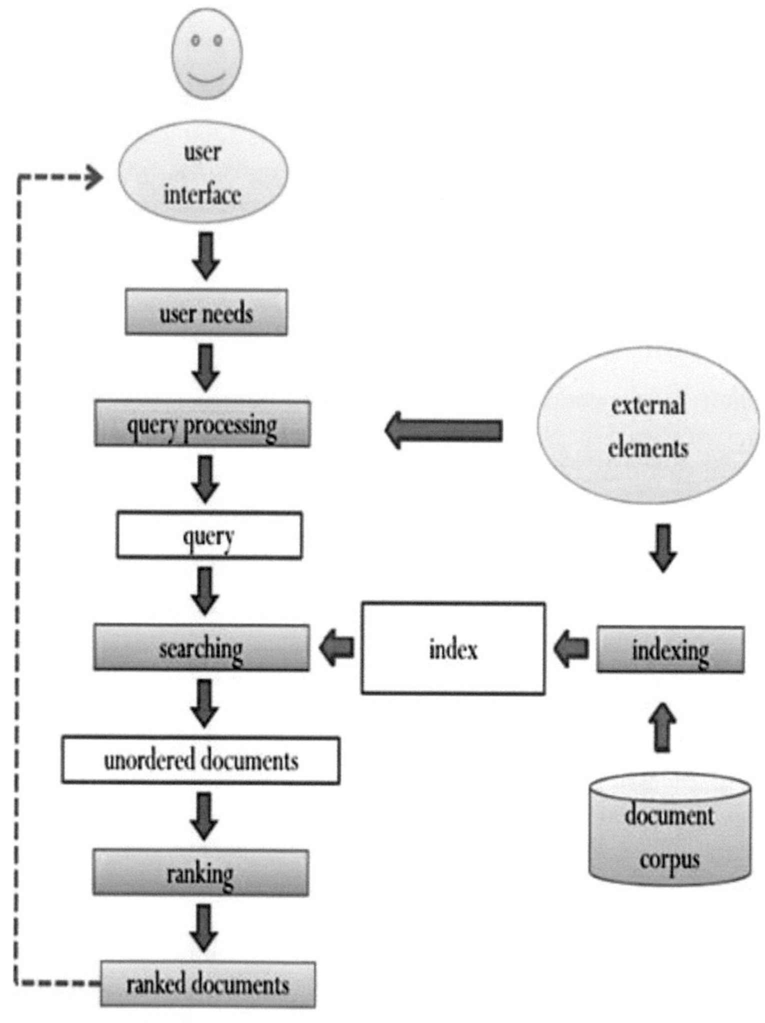

Figure 1.1 Information retrieval process [6].

The external elements play a critical role as they are used to represent, extract, and process user needs and content meanings as well as indexing. The understanding of the meaning behind information items and users' queries enhances the precision of the retrieval process, thus increasing users' satisfaction.

ii. **Indexing**: Indices are data structures constructed to speed up access and retrieval. Perceptibly, not all pieces of information items are equally significant as a representative of the information; thus, only the very significant information elements constitute an index for the information. Consequently, it is imperative to pre-process the information items to ensure proper selection of the elements to be used as index objects. Also, the construction and maintenance of index is particularly worthwhile for large and semi-static information. Besides, the inverted file is the most common indexing structure and consists of two elements: (i) a dictionary – which contains all the words in the document collection and (ii) a posting list – which, for each word in the dictionary, gives a list of all the positions where the word appears in the document collection [6, 7].

iii. **Matching**: This sub-process consists of searching and ranking activities.

   a. **Searching**: This operation matches user queries against information items, and a set of information items that meets the user needs is returned. The way this is achieved may vary depending on the format of information (text, audio, and video); usually, some form of simplification is done in the information model for tractability purposes. For instance, text retrieval commonly builds on the assumption that the matching between user query string and the documents can be based on a set of index terms.

   b. **Ranking**: The set of information items returned by the matching process generally constitutes an inexact and, by nature, approximate answer to the information need. Ranking puts the information items in the decreasing order of estimated relevance to query so that the more significant items appear topmost.

There are four fundamental development models for information retrieval system – Boolean, vector-space, probabilistic, and language model – with the first three often referred to as the classical models. It is not uncommon to see other models like fuzzy, neural network, extended Boolean, and extended vector [5, 6, 9, 10], which are offshoots of the classical models. A brief description of these fundamental models, techniques, and flaws are given subsequently.

i. **Boolean model**: This is the foremost and the most dominant model in the 1960s through 1990s. In this model, documents (*D*) are represented using bag of words – set of words or terms, and queries (*Q*) are represented as Boolean expressions of terms using three basic operators, AND, OR, and NOT, denoting intersection, union, and exclusion, respectively. The retrieval criterion of the Boolean model is usually binary (0 or 1, i.e., irrelevant or relevant). Effort to further classify the Boolean retrieval criteria based on the degree of relevance between relevant and irrelevant gave birth to the fuzzy model. Generally, the Boolean model is: (i) not scalable – as precision weakens with increasing search space; (ii) not user friendly – most users can hardly express their information needs in Boolean form; and (iii) often does not have facility to support ranking of search results.

ii. **Vector-space model:** This model represents documents, queries, and their terms as vectors in a common vector-space. A term's vector component usually consists of the term and some set of weights. Term weights are used to compute the degree of similarity used in ranking search results. These weights are assigned to define the importance or relevance of the term in a document/query vector, and the measures to weight terms simply based on statistics of term occurrences. Some of the measures include: term frequency (TF), inverse term frequency (ITF), inverse document frequency (IDF), and a hybrid such as term frequency and inverse document frequency (TF-IDF). The vector-space model uses partial matching and cosine retrieval function to retrieve and rank documents. The snag with the vector-space model is in the assumption that index terms (searchable terms) are mutually independent when, obviously, the model does not support term dependencies – a problem that is also evident in both Boolean and probabilistic models [6]. It is important to mention that the vector model according to [11] remains one of the most implemented models in commercial search engines.

iii. **Probabilistic model**: The fundamental idea behind the probabilistic model is that given a set of document *D* and query *Q*, there exists a subset *R* of the documents (*D*) that are relevant to *Q*; the model ranks *R* in the decreasing order of probability of relevance to *Q*. There are several variations of the probabilistic model, and popular among these is the Okapi BM25 – an approximation of two Poisson models [12]. Besides the fact that, in this model, index terms are assumed to be mutually independent as mentioned earlier, the model is also restricted by its representation of documents, queries, and relevance with Boolean value. Besides, the model does not consider term frequency factors.

iv. **Language model**: This model is often described as a formal approach to information retrieval and treats the generation of queries as a random process. It functions by creating a language model for each document, estimates the probability of generating the query based on each of the document's language model, and ranks the documents according to these probabilities. The language model is implemented as unigram or *n*-gram (bigram, trigram, ..., *n*-gram). The unigram model is the simplest and the most used one in information retrieval [9, 13]. The drawback of LM in information retrieval is the sparse data problem [9, 13]; some possible words that meet the information need of a user as specified in the query will not have appeared in the document at all. Though smoothening is a panacea to this problem, issues as to how to smoothen effectively still persists [9, 13].

It is worthwhile to mention that the proliferation and the high competitiveness among the creators of retrieval systems to oust one another and the quest by the users for such systems to provide the best experience (satisfaction) necessitated the emphasis on the evaluation of such systems. Consequently, there exist several retrieval evaluation methods. According to [14], the evaluation method can be broadly categorized as follows: (i) the functional evaluation – which entails testing specified systems' component for functionalities and operations' accuracy; (ii) the performance evaluation – which is a measure of the system's adeptness mainly in terms of execution time and space; (iii) the retrieval performance evaluation – which assesses how well the information retrieval system satisfies the information need of its users, which according to [6] can be either a user-based retrieval performance evaluation or system-based retrieval performance evaluation.

Though the user-based retrieval performance evaluation is, in principle, more informative, it is very expensive and difficult to explore; thus, the system-based retrieval performance approach has found more application and is as such more popular. Some of the prevalent metrics associated with the system-based approach include: precision – the ratio of documents retrieved by the system that are relevant to the query and to the total number of documents retrieved; recall – the ratio of relevant documents retrieved by the system to the total number of relevant documents in the corpus; and *f*-measure – simply the harmonic mean of precision and recall [15, 16]. Following the definitions of the metrics, it is obvious that measuring recall is arduous as it requires the knowledge of all the relevant documents to a query in the corpus, which are usually manually determined. Besides, it is pertinent to note that the values of these metrics are between 0 and 1, which can also be expressed in percentage.

1.3 Search Engine

Search engine, as mentioned earlier, is an implementation of an information retrieval technique; thus, virtually, all processes exhibited by an information retrieval system are evident in a search engine. The first search engine, Archie, was born around 1990 as a result of explosion of information on the internet – turning a global repository – to replace the traditional file transfer protocol (FTP) approach to sourcing information. This event was then followed by the advent of the first webcrawler, wandex, at about 1994 to aggregate information from various sources on the web for better experience by users. Currently, there are a myriad of search engines with these engines differing in terms of purpose and underlining approach or technique.

Purpose here refers to the type of data/information the engine can manipulate (search and retrieve) effectively; hence, the engine can be a dedicated search engine (i.e., for text only and video only) or a generic search engine (i.e., for text and video), while the approach or technique is used to imply the stratagem employed by the engine for search and retrieval, which largely culminates in either a syntactic or semantic scheme. Thus, irrespective of purpose, search engines can be branded as either syntactic or semantic. It is apposite to mention that the earliest search engines were basically syntactic in nature and that the evolution of the web from web 1.0 through web 2.0 to web 3.0 compelled the mandatory evolution of search engines from its traditional form (syntactic) to an emergent form (semantic) to help shrink the lacuna that exist between queries and the information space (repository) particularly with the current problem of information overload, which definitely will get worse. A description of search engines along these two broad categories is detailed in the subsequent subsection.

1.3.1 Traditional Search engine

A search mechanism whose strategy supports retrieving data/knowledge based on only keywords from queries irrespective of its underlining model – e.g., Boolean, vector, and probabilistic – is referred to as a syntactic or traditional search engine. A myriad of these engines exist in literature both from the industry and academia [17, 18], but very few have been successful [17]. The syntactic search engines are usually bedeviled by imprecise results hugely due to their inability to handle the problem of synonymy (several words with same meaning) and polysemy (a word having multiple meanings), with several other challenges including difficulty in gathering documents, heterogeneity of information format (txt, pdf, rtf, html, etc.) and type (textual or multimedia) [17]. Examples of some successful popular free and open source search engines both from industry and academia include: Google, Yahoo, Microsoft Bing, Apache Lucene, Terrier, and Indri.

Over the years, the Google search engine has become the standard for traditional search engines distinctively due to its level of acceptance and use; hence, its architecture, though not free from challenges, has been widely accepted as a de facto architecture for describing the traditional search engines, and it is as shown in Figure 1.2 and is described in the following.

From Figure 1.2, the URL server sends a list of URLs to be fetched to the crawler, which traverses the web to gather web pages corresponding to these URLs and sends the same to the store server, which compresses it and stores it in a repository (a process referred to as crawling). These stored web pages are then indexed by an indexer using word occurrences generated from the web pages often referred to as hits, which are then distributed into barrels (a process known as indexing). Also, important information about all the links from these web pages is stored in the anchor file. The URLResolver generates a database of links used for the computation of page rank for all documents. When a user launches a request at a terminal, the keywords from the user's need is transformed into query (query processing) which the searcher then uses the lexicon, stored indexes to answer (a process known as searching), and matches obtained are then ranked (a process known as ranking).

Furthermore, there are several tools/frameworks that one can leverage on in creating search engines instead of building from scratch – a process that is usually very onerous. Some of the tools are often seen as search engines, but, technically, they are more of a framework and as such can be leveraged for the creation of a complete and sophisticated search engine. A notable example of such framework is Apache Lucene. Additionally, the quest by businesses with online presence to be continually visible to intended customers led to their demand for search engines that will always rank them high in search results. This demand has made it incumbent for search engines to always consider optimization – search engine optimization.

There are several techniques for making a search engine perform optimally; these techniques, collectively referred to as optimization techniques,

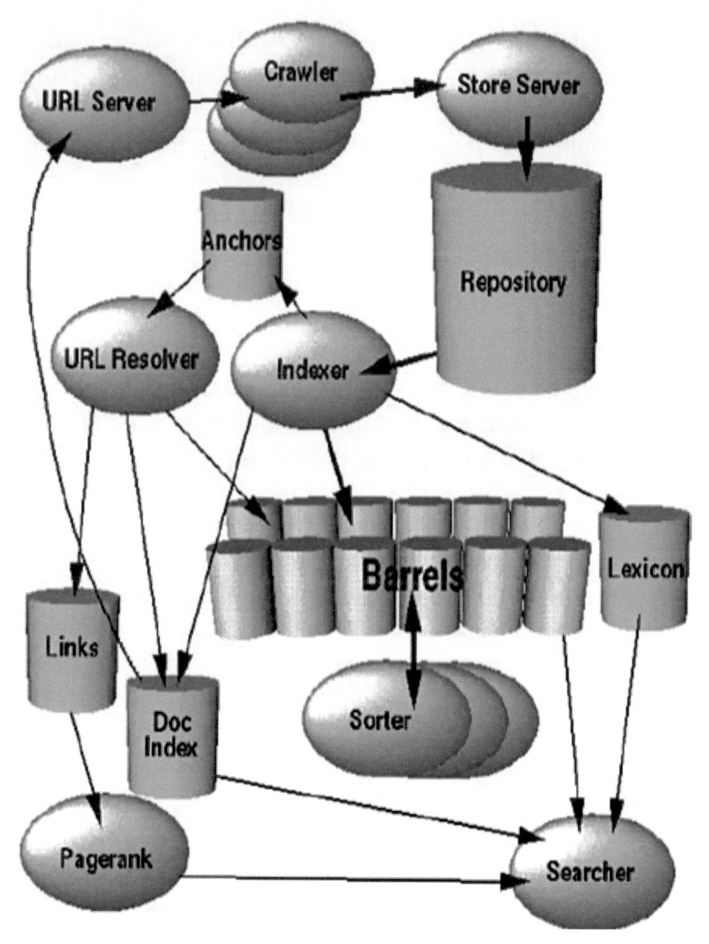

Figure 1.2 Google search engine architecture [19].

include: white hat, black hat, and gray hat. The white hat, which is the ethical and the most acceptable way of optimizing search engines, entails following the standard procedures laid down for optimization. The black hat is a way of optimizing search engines by disregarding the standard procedures for optimization. The gray hat, on the other hand, sits between the white hat and the black hat, i.e., a mix of the white hat and the black hat techniques.

Generally, the expedition to addressing the problems and challenges of the traditional search engines including their inability to penetrate the dark web, which, according to experts, holds much more information than the visible web gave birth to the semantic search engine.

1.3.2 Semantic search engine

The semantic web also referred to as web 3.0, which allows for dynamism, collaboration, and machine participation, became inescapable owing to the need to efficiently and effectively exploit the massive information (largely unstructured) and services on the web [20]. Following this, the concept has metamorphosed from being a vision [21] to a research field [22] and has found application in many expanses including areas that require repetitive tasks being carried out by machines on behalf of humans. Furthermore, the emergence of the semantic web also requires compliance of mechanisms used in its exploration; for example, a search mechanism that must thrive in semantic web space must necessarily be semantically enabled. Also, technologies that support the semantic web are referred to as the semantic web technologies. Consequently, a search engine that deploys the semantic web technologies – standards, methods, and tools at any level of its critical processes, i.e., at querying, indexing, and matching – is referred to as a semantic search engine. A typical semantic retrieval framework is as depicted in Figure 1.3.

Noticeably, Figures 1.1 and 1.3 are alike in every respect except at query and index stages where semantic web technologies were incorporated to make the system more intelligent, thereby improving retrieval precision and recall. There are several approaches and categorizations of semantic search as evident in [23]–[26], and these are discussed in the following subsections.

1.3.3 Approaches and categorization of semantic search

The deployment of semantic web methods in search engines defines its approach. These methods include: contextual analysis, reasoning engine, natural language understanding, ontology, knowledge graph, and linked data [20]. The authors [23] identified contextual analysis, reasoning engine, natural language understanding, and ontology as the basic approaches to semantic search engine.

Contextual analysis emphasizes disambiguation of data and knowledge to mitigate the void that exists between queries and the web by considering the context/sense in which the words are being used, thus helping to address the issue of polysemy and synonymy. This can be achieved by providing formal and explicit representations for contexts [20], for example, the use of lexicons like WordNet in search engines.

Reasoning in search engines emphasizes building cognitive abilities into search mechanism by embellishing it with requisite competence to enable

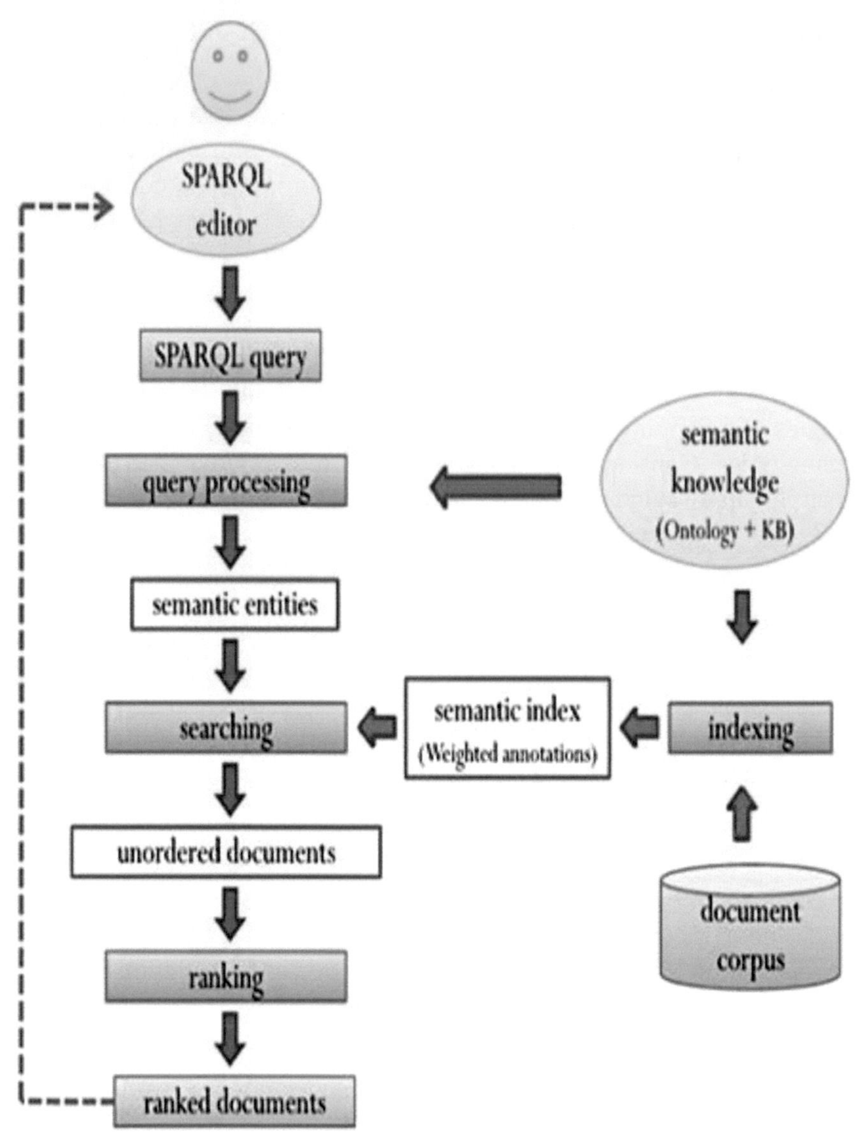

Figure 1.3 A typical semantic retrieval framework [6].

it carry out tasks, particularly repetitive ones on behalf of the users. The codification of intelligence in search engines is usually achieved by deploying logic – the foundation of reasoning; thus, search engines with such capability can easily make inference from a set of known premises.

Natural language understanding, also referred to as semantic analysis, aims to represent the meaning of texts toward enabling machine readability and comprehension. Making search engines understand every aspect of a phrase/sentence being submitted as information need will undoubtedly enhance the retrieval of more relevant results and, thus, the exploitation and popularity of this approach in several aspects of semantic search engines.

Ontology has found a great use in semantic search engines for several reasons, including its ability to separate domain knowledge from operational knowledge, and also make implicit knowledge explicit. Due to the many benefits accruing to the use of ontology as an approach to semantic web, it has been described by many experts as the corner stone of semantic web. It is important to note that each of the approaches highlighted can be used alone or can be combined where necessary and appropriately to achieve better results in search engines.

The work of [24] identified six categories of semantic search engines in research based on three dimensions, namely objectives, methodologies, and functionalities of such engines. The exposed categorizations are document-oriented search, entity and knowledge oriented, multimedia information search, relation-centered search, semantic analytics, and mining-based search. Interestingly, all the identified categories employed either one of ontological, contextual, graph based, and reasoning approaches or a combination of these approaches at either documents (text, image, etc.) or query level.

The author in [25] also exposed several categories of semantic search engines using criteria such as architecture, coupling, transparency, user context, query modification, ontology structure, and ontology technology with focus on the World Wide Web. The work also gave direction on which ideas are most semantically effective by the classification style employed. It is worthwhile to mention that the work observably propped toward search engines that exploited basically the ontological approach either at the documents or query level.

The view of Lei *et al.* [26] to the approaches of semantic search is from the perspective of the support of such engines to users – that is, friendliness of interface; hence, the following categories of such engines were identified: (i) form-based engines – which offer web forms that allow users' description of queries in a format that supports selecting ontologies, classes, properties, and values, and an example of a search engine with such contraption is the simple HTML ontology extensions (SHOE); (ii) RDF-based querying language fronted engines – which offer sophisticated querying languages to support semantic search, and an example of a search engine with such contraption is CS AKtive; (iii) semantic-based keyword engines – employ a search mechanism that enhances the performance of traditional keyword search techniques by capitalizing on the use of available semantic data as evident in TAP search engine; (iv) question answering tools – which exploits available semantic mark-up to answer questions asked in natural language format as evident in AquaLog search engine.

Also worthy of note is the fact that the described approaches are realized using several standards – formalized technologies that manifest in the form of format, vocabularies, or language [20]. These standards include uniform resource identifier (URI)/internationalized resource identifier (IRI), extensible markup language (XML), resource description framework (RDF)/resource description framework schema (RDFS), simple knowledge organization system (SKOS), web ontology language (OWL), SPARQL, and rule interchange format (RIF). For a quick acquaintance with these technologies, see [20]. Besides, there are several tools in existence that support the rapid development of semantic applications using these methods and standards. These tools include knowledge annotation tools, e.g., Annotea, Knowtator, and GATE, and knowledge acquisition and representation tools, e.g., Protégé, OntoEdit, and WebODE [20].

Overall, search engines can be broadly categorized as either syntactic or semantic when viewed from the perspective of the stratagem employed in its workings. Furthermore, it is imperative to mention that, irrespective of the view of categorizations' perspective employed in semantic search engines, the semantic web methods were evident at the query level, document level, or both in the search engines explored in these categorizations. Nonetheless, some of the popular semantic search engines that were examined in the explored categorizations comprise SHOE, CS AKtive, TAP, AquaLog, and the semantic web search engine (SWSE).

Very importantly, it must be noted that not all search engines are with crawlers as often misconstrued by many, particularly those that are new to the domain of information search and retrieval as there are crawler-based and human-powered directories and hybrid search engines; thus, the design of a semantic search engine should necessarily be influenced by several factors, including focus and level of desired "semantization" – which is a determinant of the semantic approach to imbibe and at what process or activity it should be deployed – that is, at query, indexing, or both [25, 27].

1.4 Semantic Search Engine in Industry 4.0

This section dwells on Industry 4.0 and the description of the nature of the data/knowledge generated from complex interactions of integrated participating entities in its ecosystem and the role of semantic search engine in this ecosystem.

1.4.1 Industry 4.0

The emergent fourth industrial revolution that accentuates radical automation of manufacturing processes has been described with several appellations, including smart manufacturing, future manufacturing, made in China 2025, and Industry 4.0 [28] – the often referred one in literature. Though several definitions abound in literature for the concept, with a number of variations arising from the several perspectives of view of the concept, what is undisputable is the fact that it is gripped by automation and digitalization of manufacturing processes.

Industry 4.0 is undoubtedly dependent on and driven by several disruptive computing technologies as observed and exposed by several authors [3, 27, 29]. These computing technologies according to authors [29] can be broadly classified as follows: (i) software architectural paradigm – driven by service-oriented architecture (SOA), and microservices; (ii) computing platform – which include edge, fog, and cloud computing, Internet of Things (IoT), and cyber–physical systems; (iii) communication – consisting of network technologies and protocols; (iv) data management – semantic data integration and digital twins. The espousal of these disruptive technologies in industrial settings can help lower the overall cost burden on participating organizations, improve efficiency, accelerate the production of goods and services, thereby increasing output, and ultimately increase the profit [29].

Of concern in this work is how to explore the data/information/knowledge generated within Industry 4.0 ecosystems that are critical to seamless monitoring, control, and coordination of participating entities for efficient and effective decision-making. Following this, it is apt to mention that the amount of data generated within the confines of Industry 4.0, particularly with the IoT and cyber–physical systems as mentioned earlier, is usually of massive volume (amount of data), high velocity (rate at which data is spawned), and variety (heterogeneity of data spawned), which pose several challenges such as integration, interoperability, and inaccessibility. Thus, for proper utilization of this data in the face of these challenges for diagnosis, monitoring, coordination, prediction, discovery, mining, recommendation, question answering, and analytic purposes, its veracity must be guaranteed, particularly with the quest for machine understandable and interpretable descriptions to obliterate human intervention. Most of the listed activities obviously require search systems to accomplish tasks that make it incumbent for search mechanism that can guarantee peak precision at all times for efficient and effective decision-making by man and machine throughout the production life cycle.

For proper realization, reliable implementation, and deployment of Industry 4.0, experts are in unanimity with the deployment of semantic web and its associated technologies – predominantly as a result of the successes that have been chronicled in the deployment of these technologies in addressing the aforementioned challenges with big data [3, 27–29], hence helping push toward achieving automation across several domains. Consequently, to attain the level of machine sovereignty demanded by Industry 4.0, the search mechanisms that must be deployed should be semantically deepened for seamless operation.

1.4.2 Role of semantic search in industry 4.0

This subsection gives descriptions of some of the core technologies of Industry 4.0, which require or incorporate search mechanisms in carrying out their tasks and the design requirements of such engines to perform optimally in achieving tasks. These technologies include IoT, internet of service (IoS), and big data and analytics.

1.4.2.1 Search engines for Internet of Things (IoT)

The IoT is one of the core technologies underpinning Industry 4.0. The concept undoubtedly predates Industry 4.0, and the successes that have been recorded in its deployment in distributed environment led to its adoption in Industry 4.0, where it is now sometimes referred to as industrial Internet of Things (IIoT) [4, 30]. A notable definition of the concept is that given by Guillemin *et al.*, which defined IoT.

"as a dynamic global network infrastructure with self-configuring capabilities based on standard and interoperable communication protocols where physical and virtual "things" have identities, physical attributes, and virtual personalities and use intelligent interfaces, and are seamlessly integrated into the information network" [31], p. 6.

Following this definition of IoT, the need for semantic descriptions of some critical processes is inevitable for its full realization. Also, we cannot help but allude to the fact that the survivability of IoT is strongly dependent on middleware technologies (for example, web services) – which are dominated and driven by service-oriented architecture (SOA). More so, the discovery of heterogeneous devices on the IoT directory requires a search mechanism [30]. This mechanism obviously relies on the syntax and semantic of information contained in data for retrieval, with greater emphasis in recent time on semantics, chiefly as the quest for machine sovereignty deepens. Thus, for activities like query processing (formulation and reformulation), data integration, and interoperability; deploying approaches like ontologies, linked data, and knowledge graph among others have gained more relevance—particularly with real-time applications [32], as demanded by Industry 4.0.

Some notable works on IoT middleware incorporating search engines include (i) the work of [32], where linked sensor middleware (LSM) is deployed for live sensor data to be published, visualized, annotated, and queried by employing several semantic web technologies, e.g., SPARQL for query purposes, (ii) GNS, which addresses the problem of data integration and distributed query processing to ensure veracity of data – see [30], (iii) Xively, which allows product connectivity and gathers data from connected devices to provide real-time monitoring, control, and data storage, and (iv) the work of [30], which developed a framework named context-based search engine (COBASEN) that allows users to search and select devices on IoT using a context-based semantic approach. Observably, all the engines employed in the enumerated middleware deployed the semantic search approach, except Xively whose underlining stratagem is keyword search.

1.4.2.2 Search engines for internet of services (IoS)

The internet of services (IoS), like the IoT, is one of the pillars of Industry 4.0. The IoS may easily be misconstrued for IoT due to the perception that IoT subsumes IoS and, as such, may lead to the temptation of using the concepts interchangeably. Technically, there is a definite fine line between these two concepts; hence, while the IoT deals with tangible objects, machines, and sensors, the IoS deals with intangibles (abstract functionalities) [33]. IoS is strongly dependent on SOA with web services being its main vehicle. One of the attributes of web services is its support for location transparency, where services have definitions and location data published in a registry – the universal description discovery and integration (UDDI) [33–35]; thus, the discovery of services – which could be static in which case the service implementation details are bounded at design time or dynamic in which case the service implementation details are bounded and unbounded at execution time [34] – for consumption inevitably requires a search mechanism.

There are several search mechanisms in literature for discovery of services, many of which are based on web services description language (WSDL), and with the emergence of semantic web services and the proliferation and diffusion of semantic web technologies, including OWL-S, WSDL-S, web services modeling ontology (WSMO), and semantic web services framework (SWSF), toward machine comprehension and processability of services, it is incumbent that the search mechanism deployed be semantically embellished. For a description of some of these semantic web services technologies, see [36]. Some notable works on semantic search engines for web services include those of [37], which offered a fuzzy matchmaking approach for semantic web services by semantically annotating data; the authors in [38] employed a semantic approach (use of ontology) to integrating data from disparate sources (open geospatial consortium specifications and UDDI standards) and easy discovery of the same using semantic reasoning. The authors in [39] developed a semantic mash-up discovery algorithm to traverse their hybridized frameworks of semantic web services discovery, UDDI repositories, and existing mash-up tools.

1.4.2.3 Search engines for big data

From the perspective of big data, the essence of search engines is to provide users with their information needs from the humongous amount of data spawned within the Industry 4.0 ecosystem [40] – an activity that becomes more complicated to handle due to further widening of the dichotomy between queries and data space as the amount of data gets larger. This provision requires that results be presented efficiently and precisely at all times. Following this, with the challenges posed by multiplicity of data from disparate sources and the fact that machines constitute a huge portion of the consumers of the data in the Industry 4.0 ecosystem, it is inescapable to deploy search engines that are semantically draped, particularly as many of the processes of Industry 4.0 are semantically compelled [4, 28]. A notable work along that path is that of [28], which developed a semantic-based visual query system for Industry 4.0 by employing ontology to ease the exploration of data and visualization of machines in 2D.

It is pertinent to categorically state that control, coordination, diagnosis, and quick and automatic fixes of failed and failing entities within the Industry 4.0 ecosystem would be stifled without data discovery, access, gathering, and harvesting – activities driven by search mechanism; thus, the importance of search engines, particularly semantically draped search engines, cannot be overemphasized in Industry 4.0. Very importantly, the roles of the semantic search engines will become more visible as semantic web technologies' applications diffuse more into Industry 4.0.

Also, despite the huge benefits accruing to the deployment of semantic web technologies in search engines and other parts of Industry 4.0 yearning for its deployment, it is germane to note that there are challenges that must be surmounted to fully key into the limitless computational capability that semantic web and its associated technologies can offer. Some of these challenges include: (i) linguistic challenges – which, on the one hand, poses difficulty in determining which words in a query should constitute the concept to be searched and, on the other hand, the multiplicity of languages, which begs for the system to comprehend multiple languages; (ii) knowledge acquisition bottleneck – a common problem that often arises when converting big data to knowledge, an activity which is usually very expensive both from the perspective of the knowledge engineer and domain experts [20]; (iii) usability – some of the systems' interfaces requires a great deal of technical expertise in query languages like SPARQL and RDQL for interaction purposes, thus making the system arduous for naïve users. Also for an expert user of such systems, there is a need to have a grasp of the system's underlining semantic technologies for effective manipulation (querying). It must be noted that even with interfaces that allow natural languages, the need to repeatedly convert such to search items query without losing meaning is practically impossible.

1.5 Conclusion

The chapter began with an overview of search mechanism opening with the concept of information retrieval, its critical processes, models, and evaluation, and reiterated that search engine is an instance of information retrieval. The work exposed the several processes and components in search mechanism and discussed the search mechanism along two broad categories – traditional search and semantic search and made clear what distinguishes one from the other is the stratagem underpinning its workings. It was also made clear that the several challenges bedeviling the traditional search engines such as problem of synonymy and polysemy among others made it incumbent for the emergence of semantic search engine – which deploys the semantic web technologies – standards (e.g., XML, RDF, OWL, and OWL-S), and methods (e.g., contextual analysis, natural language understanding, reasoning, and ontology), for automatic discovery, access, and harvesting of data/knowledge and services from repositories with peak precision to ensure data/knowledge and services presented for consumption are of required veracity for efficient and effective decision making.

The work further exposed some of the categorization of semantic search engines along some critical dimensions, including the semantic web technologies deployed and the objective, methodologies, and functionality, among others. Subsequently, focus was shifted to the exposition of Industry 4.0 – an advanced manufacturing model, which, at provenance, is used to refer to modifications linked to automation fields fused with information technology, with its emergence driven by three fundamental players: knowledge, experimentation, and innovation, bringing to bare a revolution in the way goods and services are handled from the point of design, through production, supply, and distribution to consumption – obviously aided by several disruptive technologies (e.g., IoT, IoS, and big data and analytics). Finally, the work exposed the essence and the roles of semantic search engine in the seamless operation of all participating entities in Industry 4.0 and made clear that despite the humongous benefits in deploying semantic search engines in Industry 4.0, there are still challenges acting as "clog in the wheel" which must be addressed for full realization of Industry 4.0.

References

[1] Carvalho, N. G. P. and Cazarini E. W. (2020) "Industry 4.0 – What Is It?," in *Industry 4.0, Current Status and Future Trend*, ed. H. Ortiz, INTECHOPEN, 3–11, http://dx.doi.org/10.5772/intechopen

[2] Nagy, J. Oláh, J., Erdei, E., Máté, D. and Popp, J. (2018) The Role and Impact of Industry 4.0 and the Internet of Things on the Business Strategy of the Value Chain—The Case of Hungary, *Journal of Sustainability*, MDPI, 10, 1–25, htpps://doi.org/10.3390/su10103491

[3] Cho, S., May, G. and Kiritsis, D. (2019) "A semantic-driven approach for Industry 4.0", in *Proceedings of the 15th International Conference on Distributed Computing in Sensors Systems 2019 (DCOSS)*, IEEE, 347–354. https://doi.org/10.1109/DCOSS.2019.00076

[4] Obitko, M. and Jirkovsky, V. (2015) "Big Data Semantics in Industry 4.0", in *HoloMAS*, eds V. Mařík, et al., LNAI 9266, 217–229. https://doi.org/10.1007/978-3-319-22867-9_19

[5] Dominich, S. (2000) Foundation of Information Retrieval, *Mathematica Pannonica*, 11, 137–153.

[6] Sanchez, M. F. (2009) Semantically Enhanced Information Retrieval: An Ontology-based Approach, Ph.D. Thesis, Escuela Politécnica Superior, Universidad Autonoma De Madrid.
[7] Kharkevich, U. (2010) Semantics Enabled Information Retrieval, PhD Thesis, International Doctorate School in Information and Communication Technologies DIT, University of Trento.
[8] Carpineto, C. and Romano, G. (2012) A survey of automatic query expansion in information retrieval", *ACM Computing Survey*, ACM, 44, 1–50. http://doi.acm.org/10.1145/2071389.2071390
[9] Manning, C. D., Raghavan, P. and Schutze, H. (2009) An Introduction to Information Retrieval, Cambridge University Press, Cambridge, England.
[10] Zuo, J., Wang, M., Wano, J. and Luo, W. (2013) "Information Retrieval Model Combining Sentence Level Retrieval", in *International Conference on Asian Language Processing*, IEEE, 37–40.
[11] Akerkar, R. and Lingras, P. (2007) *Building an Intelligent Web: Theory and Practice*, Jones and Bartlett Publishers, 1–325.
[12] Zhai, C. (2008) Statistical Language Models for Information Retrieval, a Critical Review, *Foundation and Trends in Information Retrieval*, 2, 137–213.
[13] Song, F. and Croft, W. B. (1999) "A General Language Model for Information Retrieval", in *Proceedings of the 22nd Annual International ACM-SIGIR Conference on Research and Development in Information Retrieval*, New York, ACM, 279–280.
[14] Baeza-Yates, R. and Ribeiro Neto, B. (1999) *Modern Information Retrieval*, Addison- Wesley, 1–518.
[15] Ebietomere, E. P. (2018) A semantic retrieval system for Nigerian case law, Doctoral Thesis, Department of Computer Science, University of Benin, Benin City.
[16] Ebietomere, E. P. and Ekuobase, G. O. (2019) A semantic retrieval system for case law, *Applied Computer Systems*, 24, 38–48, 2019. https://doi.org/10.2478/acss-2019-0006
[17] Khabsa, M., Carman, S., Choudhury, S. R. and Giles, C. L. (2012) "A framework for bridging the gap between open source search tools", in *SIGIR 2012 Workshop on Open Source Information Retrieval*, 32–39.
[18] Trotman, A. Clarke, C. L. A., Ounis, I., Culpepper, S., Cartright, M. and Gera, S. (2012) "Open Source Information Retrieval: A Report on the SIGIR 2012 Workshop", ACM, 46, 95–101.

[19] Brin, S. and Page, L. (1998) The anatomy of a large-scale hypertextual Web search engine, *Computer Networks and ISDN Systems*, 30, 107–117.

[20] Ebietomere, E. P. and Ekuobase, G. O. (2022) "Semantic Web Technologies," in *Semantic Web Technologies: Research and Applications* , eds A. Patel, N. C. Debnath and B. Bhusan, CRC Press, Taylor & Francis Group, 1–23

[21] Berners-Lee, T. (1998) *Semantic Web Road Map*, https://www.w3.org/DesignIssues/Semantic.html

[22] Hitzler, P., (2021) A Review of the Semantic Web Field, *Communications of the ACM*, 64, 76–83. https://doi.org/10.1145/3397512

[23] Sudeepthi, G., Anuradha, G. and Babu, M. S. P. (2012) A Survey on Semantic Web Search Engine, *International Journal of Computer Science Issues (IJCSI)*, 9, 241–245.

[24] Wei, W., Barnaghi, P. M. and Bargeila, A. (2008) Search with Meanings: An Overview of Semantic Search Systems, Tech. Rep., School of Computer Science, University of Nottingham Malaysia Campus.

[25] Mangold, C. (2007) A Survey and Classification of Semantic Search Approaches, *Int. J. Metadata, Semantics and Ontology*, 2, 23–34.

[26] Lei, Y., Uren, V. and Motta, E. (2006) "Semsearch: A search engine for the semantic web," in *EKAW 2006*, eds. S. Staab, and V. Svátek, LNCS (LNAI), Springer, 238–245.

[27] Ezhilarasi, K. and Kalavathy, G. M. (2018) "Literature Survey: Analysis on Semantic Web Information Retrieval Methodologies", in *Proceedings of the International Conference for Phoenixes on Emerging Current Trends in Engineering and Management (PECTEAM 2018), Advances in Engineering Research (AER)*, Atlantis Press, 142, 93–108.

[28] Berges, I., Ramírez-Durán, V. J. and Illarramendi, A. (2021) A Semantic Approach to Big Data in Industry 4.0, *Journal of Big Data Research*, Elsevier, 25, 11–14. https://doi.org/10.1016/j.bdr.2021.100222

[29] Siqueira, F. and Davis, J. G. (2021) Service Computing for Industry 4.0: State of the Art, Challenges, and Research Opportunities. *ACM Computing Survey*, ACM, 54, 1–38. https://doi.org/10.1145/3478680

[30] Lunardi, W. T., Matos, E., Tiburski, R. L., Amaral, A., Marczak, S. and Hessel, F. (2015) Context-based Search Engine for Industrial IoT: Discovery, Search, Selection, and Usage of Devices, IEEE, 1–8.

[31] Guillemin, P. Friess et al., P. (2009) Internet of things strategic research roadmap, The Cluster of European Research Projects, Technical Report.

[32] Le-Phuoc, D. Quoc, H. N. M., Parreira, J. X. and Hauswirth, M. (2011) "The linked sensor middleware–connecting the real world and the semantic web," in *Proceedings of the Semantic Web Challenge*.

[33] Reis, J. Z. and Gonçalves, R. F. (2018) "The role of Internet of Services (IoS) on Industry 4.0 through the Service Oriented Architecture (SOA)", in *Proceedings of the IFIP WG 5.7 International Conference, Part II, APMS 2018*, Seoul, Korea, 1–7. https://doi.org/10.1007/978-3-319-99707-0_3

[34] Papazoglou, M. P. (2008) *Web Services*, 1st Edition, © Pearson Education Limited.

[35] Karuppiah, T. and Ramakrishnan, M. (2014) Intelligent Search Engine-Based Universal Description, Discovery and Integration for Web Service Discovery, *Journal of Computer Science*, 10, 1798–1810. https://doi.org/10.3844/jcssp.2014.1798.1810

[36] Stollberg, M., Feier, C., Roman, D. and Fensel, D. (2006) "Semantic Web Services - Concepts and Technology," in *Language Technology, Ontologies, and the Semantic Web*, eds. N. Cristea, and D. Tufis, Kluwer Publishers, 1–25.

[37] Liu, M., Shen, W., Hao, Q., Yan, J. and Bai, L. (2012) A fuzzy matchmaking approach for semantic web services with application to collaborative material selection, *J. Comput. Indus.*, 63, 193–209.

[38] Tian, Y. and Huang, M. (2012) Enhance discovery and retrieval of geospatial data using SOA and Semantic Web technologies, *Expert Systems Applications*, 39, 12522–12535. https://doi.org/10.1016/j.eswa.2012.04.061

[39] Meditskos, G. and Bassiliades, N. (2011) A combinatory framework of Web 2.0 mashup tools, OWL-S and UDDI, *Expert Systems Applications*, Elsevier, 38, 6657–6668. https://doi.org/10.1016/j.eswa.2010.11.072

[40] Drivas, I. C., Sakas, D. P., Giannakopoulos, G. A. and Kyriaki-Manessi, D. (2020) Big Data Analytics for Search Engine Optimization, *Big Data and Cognitive Computing*, MDPI, 4, 1–22. https://doi.org/10.3390/bdcc4020005

2

Semantic Web Services: The Interoperable Middleware Technology for Industry 4.0

Godspower Osaretin Ekuobase and Esingbemi Princewill Ebietomere

Department of Computer Science, University of Benin, Nigeria
godspower.ekuobase@uniben.edu; princewill.ebietomere@uniben.edu

Abstract

Industry 4.0 furthers the diminution of human labor in the manufacturing industry, digitally transforming goods manufacturing factories into cyber–physical manufacturing, logistics, and value-addition systems for optimal customer satisfaction and organizational profit. These cyber–physical systems are inherently heterogeneous and necessarily require a standard interoperable infrastructure for seamless interaction. Web services are a natural infrastructure for maintaining a rich structured form of interoperability, but its discovery and consumption data are not machine comprehensible and thus unsuitable for machine sovereignty as required by Industry 4.0. However, semantic web services (SWS) – web services with its description aspects semantically enriched – will definitely guarantee the seamless interoperability demands of Industry 4.0. This chapter explicates the concepts, features, prospects, and challenges of Industry 4.0, establishes interoperability as the oxygen of Industry 4.0, and exposes SWS as the interoperable middleware technology indispensable for efficient and effective deployment of Industry 4.0. At the end of this chapter, readers, in addition to appreciating the concepts and technologies underpinning Industry 4.0, will also appreciate its beauty and challenges as an offshoot of the semantic web vision in the manufacturing industry and also the role of SWS in the realization of the full potential of Industry 4.0.

Keywords: Cyber–physical system, Industry 4.0, interoperability, semantic web, manufacturing industry, smart factory, web services.

2.1 Introduction

Industry 4.0 is the manifestation of the semantic web vision in the manufacturing industry. The human species have consistently improved production and convenience in manufacturing, taking advantage of a new disruptive ideology as human demands begin to overwhelm the incumbent ideology. Thus, we had the first industrial revolution (Industry 1.0), the second industrial revolution (Industry 2.0), and the third industrial revolution (Industry 3.0) triggered by mechanization, electrification, and codification, respectively [1–3]. Each of these manufacturing ideologies or drives refines and subsumes preceding drives toward increased production, human convenience, organizational profit, and customer satisfaction. However, the triggered revolutions only saw the wane of human physical labor in manufacturing, not the human cognitive labor, because the machine remained a dunce – unable to understand fresh realities and respond appropriately in real time. The fourth industrial revolution (Industry 4.0) triggered by "semantization" – the idea of making machines comprehend data, seamlessly interoperate, and take cognitive and operational decisions in real time unaided, which, besides furthering the diminution of human physical labor, will also see to the waning of human cognitive labor in manufacturing – has begun [1, 4].

Industry 4.0 is a manufacturing ideology whose production and value addition chains are autonomously driven by smart machines that are necessarily heterogeneous and distributed. This manufacturing ideology subsumes the primary and tertiary industrial activities in the line of supply, production, distribution, and value addition of manufactured goods into a single manufacturing circle operated and managed by smart machines, relegating human roles in manufacturing to auxiliary. Despite the social odds (e.g., mass unemployment) against Industry 4.0 [4], the increasing human demands, quality stakes, willingness to pay less, and personalized and rapidly changing human requirements in this ever-emerging and exhilaratingly competitive world apparently attenuated by the cyber-space make Industry 4.0 imminent. This is indeed a revolution against human beings in the manufacturing industry by smart machines.

Smart machines, besides being intelligent, seamlessly comprehend possibly fresh data, autonomously grow experience (self-learning), and take possibly fresh cognitive decision or action accurately in real time based on the prevailing reality (context or situation) and prior knowledge. For machines to be smart, they must therefore be semantically enabled to comprehend data and take cognitive decisions in real time unaided. This is the vision of semantic web – making machines smart – in contrast with artificial intelligence that has the vision of making machines intelligent[1]. Data consumed by smart machines, particularly in Industry 4.0, are usually from multiple sources that are necessarily heterogeneous and distributed and generate possibly large data in real time. The smart machine and data source configuration in Industry 4.0 makes the following technologies mandatory for the realization of Industry 4.0: (i) broadband, (ii) cyber–physical systems, (iii) Internet of Things, (iv) smart applications, (v) internet of services, (vi) big data and analytics, (vii) cloud computing, (viii) fog/edge computing, (ix) additive manufacturing, (x) digital twin, (xi) augmented reality, (xii) cyber security, and (xiii) systems integration. These technologies and their germaneness to Industry 4.0 [5–7] are briefly discussed as follows.

(i) **Broadband:** Broadband is a transmission technique that enables constant and instantaneous exchange of large messages among interacting entities on the internet[2]. Industry 4.0 is heavily dependent on steady high-speed internet connectivity among its many components due to its distributed nature, heavy reliance on high volume and velocity data exchange, as well as the instantaneous response or low latency requirement of the manufacturing ideology. Thus, the most basic infrastructure for Industry 4.0 is typically broadband. Common types of broadband[3] are fiber broadband, satellite broadband, wireless broadband, cable broadband, digital subscriber line (DSL) broadband, and broadband over power lines (BPL).

(ii) **Cyber–physical systems:** Cyber–physical systems are the core components of Industry 4.0. These are internetworked programmable mechatronic devices that seamlessly interact in real time with one another, computers (for control and coordination), and humans (for auxiliary services) to essentially perform a given physical task(s) safely and

[1] Intelligent entities are not necessarily smart, but smart entities are inherently intelligent.

[2] The term internet in this chapter includes extranet or intranet or any distributed internetwork.

[3] The fcc.gov site contains a succinct description of the common broadband types.

efficiently. The term "physical task" has to do with physical motion and sensory. When the ultimate task is sensory only, the term sensor device is preferred. Network of sensor devices (embedded or not embedded) is termed Internet of Things (IoT). The principal contrast between cyber–physical systems and IoT, therefore, is in physical motion and sensory tasks. Cyber–physical devices include robots, smart factories, self-driving cars, and drones. That a smart phone has sensors embedded in it does not qualify it to be a cyber–physical device. However, such a smart phone may only qualify as an IoT device. That a printer has a sensor embedded in it to detect the state of its toner cartridge does not qualify it as an IoT device – it is not programmable though mechatronic and its task involves physical motion; it is a computer peripheral (output) device; it is useless without the computer. It is advised on grounds of pedagogy for modern electronic devices to be classified based on their operational characteristics not only to avoid conflicts of terms but also to enable designers, managers, and users to see these devices from the same stand point; for our stand points determine what we see.

(iii) **Internet of Things:** IoT is a network of sensory devices that consistently, accurately, and autonomously captures one or more states of their neighbors or environment for real-time display, cognitive response, or transmission as digital signal (data) across its internetwork for storage or further processing by designated entities in the internetwork. IoT is a critical component of Industry 4.0, a super network, as IoT serves as the "sensory organs" of Industry 4.0. These sensory devices, each with a unique identifier (UID), are usually embedded in animate or inanimate object called "thing." Many of our smart phones are now IoT devices. In Industry 4.0, most cyber–physical devices are things and this is important for self-healing or predictive maintenance of the cyber–physical systems.

(iv) **Smart applications:** Computing or cyber–physical devices are smart only to the degree permitted by the software that runs in them. Smart applications are basically intelligent, autonomous, real-time, data-driven, and context-aware software applications. Smart applications are heavily dependent on artificial intelligence, machine learning, and semantic web technologies for intelligence, learning, data comprehension, and contextual decisions. Industry 4.0, therefore, necessarily requires its applications to be smart.

(v) **Internet of services:** Internet of services (IoS), unlike IoT, is not a network but an open service platform that enables public publication and consumption of data and service applications (soft-services) across the internet. IoS is highly cost effective and reliable. A smart factory usually on subscription basis can focus on the business of manufacturing without getting engulfed in the hectic tasks of software development, maintenance, and sustained operation. IoS encourages professionalism, enhances software (functional and non-functional) qualities, and provides the needed competition enabling subscribers to switch to a preferred soft-service provider and back in real time, say during downtime or failure of subscribed soft-service. A given smart factory, therefore, can subscribe to multiple soft-service providers for varying soft-services based on varying factors for enhanced productivity and profit. Soft-service providers are heavily encouraged in Industry 4.0, though already sprouting, to professionally make the needed smart applications for running Industry 4.0 publicly consumable at low-cost and prevent Industry 4.0 from collapsing into a pseudo-software house.

(vi) **Big data and analytics:** A key feature of Industry 4.0 is its heavy reliance on fresh data. These data can only be consumed by both human and machines after available data has been computationally streamlined into valuable insights since the data were originally heterogeneous, of large volume and from multiple sources. The computational streamlining of data is the "data analytics" while the original nature of the data (volume, variety, velocity, veracity, and value) before processing into useful insights is what is meant as "big data." The insights from big data and analytics underpin the personalized product manufacturing and other smart cognitive decisions of Industry 4.0.

(vii) **Cloud computing:** Cloud computing is a service mechanism that enables the extension of machine's capabilities (infrastructure, software, or platforms) in real time with those of shared ultra-powerful and sophisticated remote machines usually on subscription. Thus, the computational capabilities of the machines of Industry 4.0 (computer, cyber–physical, and IoT devices), however operationally sophisticated, can be computationally thin while leveraging cloud computing for data, computational resources, and capabilities.

(viii) **Fog/edge computing:** Fog computing and edge computing are computational mechanisms meant to decongest the global network and attenuate latency. They, therefore, bring specific computation close to the data sources and similarly bring specific computation and data insights close to machines that regularly consume such. The relationship between cloud and fog/edge is akin to that of servers and proxy-servers. However, while edge is closer to the devices in Industry 4.0, fog is closer to the cloud. Also, while several edges can be served by a fog, several fogs can be served by a cloud. Fog/edge computing is particularly useful to Industry 4.0 as it mitigates latency and improves cyber security. This is crucial for improved manufacturing efficiency and reliability of Industry 4.0.

(ix) **Additive manufacturing:** Additive manufacturing (AM) is an autonomous layer-by-layer artifact production process based on 3D data. AM, also referred to as 3D printing, enables efficient, cost effective, and high quality personalized manufacturing on demand. AM is also critical to the predictive maintenance of Industry 4.0 – data can be exchanged in real time with part manufacturers for on-demand manufacture and timely supply of parts that need replacement.

(x) **Digital twin:** Digital twin (DT) is a virtual replica of a system (object or process) state and behavior in real time. DT is an advanced form of simulation additionally driven by real-life data in real time for status monitoring, failure or fault detection, and improvements of the physical twin [8, 9].

(xi) **Augmented reality:** Augmented reality (AR) uses real-life, and real-time data analytics to give deeper sensory and cognitive insights to humans interacting with physical systems, products, and processes, enabling seamless interaction with humans, and decisions hitherto herculean. AR is particularly useful in Industry 4.0 as it enables seamless interaction with humans, and smart factory components, products, and processes aiding in accurate information exchange, inspection, detection, direction, operation, maintenance, and quality assurance.

(xii) **Cyber security:** The authors in [10] described cyber security as "the protection of the cyber space – programs and data – from false or forced failure, mal-function, acquisition, deletion, alteration, destruction, comprehension, access or use as well as the protection of their environment – computers and its peripheral devices – from forced failure or mal-function." Cyber security is critical to Industry 4.0, being a distributed system particularly heavily dependent on fresh large data, real-time data processing activities, and on-demand autonomous operational technologies (OTs), making cyber security a necessity for sustained and effective operations of Industry 4.0.

(xiii) **System integration:** System integration is a task mechanism that enables seamless and sustained interoperability of independent but connected information and operational components or technologies with the sub-task data or capabilities. This is particularly necessary in Industry 4.0 as it is heavily dependent on real-time data exchanges, autonomous processes, heterogeneous data, and technologies. Invariably, Industry 4.0 is not even possible in the first place without efficient system integration mechanism. Thus, the integration efficiency of Industry 4.0 defines its existence, sustenance, and efficiency, making interoperability the Achilles' heel of Industry 4.0. A natural infrastructure for system integration and interoperability is web services. However, web services are semantically impoverished and unsuitable for the characteristic autonomous nature of Industry 4.0 machines. As a consequence, semantic web services (SWS) – semantically enriched web services – have remained an indispensible system integration technology in Industry 4.0. This book chapter explicates SWS in the context of Industry 4.0 – a context that best illuminates its capabilities as an interoperable middleware technology.

2.2 Semantic Web Services

Semantic web services (SWS) is a web service platform semantically enriched for autonomous discovery and consumption of web service. A web service is a distributed functionality with global web accessibility. Web service description must necessarily be machine understandable for automatic discovery and consumption. A web service, however, is completely different from web services – an interoperable middleware platform that enables the creation, description, publication, discovery, and consumption of possibly composite web service [11, 2]. Web services are, therefore, an integration infrastructure that, in its original form, syntactically enables the description, discovery, and consumption of web service. This unarguably, as Industry 4.0 demands, makes web services unsuitable for autonomous discovery and consumption without it being compliant with the semantic web (web 3.0) vision of machine comprehension of data. When web services supports and enables automatic discovery and consumption of web service, it is termed SWS. The following sub-sections detail web services, web services potentials, and limitations that necessitated SWS as well as the SWS features and capabilities that guarantee seamless integration of machines and processes in Industry 4.0.

2.2.1 Concepts of [web] service

A service is a value co-production mechanism that enriches the state or asset (tangible or intangible) of participating entities (natural or juristic). This mechanism can bc social, socio-technical, or automatized. When the mechanism is drenched in human physical labor and interactions, it is a social mechanism; when it is drenched in human cognitive labor and interactions, it is a socio-technical mechanism – otherwise, it is an automatized mechanism, drenched in machine intelligence and interactions. In the service concept, there is this illusion that a set of participating entities (service provider) provides the service and the other set (service consumer) consumes or enjoys the service. The reality is that service is co-produced and co-consumed by participating entities. The product of a service that causes state or asset enrichment in participating entities is called services – an intangible and time perishable experience. Goods are simply carriers of services [12] – they are "things" that, when properly interacted with by suitable entities, result in the entities' state or asset enrichment. Thus, Industry 4.0 is a typical automatized service mechanism that produces service-encapsulated things called goods and ensures that the things are properly interacted with by suitable entities and continuously enhances the value of participating entities on both sides of the service divide – producers and consumers. The automatized service mechanisms as Industry 4.0 are characterized by increased productivity, quality, efficiency, profit, human convenience and safety, globalization, and personalization [13]. They may, however, be capital intensive and result in socio-economic disruptions [13]. Irrespective of the modes of service, if the interaction is enabled by web technologies, the term web service is preferred. Typically, Industry 4.0 production interactions are driven by SWS. The SWS is often described as the marriage between web services and semantic web [14].

2.2.2 Web services

Web services is "an infrastructure for maintaining a rich structured form of interoperability between clients and servers, allowing complex applications to be developed by providing services that integrate other services" [15]. The authors in [16] described web services as a means of connecting service functionalities to each other using service-oriented architecture (SOA) – allowing a collection of service to communicate, interact, and coordinate their execution by message passing. Web services allows the seamless integration of software components and business processes and provide support for synchronous and asynchronous transaction by means of service interface [17]. Web services have become the de facto interoperable middleware technology for realizing end-to-end business transactions in distributed systems [12]. Basically, web services has capabilities that enable remote publication, discovery, and seamless data exchange (consumption) among heterogeneous service functionalities via open web standards as captured in Figure 2.1.

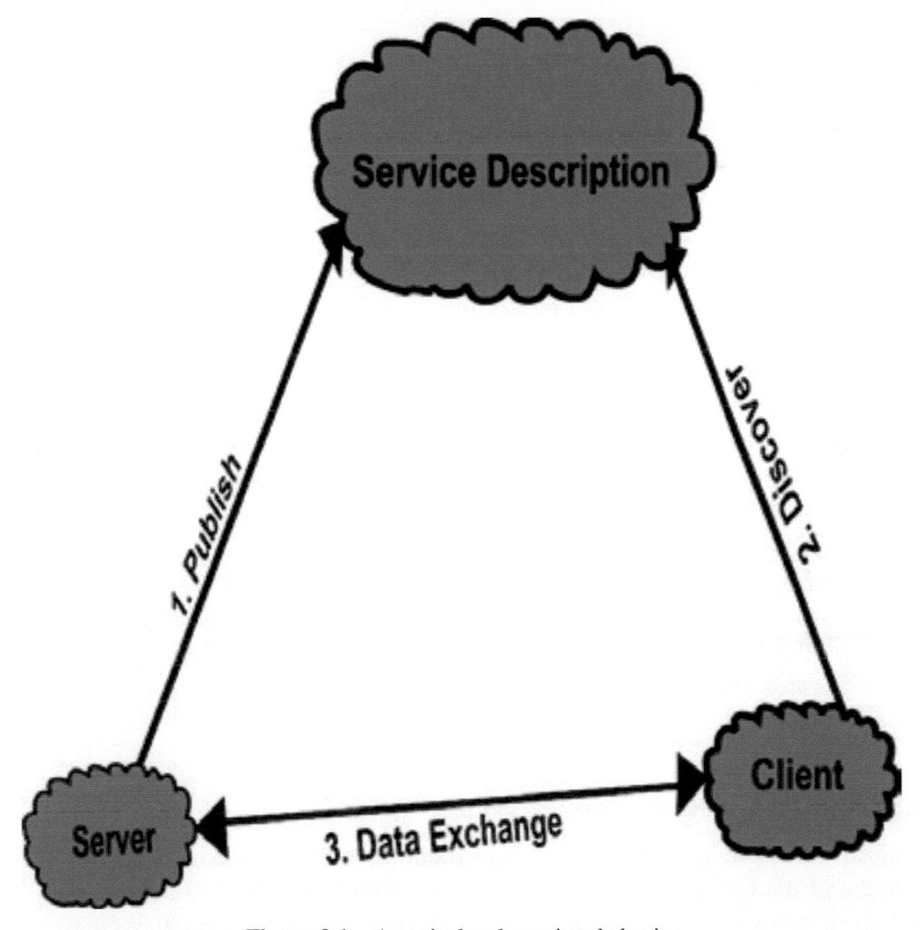

Figure 2.1 A typical web services behavior

Figure 2.1 describes the behavior of a typical web services: support for server hosting of web service functionality and constraints (server) with the mechanisms for the description (service description) and exposition of the service interface in the open web (publishing), a client application (client) with mechanism for the discovery of the published service interface, and the mechanism for the consumption of the service by client via data exchange between the client and the server. Consequently, we identified three standard web services named after their remote data exchange mechanism [18]: (i) SOAP web services, (ii) REST web services, and (iii) GraphQL web services.

a. **SOAP web services:** SOAP (initially an acronym for simple object access protocol but not anymore) is an extensible markup language (XML) based remote procedure call (RPC) style implementation of web service. SOAP, despite its support for secured asynchronous and state-sensitive machine interactions, is bedeviled with the problem of implementation complexity and high bandwidth consumption that has resulted in the emergence and embrace of REST for web service implementation.
b. **REST web services:** REST is an acronym for representational state transfer. It is an application programming interface (API) architectural style implementation of web service that views data and functionalities as web resources that it accesses using URIs. It can support plain text, HyperText Markup Language (HTML), and JavaScript Object Notation (JSON) in addition to XML. Thus, REST can handle SOAP, but SOAP cannot handle REST. Although cacheable, stateless, and simple to implement and maintain, REST web services have the problem of poor data fetching, particularly from multiple sources.
c. **GraphQL web services:** GraphQL is simply an API query language that enables API architectural style implementation of web services like REST but furthers the agility and bandwidth savings associated with REST as it supports self-documentation, client/server decoupling, and tailored data fetching. Although GraphQL implementation is simpler than "the simple" SOAP, it is not as simple as REST.

2.2.3 Semantic web and web services

Semantic web (or web 3.0) is a vision dedicated to making machines comprehend data and autonomously perform cognitive tasks [10]. A manifestation of this vision in manufacturing is Industry 4.0, which is necessarily distributed and heterogeneous and, thus, requires semantically enriched web services for interoperability and sustenance. The web services discussed in Section 2.2.2, however efficient, flexible, or light weight, are semantically malnourished for effective automatic machine discovery and consumption. To realize a semantically enriched web services, or SWS, for autonomous machines interoperability – without which there is no Industry 4.0 – ontology remains the underlining data model for SWS languages and protocols. For concise introduction to ontologies and other semantic web technologies critical to the realization of SWS, see [19]. These semantic web technologies may not be given any special attention in this work. However, attention will be given to SWS languages and protocols that help enable automatic integration and interoperability of heterogeneous data and systems in Industry 4.0. Common SWS languages and protocols include: OWL-S, WSDL-S, WSMO, WSML, WSMX, SWSF, SAWSDL, SAWSDL, WSMO-Lite, hRESTS, SA-REST, and MicroWSMO. Specifically, for Industry 4.0, we have SAREF, MPMO, DPWS, CoAP, MQTT, and AMQP, to mention a few. These protocols and languages are highlighted as follows.

a. **OWL-S:** OWL-S is an acronym for web ontology language (OWL) services. OWL-S is ontology of services expressed in OWL-DL, which enables human and program automatic discovery, consumption, and monitoring of service functionalities on the web. It consists of three main parts [20]: (i) service profile used to describe service publication and discovery; (ii) process model used to describe a service operation; and (iii) grounding used to describe how to interoperate with a web service. Although OWL-S has rich and flexible semantic expressivity, it is however restricted to the OWL ontologies, results in duplicate description of web service, and has ambiguous formal language properties.
b. **WSDL-S:** WSDL-S is an acronym for web service description language (WSDL) with semantics. WSDL-S is a mechanism that enables the association of WSDL with semantic annotations to support automatic discovery, consumption, and monitoring of service functionalities on the web [21]. Its main advantage is its support for several semantic description languages and retention of existing industrial standards for web services. It is, however, not as rich as OWL-S in specifying semantics for web service.
c. **WSMO:** WSMO is an acronym for web service modeling ontology. WSMO is a conceptual framework based on established web services modeling language (WSML) for semantic description of the important

aspects of web services for automatic discovery and consumption. It consists of four main parts [22]: (i) ontologies for specifying concepts used by other WSMO elements; (ii) web service description that describes a service functionality and behavior; (iii) goals for describing possible user objectives or desires; and (iv) mediators for resolving interoperability issues between WSMO elements.

d. **WSML:** WSML is the language, a formal language, for specifying WSMO elements. Several variants of WSML with varying expressivity exist depending on the adopted logical formalism(s) (e.g., descriptive logic, horn logic, logic programming, and first-order logic), namely: WSML-Core, WSML-DL, WSML-Flight, WSML-Rule, and WSML-Full [23]. WSML supports both human readable syntax and machine interpretable syntax in the form of XML and RDF syntax.

e. **WSMX:** WSMX is an acronym for web service execution environment. WSMX is an execution environment that supports a complete semantic web services implementation based on WSMO. WSMX also serves as a reference implementation for WSMO. WSMX consists of the following components: "core component, resource manager, service discovery, service selection, data and process mediator, communication manager, choreography engine, web service modeling toolkit, and reasoner" [24].

f. **SWSF:** SWSF is an acronym for semantic web service framework and it consists of the semantic web services language (SWSL) and semantic web services ontology (SWSO) [25]. SWSL is used to specify SWSO and web service. There are two parts to SWSL: (i) SWSL-FOL based on first-order logic; and (ii) SWSL-rules, which is rule-based. While the former specifically provides horizontal interoperability, the latter basically settles for only service functionality and behavior [26]. SWSOs are of two types: (i) first-order logic ontology for web services (FLOWS or SWSO-FOL); and (ii) rule-based ontology for web services (ROWS or SWSO-rules). While FLOWS was specified using SWSL-FOL, ROWS is simply the rule-based version of FLOWS created by the systematic transformation of FLOWS using SWSL-rules language [27]. FLOWS enables horizontal interoperability and complex service composition in addition to specifying web service for automatic discovery and consumption.

g. **SAWSDL:** SAWSDL is an acronym for semantic annotations for WSDL and XML schema. SAWSDL enables semantic annotations of the various aspects of WSDL and XML schema documents, thus semantically enriching web services for machine interpretation, discovery, and consumption [28]. SAWSDL defines three provisions for annotation: (i) modelReference, (ii) liftingSchemaMapping, and (iii) loweringSchemaMapping [28]. The annotation mechanism of SAWSDL is neutral to any semantic technology.

h. **WSMO-Lite:** WSMO-Lite is a lightweight semantic description of web service in RDFS for semantic annotation of the various aspects of WSDL using the SAWSDL annotation mechanism. WSMO-Lite defines a limited extension of SAWSDL and was inspired by the WSMO ontology as it used a small subset of WSMO to extend SAWSDL, hence the name WSMO-Lite. The extension of SAWSDL (or WSMO-Lite) strengthens its annotation with palpable semantic service description [29]. WSMO-Lite provides ontology for describing the four aspects of service semantics: information model, functional, non-functional, and behavioral semantics. WSMO-Lite neutrality, lightweight, and semantic richness make it a SWS technology of choice for automatic and intelligent discovery, negotiation, invocation, composition, and integration of different kinds of services (SOAP or RESTful web services) on the web [29], thus fostering excellent web services interoperability as Industry 4.0 demands.

i. **hRESTS:** hRESTS is an acronym for HTML for RESTful services. hRESTS is a microformat that enables machine comprehensible description of web APIs built on the REST architectural style [30, 31]. hRESTS supports annotation of textual descriptions, hyperlinks, and forms. Unlike previously discussed SWS technologies oriented toward SOAP web services, hRESTS, and its extensions, SA-REST and MicroWSMO are oriented toward the REST web services. hRESTS microformat definitions are service, operation, address, method, input/output, label, and default classes [30].

j. **SA-REST:** SA-REST is an acronym for semantically annotated RESTful web services. SA-REST defines classes for data format (data-format class) and programming languages binding (p-lang-binding class) properties of web APIs [30, 32]. It is a cross-platform extension of hRESTS, enabling seamless mash-up of public and usually heterogeneous RESTful services on the web.

k. **MicroWSMO:** MicroWSMO is a SAWSDL-like extension of hRESTS. MicroWSMO, like SAWSDL for WSDL, extends hRESTS using three link relations: model, lifting, and lowering links [30, 31], thus enabling support for automatic discovery and consumption of RESTful web services. MicroWSMO lays the requisite foundation for full semantic enrichment and autonomous discovery, composition, integration, and consumption of RESTful APIs alongside SOAP web services using the WSMO-Lite SWS technology [31].

l. **SAREF:** SAREF is an acronym for smart application reference ontology. SAREF originally intended to foster the interoperability of IoT systems has been extended to support the smart industry and manufacturing domain and named SAREF4INMA [33]. The main focus of SAREF is device – a tangible entity (thing) that can accomplish a task [34]. SAREF4INMA enables horizontal and vertical interoperability among industrial assets (including production items) and services in the entire production value chain of Industry 4.0.

m. **MPMO:** MPMO is an acronym for manufacturing predictive maintenance ontology. MPMO is a notable ontology built on OWL for condition-based maintenance of manufacturing devices in the industry – covering manufacturing, context, and condition monitoring [35].

n. **DPWS:** DPWS is an acronym for device profile for web services, an OASIS standard [36]. DPWS enables secure integration and messaging, dynamic discovery, and eventing among resource constrained internet devices (e.g., IoT devices), enterprise infrastructures, and the web [37]. It is based on SOAP web services and can be extended using REST proxy to support the REST architectural style to make it lighter [37].

o. **CoAP:** CoAP is an acronym for constrained application protocol. CoAP enables efficient and autonomous discovery and interoperability among resource-constrained devices, enterprise infrastructures, and the web [38]. CoAP is a low latent lightweight binary protocol based on the REST architectural style and operates over the user datagram protocol (UDP) [39].

p. **MQTT:** MQTT is an acronym for message queue telemetry transport. MQTT is an OASIS open standard that enables scalable integration and interoperability of IoT devices, enterprise infrastructure, and the web on publish and subscribe model. MQTT is lightweight, though heavier than CoAP, highly latent, and operates over the transmission control protocol/internet protocol (TCP/IP) and secured socket layer/transport layer security (SSL/TLS) [40]. For the latest version of MQTT standard, see [41].

q. **AMQP:** AMQP is an acronym for advanced message queuing protocol, an open OASIS standard. AMQP enables scalable vertical and horizontal integration and interoperability of IoT devices, enterprise infrastructure, and the web. Like CoAP, AMQP is a low latent binary protocol that can operate on either the publish/subscribe model or the request/response model. Although a lightweight protocol, AMQP is heavier than MQTT. Like MQTT, AMQP is TCP/IP based, reliable, and even more secured [40].

2.3 Challenges and Prospects of Industry 4.0

This section attempts to promote Industry 4.0 but not blindly. We, therefore, begin this section with the exposition of some disturbing challenges of Industry 4.0 so as to end this chapter on a high note.

2.3.1 Challenges of industry 4.0

Industry 4.0 may continue to struggle with the following issues: adoption and large-scaling, integration and standardization, semantic interoperability conflicts, internationalization and legal frameworks, security, social–cultural conflicts, and sustainability, which are discussed as follows.

a. **Adoption and large-scaling of Industry 4.0:** The dragging adoption of Industry 4.0 across industries and the regions of the world is evident with about three of every five of the less than 30% global manufacturing companies that have adopted Industry 4.0 trapped in "pilot purgatory" [42, 43]. Pilot purgatory is a phenomenon "where technology is deployed experimentally at reduced scale for an extended period due to inability or lack of conviction to roll it out at production system scale" [42]. This phenomenon is more terrifying for small- and medium-sized production enterprises [42]. Palpable culprits of this phenomenon include: infrastructure, knowledge, resource, cost, initial disruption, unclear value, use cases assurance, competing platforms, platform trust, and leadership support and attention [43]. Generally, Industry 4.0 adoption is still alien to most manufacturing companies in the continents of the world with the exception of a few North American, European, and Asian companies [43].

b. **Integration and standardization in Industry 4.0:** Industry 4.0 characterized by increasing device, platform, architecture, framework, and data varieties is becoming overwhelmed also by the varieties of integration standards meant to bring about the seamless integration of these industrial assets to realize the open global internetworking and dynamic automation of Industry 4.0. Thus, Industry 4.0 is presently grappling with the burden of setting up a global standard for seamless interoperability of its increasing assets that include production items and existing integration standards [44]. To this end, semantic interoperability will remain a critical focus of Industry 4.0. However, semantic interoperability in Industry 4.0 is not immune to the awkward conflicts it is struggling with [45].

c. **Semantic interoperability conflicts in Industry 4.0:** Semantic interoperability is the seamless exchange of data with unambiguous and shared meaning between computational entities. Semantic interoperability conflicts (SICs) "denote differences in the modeling of different and/or equivalent concepts and how these concepts are expressed" [45]. The author [45] described seven general SICs: structuredness, schematic, domain, representation, language, granularity, and missing item. The author in [45] further identified three categories of Industry 4.0 SICs, CPS-related, standardization framework, and standard, and exposed their relationships with the general SICs as depicted in Figure 2.2. The red, brown, and black rectangular boxes in Figure 2.2 hold the general SICs, SICs, and Industry 4.0 SIC categories, respectively, while the blue line depicts relation between general SICs and CPS-related SICs or Industry 4.0 SIC categories. Figure 2.2 shows that the CPS-related SICs of Industry 4.0, value processing, granularity, schematic difference, conditional mapping, bi-directional mapping, grouping and aggregation, and restriction on value, are only related to three of the general SICs, namely schematic, representation, and granularity. The other two Industry 4.0 SIC categories, standard and standardization, framework are shown to be related to the same set of general SICs, namely structuredness, schematic, and domain. The efficient, dynamic, and automatic conciliation of these SICs is a necessity in Industry 4.0 but non-trivial.
d. **Internationalization and legal frameworks in Industry 4.0:** Industry 4.0 in its full manifestation is an open and global autonomous socio-technical system necessitating global legal frameworks in the area of risks and liabilities. Can the world ever come to a consensus on issues that border on culture, human-right, cyber security and privacy, levies and taxes, liability, data ownership, and transparency? No doubt,

Industry 4.0 needs globally harmonized legal and regulatory processes that will not stifle its interoperability and innovation but minimize risks. The present silos of legal and regulatory frameworks of nations, regions, or individual technology may frustrate Industry 4.0 adoption and efficiency. The legal frameworks in Industry 4.0 should not only be holistic but internationalized.

e. **Security in Industry 4.0:** The open, dynamic, and autonomous nature of Industry 4.0 coupled with its heavy reliance on fresh data and on-demand OTs complicates its cyber vulnerability. Data integrity has become a stronger security issue in Industry 4.0 than known cyber-security threats. Additionally, security in Industry 4.0 includes human and environmental safety as CPS and cobots become colleagues to humans. The security of internet-driven systems including Industry 4.0, needless to say, is an endless challenge that must be passionately kept deprecated.

f. **Social-cultural conflicts in Industry 4.0:** Industry 4.0 may increase unemployment (or minimally, result in massive job up-skilling) for it is capable of replacing humans in the production chain and eliminating middle-men from the supply chain. Thus, massive deployment of Industry 4.0 may result in social tension, at least in the short term, and unprecedented shift in human living and interaction. Industry 4.0 will obviously bring about unprecedented social–cultural disruptions that may garner human resistance and result in weak leadership support and attention.

g. **Sustainability of Industry 4.0:** An impending challenge of Industry 4.0 is intellectual sustainability. The gains from Industry 4.0 will lure humans into building ultra-intelligent machines for more gains – making humans indolent and doltish – and, ultimately, drive humanity into singularity [46].

2.3.2 Prospects of industry 4.0

Industry 4.0 has the capability in manufacturing and is already showing strong indications, for improved productivity, production efficiency, process optimization, product quality, product personalization, innovation,

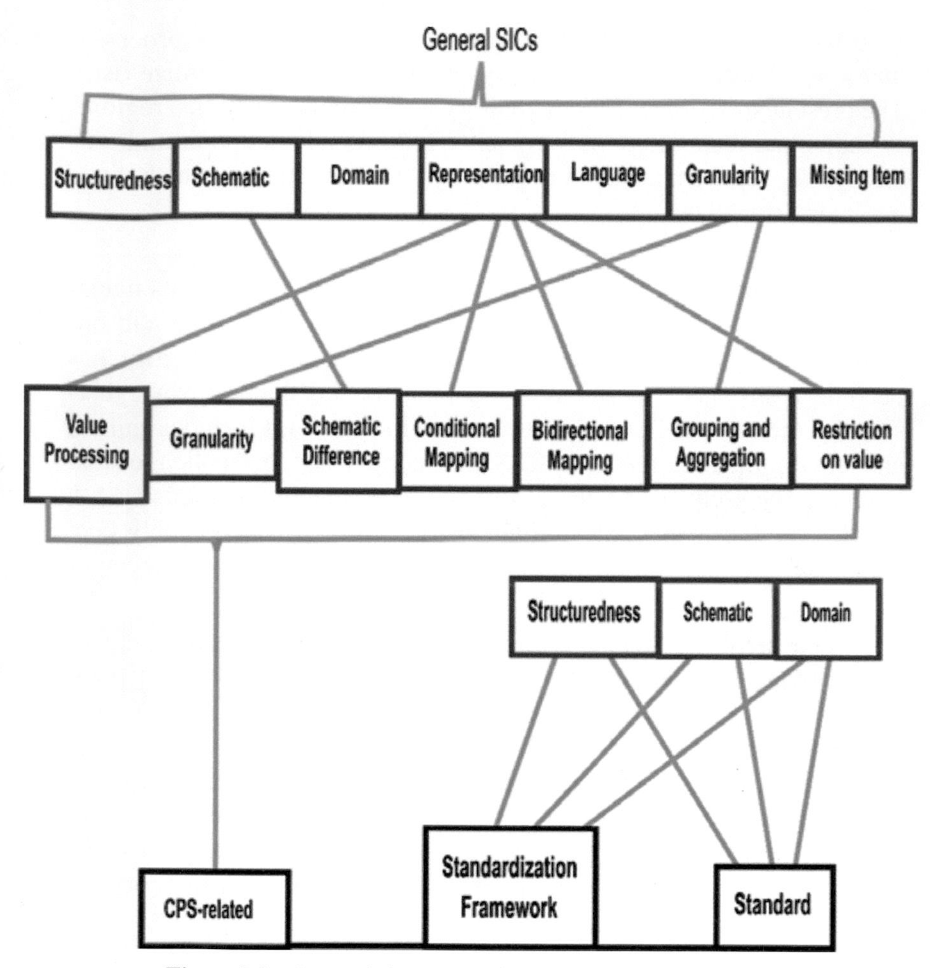

Figure 2.2 Semantic interoperability conflicts in Industry 4.0.

production monitoring, vertical and horizontal collaboration, customer experience and feedback, profitability, and human convenience and safety as well as reduced risks and cost of manufactured goods [13]. These benefits are incident on the seamless and autonomous machine integration and comprehension of (fresh) data, sound real-time cognitive and operational decisions, and semantic interoperability of Industry 4.0 occasioned by SWS and other semantic web technologies.

2.4 Conclusion

Human quest for efficient production and convenience in manufacturing has taken humans through disruptive manufacturing ideologies – which saw the diminution of human physical labor in manufacturing – to an emerging disruptive manufacturing ideology called Industry 4.0 that will further the diminution and also wane human cognitive labor in manufacturing. The chapter made evident that this latest manufacturing ideology is triggered by semantization. The chapter identified and discussed the key features of Industry 4.0 (internet broadband, CPS, IoT, IoS, smart applications, big data and analytics, cloud computing, fog/edge computing, AM, DT, AR, cyber security, and systems integration) that semantically work in concert and subsumes logistics and customer services into manufacturing, resulting in unprecedented automation that relegates human roles in manufacturing to auxiliary – semantization makes machines smart if not smarter than humans. Industry 4.0 is the "work in concert" of its constituents toward sustained production of quality goods. Industry 4.0, therefore, only exists and succeeds to the extent to which its constituents "work in concert" or interoperate as a whole. The technology, a middleware technology, that makes Industry 4.0 constituents interoperate seamlessly – web services, SWS in particular – was discussed under concepts, types, languages, protocols, and frameworks particularly as they support or relate to Industry 4.0. The chapter drew its curtain reiterating the prospects, with successes already, of Industry 4.0 not however without exposing some of its serious challenges.

This work exposed Industry 4.0 as the fourth industrial revolution in history that subsumes prior industrial revolutions triggered by mechanization, electrification, and codification and their most successful technologies and organized them in concert using semantic web technologies for real-time autonomous goods manufacturing, expanded to now include logistics and customer services. Like its predecessors, Industry 4.0 will survive and transform civilization but to a fearful end of singularity. This warning is necessary to make singularity a known threat of Industry 4.0 to human universal dominance and existence, for humans are capable of handling known fears if given sufficient time. Other warnings necessary for the success of Industry 4.0, as made evident in this chapter, are global digital-infrastructure deployment, dissemination of technical know-how, standardization, and legal frameworks for Industry 4.0. Also critical is the efficient, dynamic, and automatic conciliation of SICs in Industry 4.0, for Industry 4.0 is simply the seamless semantic interoperability of technologies for sustained autonomous manufacturing of quality goods by machines for improved convenience and prosperity of humans and the manufacturing industry – some of these technologies are known and have been discussed, but more will unfold in future. Whatever the technologies that may unfold, Industry 4.0 remain an offshoot of the semantic web vision in manufacturing, with SWS being the specific underpinning semantic web technology.

References

[1] Pozdnyakova, U. A., Golikov, V. V., Peters, I. A. and Morozova, I. A. (2019) "Genesis of the Revolutionary Transition to Industry 4.0 in the 21st Century and Overview of Previous Industrial Revolutions," In *Industry 4.0: Industrial Revolution of the 21st Century*, eds. E. G. Popkova, Y. V. Regulina, and A. V. Bogoviz, Springer, 11–20.

[2] Oliveira, M. and Afonso, D. (2019) "Industry Focused in Data Collection: How Industry 4.0 is Handled by BigData," In *Proceedings of 2019 2nd International Conference on Data Science and Information Technology (DSIT'19)*, Seoul, Republic of Korea, ACM, 1–7. https://doi.org/10.1145/3352411.3352414

[3] Tay, S. I., Lee, T. C., Hamid, N. A. A. and Ahmad, A. N. A. (2018) An Overview of Industry 4.0: Definition, Components, and Government Initiatives. *Journal of Advanced Research in Dynamical and Control Systems*, 10, 1379–1387.

[4] Sukhodolov, Y. A. (2019) "The Notion, Essence, and Peculiarities of Industry 4.0 as a Sphere of Industry," In *Industry 4.0: Industrial Revolution of the 21st Century*, eds. E. G. Popkova, Y. V. Regulina and A. V. Bogoviz, Switzerland: Springer, 3–10.

[5] Gilchrist, A. (2016) *Industry 4.0: The Industrial Internet of Things*, Apress. https://doi.org/ 10.1007/978-1-4842-2047-4

[6] Duan, L. and Xu, L. D. (2021) Data Analytics in Industry 4.0: A Survey. *Information Systems Frontiers*, Springer, 1–17. https://doi.org/10.1007/s10796-021-10190-0

[7] Oztemel E. and Gursev, S. (2020) Literature review of Industry 4.0 and related technologies. *Journal of Intelligent Manufacturing*, Springer, 31, 127–182. https://doi.org/10.1007/s10845-018-1433-8

[8] Fuller, A., Fan, Z., Day C. and Barlow, C. (2020) Digital Twin: Enabling Technologies, Challenges and Open Research. IEEE Access, 1–21. https://doi.org/ 10.1109/ACCESS.2020.2998358

[9] Parrot, A. and Warshaw, S. (2017) "Industry 4.0 and The Digital Twin: Manufacturing Meets its Match," In *Deloitte series on Industry 4.0, Digital Manufacturing Enterprises, and Digital Supply Networks*, Deloitte University Press, 1–20.

[10] Ekuobase G. O. and Ebietomere, E. P. (2022) "Latest Applications of Semantic Web Technologies for Service Industry," *In Semantic Web Technologies: Research and Applications*, eds. A. Patel, N. C. Debnath and B. Bhusan, CRC Press, Taylor & Francis Group, pp. 73–103.

[11] Ekuobase, G. O. and Onibere E. A. (2010) "Understanding Web Services: The Platform Technology for Next-generation Computing Systems," *International Journal of Natural and Applied Sciences (IJONAS)*, 6, 382–390.

[12] Ekuobase, G. O. (2020) *Service Innovation Computing – Nigeria's Pilot to El-Dorado, 232nd Inaugural Lecture Series*, University of Benin Press.

[13] Sony, M. (2020) "Pros and Cons of Implementing Industry 4.0 for the Organizations: A Review and Synthesis of Evidence," *Production & Manufacturing Research*, Taylor & Francis, 8, 244–272. https://doi.org/10.1080/21693277.2020.1781705

[14] Cardoso, J. and Sheth, A. P. (2005) Introduction to Semantic Web Services and Web Process Composition, *Semantic Web Process: Powering Next Generation of Processes with Semantics and Web Services*, LNCS, Springer-Verlag, 3387, 1–13.

[15] Coulouris, G., Dollimore, J., Kindberg, T. and Blair, G. (2012) *Distributed System: Concepts and Design*, Addison-Wesley, 1047pp, 2012.

[16] Barry D. K. and Dick, D. (2013) *Web Services, Service-Oriented Architectures, and Cloud Computing, The Savvy Manager's Guide, 2nd Edition*, Elsevier Publishers, 1–244.

[17] Erl, T. (2008) *SOA: Principles of Services Design, SOA Systems*, Prentice Hall Publishers, 573pp.

[18] Brito, G. and Valenta, M. T. (2020) "REST vs GraphQL: A Controlled Experiment", IEEE International Conference on Software Architecture (ICSA), IEEE, 11pp. 2020. https://doi.org/10.1109/ICSA47634.2020.00016

[19] Ebietomere, E. P. and Ekuobase, G. O. (2022) "Semantic Web Technologies," *In Semantic Web Technologies: Research and Applications*, eds. A. Patel, N. C. Debnath and B. Bhusan CRC Press, Taylor & Francis Group, 1–23.

[20] Martin, D. Burstein M. et al. (2004), *OWL-S: Semantic Markup for Web Services*, W3C, http://www.w3.org/Submission/OWL-S_
[21] Akkiraju, R. Farrell J. et al. (2005) *Web Service Semantics – WSDL-S*, W3C, http://www.w3.org/Submission/WSDL-S/_
[22] Lausen, H., Polleres A. et al. (2005), *Web Service Modeling Ontology (WSMO)*, W3C, http://www.w3.org/Submission/WSMO/_
[23] De Bruijn, J. Lausen H. et al. (2005), "Web Service Modeling Language (WSML)," W3C, http://www.w3.org/Submission/WSML/_
[24] Cimpian, E. Zaremba M. et al. (2005), *Web Service Execution Environment (WSMX)*, W3C, http://www.w3.org/Submission/WSMX/_
[25] Battle, S. Bernstein A. et al. (2005), *Semantic Web Services Framework (SWSF) Overview*, W3C, http://www.w3.org/Submission/SWSF/_
[26] Battle, S. Bernstein A. et al. (2005) *Semantic Web Services Language (SWSL)*, W3C, http://www.w3.org/Submission/SWSF-SWSL/_
[27] Battle, S. Bernstein A. et al. (2005), *Semantic Web Service Ontology (SWSO)*, W3C, http://www.w3.org/Submission/SWSO/_
[28] Farrell J. and Lausen, H. (2007) *Semantic Annotations for WSDL and XML Schema*, W3C, http://www.w3.org/TR/sawsdl/_
[29] Fensel, D. Fischer, F. Kopecky, J. Krummenacher, R. Lambert D. and Vitvar, T. (2010) "WSMO-Lite: Lightweight Semantic Descriptions for Services on the Web," W3C, http://www.w3.org/Submission/WSMO-Lite/_
[30] Kopecky, J., Gomadem, K. and Vitvar, T. (2008) "hRESTS: An HTML Microformat for Describing RESTful Web Services", *In Proceedings of the IEEE/WIC/ACM International Conference on Web Intelligence*, https://corescholar.libraries.wright.edu/knoesis/259
[31] Roman, D., Kopecky, J., Vitvar, T., Domingue J. and Fensel, D. (2014) WSMO-Lite and hRESTS: Lightweight Semantic Annotations for Web Services and RESTful APIs, *Web Semantics: Science, Services and Agents on the World Wide Web*, 1–20. http://dx.doi.org/10.1016/j.websem.2014.11.006
[32] Gomadam, K., Ranabahu, A. and Shet, A. (2010) *SA-REST: Semantic Annotation of Web Resources*, W3C Member, http://www.w3.org/Submission/SA-REST/
[33] Daniele, L., Izquierdo, A. F., Garcia-Castro, R. and De Roode, M. (2019) *SAREF4INMA: an extension of SAREF for the industry and manufacturing domain*. https://saref.etsi.org/saref4inma/

[34] Daniele, L., Garcia-Castro, R,. Lefrancois, M. and Poveda-Villalon, M. (2020) *SAREF: the Smart Applications REFerence ontology*. https://saref.etsi.org/core/_

[35] Cao, Q., Samet, A., Zanni-Merk, C., de Beuvron, F. B. and Reich, C. (2020) Combinning Chronicle Mining and Semantics for Predictive Maintenance in Manufacturing Processes, *Semantic Web*, IOS Press, 11, 927-948. https://doi.org/10.3233/SW-200406.

[36] Driscoll, D. and Mensch, A. (2009) *Devices Profile for Web Services Version 1.1: OASIS Standard*, OASIS Open. http://docs.oasis-open.org/ws-dd/dpws/wsdd-dpws-1.1-spec.thml

[37] Han, N. S., Park, S., Lee G. M. and Crespi, N. (2015) Extending the Devices Profile for Web Services (DPWS) Standard using REST Proxy, *IEEE Internet Computing*, IEEE, 19, 10 – 17. https://doi.org/10.1109/MIC.2014.44

[38] Shelby, Z., Hartke, K. and Bormann, C. (2014) *The Constrained Application Protocol (CoAP)*. http://www.rfc-editor.org/info/rfc7252

[39] Chen, X. (2014) *Constrained Application Protocol for Internet of Things*. http://www.csc.wustl.edu/~jain/csc574-14/ftp/coap/index.html.

[40] Solovev, A. and Petrova, A. (2020) *Reasons and Peculiarity of Choosing MQTT Protocol for IoT Devices*. http://www.integrasources.com.

[41] Banks, A., Briggs, E., Borgendale, K. and Gupta, R. (2019) *MQTT Version 5.0: OASIS Standard*. http://docs.oasis-open.org/mqtt/mqtt/v5.0/mqtt-v5.0.thml.

[42] World Economic Forum (2018) *The Next Economic Growth Engine: Scaling Fourth Industrial Revolution Technologies in Production*, WEF.

[43] Wopata, M. (2020) *Industry 4.0 Adoption 2020 – Who is Ahead?*. http://www.iot-analytics.com _

[44] Burns, T. Cosgrove, D. J. and Doyle, F. (2019) A Review of Interoperability Standards for Industry 4.0, *Procedia Manufacturing*, 38, 646–653.

[45] Grangel-Gonzalez, I. (2019) A Knowledge Graph Based Integration Approach for Industry 4.0. PhD Thesis, University of Bonn (Rheinischen Frieddrich-Williams-Universitat Bonn), Bonn, Germany.

[46] Chalmers, D. J. (2010) The Singularity: A Philosophical Analysis, *Journal of Consciousness Studies*, 17, 7–65.

3

Semantic Web of Things for Healthcare Interoperability using IoMT Technologies

Rajani Reddy Gorrepati[1], Prathiba Jonnala[2], Sitaramanjaneya Reddy Guntur[3], and Do-Hyeun Kim[4]

[1]Department of Computer Science and Engineering, Koneru Lakshmaiah Education Foundation, Vaddeswaram, Guntur, India
[2]Department of Electronics and Communication Engineering, Vignan's Foundation for Science, Technology, and Research, India
[3]Department of Biomedical Engineering, Vignan's Foundation for Science, Technology, and Research, India
[4]Department of Computer Science and Engineering, Jeju National University, Republic of Korea
E-mail: rajani.cse20@gmail.com; pratibha.j@gmail.com; gunturu.ram@gmail.com; kimdh@jejunu.ac.kr

Abstract

The semantic web of things (SWT) is useful for the remote-based automatic assessment of smart healthcare devices connected via the internet at each level. The SWT's interoperability across many platforms and systems based on the Internet of Things (IoT) represents a significant and highly complex challenge. However, various healthcare applications can result in tremendous benefits. Current SWT usage trends in IoT healthcare applications are remote patient monitoring, drug delivery traceability, and patient diagnosis. However, the current technology does not address reusability and interoperability in SWT in IoT applications. To overcome the limitations of an existing system, an ontology-based modeling framework and an IoT-based semantic M2M platform are proposed. This framework employs ontologies used in the IoT healthcare domain to design knowledge frameworks, automated sequencing, and processing and offer semantic descriptions of various concepts and their relationships. The SWRL rule addresses the issue of insufficient ontology expressivity in property association and operation to facilitate spatial connection reasoning. Using Protégé and its plug-ins, a system based on ontologies and rules was constructed. The proposed IoT-based SWT framework facilitates hospital device interoperability and semantic annotation. This chapter focuses on a collaborative approach to developing a unique framework for semantic interoperability in healthcare applications employing IoT.

Keywords: IoT, interoperability, M2M, ontology, Protégé, RFD, Semantic web of things.

3.1 Introduction

Several features of the IoT connected to smart devices, which are embedded technologies to perceive, communicate, interact, and collaborate with other things that are establishing a network of physical items, have emerged in this decade. As a result of today's technology, the concept has exploded in popularity at present. Semantic models have been developed for IoT applications. The innovation of the IoT application development tool enables node connectivity, data transfer, and orchestrations for complex industrial applications. The semantic sensor network (SSN) ontology is one of the most important and widely used knowledge models for characteristic IoT domain systems. This model represents sensors semantically, including their attributes, stimuli, and observations. However, several IoT items, such as actuators, identifiers, and processing services, are not included in the SSN ontology.

More effort and vocabulary description frameworks are required to represent IoT health domain resources, data, and services. The querying ontology is used to retrieve information to improve the performance and usage of semantic data in the IoT healthcare sector. Given a large amount of information and knowledge representation, IoT health devices are easily connected to various sensors to obtain information. The data can access and provide a connection to big data storage and data collection from other devices depending on the semantic query requests. The typical query systems have proposed ontology SPARQL, which requests information from ontologies written in the resource description framework (RDF) and OWL. The increased and huge number of information and knowledge representation extraction and exploration aim to enhance the semantic data processing in the IoT health domain. The data can be accessed by using semantic query requests, which provide connections to big data storage for data collected from smart health devices. The typical querying system for SPARQL has been proposed, which requests information from ontologies RDF and OWL. SQL querying for relational databases has inspired it.

3.1.1 Overview of industrial internet of things (IIoT)

One of the many benefits of industrial Internet of Things (IIoT) platforms is the ability to use data to improve industrial environments and process production. However, the challenge of collecting and processing data to implement the IIoT application remains. Figure 3.1 depicts the semantic industry for Internet of Things. IoT platforms must rely on IoT infrastructure and the number of actuators and sensors required to provide appropriate information by an IoT platform. However, such systems typically use IoT hardware from specific vendors. The IoT is a technological advancement that links things, thus providing a foundation for next-generation basic structure, everyday services, and healthcare domain commercial applications. However, the number of Internet-connected gadgets has surpassed the world's total population [1–3]. Researchers estimate that IoT devices will have billions of connected devices, and the convergence of which will evolve into the internet of everything.

Industry 4.0 refers to the recent revolution in the automation of trends and data exchange in medicare manufacturing technologies connected via smart manufacturing assets factory, which has horizontal integration via connectivity throughout the entire healthcare supply chain. Web of things (WoT) and SWT technical innovations have a promising role to play in addressing industry concerns in the 4.0 version [4, 5]. Communication between heterogeneous industrial assets is enabled by integrating SWT and WoT. Manufacturing knowledge can also be represented in a machine readable fashion using the semantic web. The semantic modeling of industrial assets and services results in clear and understandable inventions in healthcare services across domains.

The IoT environment provides interdisciplinary services in health sectors, such as building, energy, consumers, and healthcare data transformation. The value of a network is determined by its ability to comprehend a situation or context to enable services and medicare applications. The variety of devices that cover various parts of the environment and have direct data as well as the ability of higher-level systems to analyze and generate process data can be constrained. A circumstance or context enables services and medicare

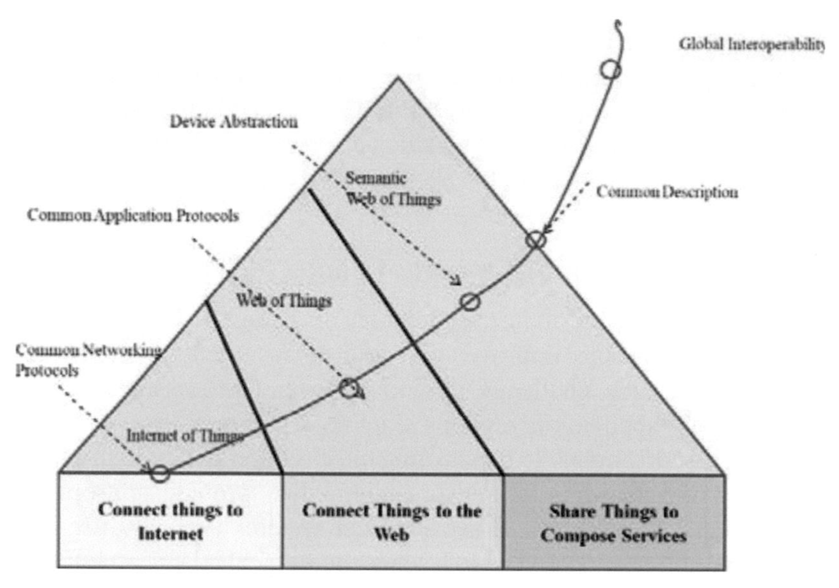

Figure 3.1 Semantic industry for Internet of Things.

applications [6, 7]. Direct data are generated by a range of devices that cover many parts of the environment, and the capacity of higher-level systems to interpret and generate process data is restricted. Data are easy to collect using smart health devices, such as sensors, scanners, smartphones, biometric devices, imaging devices, cameras, and video surveillance [8]. Most of these devices are in the remote area network of healthcare smart devices in the healthcare industry sectors. The primary goal of the health-field sector is to provide high-quality healthcare services by storing a large amount of data about people's health records and allowing wearable devices to collect real-time applications about their health [9, 10]. The data are then transmitted to transportation agencies or healthcare providers for analysis and decision-making. All these issues create communication challenges when reading different data formats and exchanging specific data contexts.

The IoT is gradually gaining significance, which is hampered by the fact that it is a ubiquitous technology. Health devices are able to transmit data clearly and unambiguously with shared W3C web of things architecture to provide semantic data exchange interoperability for healthcare domains. To reach higher levels of interoperability of semantic sensor technologies for IoT, designers develop solutions that enhance the web of things (WoT) architecture [11, 12]. Information models and computational IoT ontologies have been developed in different health sector organizations, thus resulting in the creation of semantic web ontologies within different IoT ecosystems [13]. According to their 300 different ontologies defined in IoT domains and knowledge, representation is inconsistent with several standards and IoT solutions caused by semantic fragmentation [14].

3.1.2 Requirements of SWT for medical devices

The first step has been to provide interconnection to all healthcare devices so that they can be connected to the internet. Interoperability among the various entities is enabled after connection with heterogeneity being realized. Therefore, the current heterogeneous islands of devices have been concentrated on IPV6. Once connectivity has been established, a common protocol for the transport and application layers is required. The worldwide web consortium is the most extensive application on the internet. Hypertext transfer protocol (HTTP) is a web-based data transfer protocol. During the last decade, web services have been capable of providing a homogeneous application protocol in HTTP. Its potential is being realized using hypertext markup language for resource representation and JavaScript for logic and intelligence [15, 16].

Science and technology have evolved into critical components of medical technology. The emergence of SWT has increased the impact of AI, blockchain, IoT, and technology on global health systems. Semantic interoperability is one of the most important levels of interoperability [17, 18]. It refers to being automatically interpreted by machines, platforms, and domains. Researchers use the semantic web because it enables the semantics of information and services. The SWT is developed on a set of standards, including RDF, RDF schema, OWL, and SPARQL for data representation and querying, respectively.

The semantic web is used in IoT architecture to achieve semantic interoperability between ontology domains. An ontology coherently represents the heterogeneous data collected. A set of SRWL rules enrich the ontology by discovering new hidden relationships, and a reasoning engine discovers new knowledge about people's health status by providing them with quality transportation services, adapting their health, and applying real-time semantic processing to the heterogeneous data collected [19, 20]. As such, SWoT challenges of interoperability and health ontology of IoT domain in healthcare transport have emerged. The emerging technology for the internet of revolution poses enormous challenges in every aspect of life for specific health domain sectors [21, 22]. The emerging technologies that are being adapted to the internet of semantic things (IoST) include several aspects of security protocols applied in artificial intelligence (AI), blockchain cyber security, mobile technologies for the survey of healthcare sectors in M2M, tracing of elderly patients, remote-based monitoring, telemedicine, tracking drug delivery management, diagnostic imaging, and biomedical equipment.

IoT refers to billions of network devices that communicate with one another and are flexible enough to exchange patient data and wealth analysis for planning, management, and decision-making in patient tracking and remote monitoring using smartwatches, biosensors, and smart sensor devices. The IoT industry has been applied to the health sector in urban hospitals. Semantic analytics have been used by various physicians in different areas. To communicate directly, they have also installed smart devices to measure daily patient temperature, heart rate, blood sugar level, oxygen level, and so on. Minimizing the interference in the IoT industry and SWT technologies in health institutions can help specialists interact better to keep track of health check-ups continuously and to make prescriptions or decisions in the ontologies in IoT health domains, requirements, security, and privacy [23, 24].

Meanwhile, SWT technologies define the enrichment of things for syntactic web information, which consists of documents as well as data on the SWT and is transformed more effectively by machines. The semantic web improves the drug supply chain management's ability to select high-quality health information service providers, as well as automated decision-making medical smart sensor devices, delivery of healthcare systems that are rule-based ontologies in health domain sector systems, interoperability, and use ontologies from various health domain sectors [25, 26].

3.1.3 Semantic interoperable healthcare industry using IoT system

The SWT automatic industry is remotely based on the evolution of smart healthcare devices at each level. Such smart healthcare devices are connected via remote-based internet. Interoperability of big biomedical data analytics for SWT across diverse IoT platforms and systems is a complicated problem, and various healthcare applications might provide huge benefits. IoT application development techniques, such as node connectivity, have advanced to the point that complicated industrial applications, such as IoT orchestrations, have been developed. At present, the developments in the use of semantic industrial technologies in the IoT domain include several healthcare applications, such as remote patient monitoring, tracking drug deliverables, and patient diagnosis [27, 28]. Furthermore, it expands on the SWT, which enables middleware, frameworks, and M2M architecture to overcome smart device and data heterogeneity constraints. The current technology is re-usability and interoperability applications that cannot be used to develop interoperability problems in IoT applications in SWT. Based on ontology modeling, a framework for an IoT semantic M2M platform has been proposed. Ontologies are utilized in this framework to define the structure, allow knowledge of automated sorting and instance detection, and give formal semantics for the description of numerous concepts and their interactions in the IoT healthcare area.

To support spatial connection reasoning, the SWRL rule can be used to address the issue of insufficient ontology expressivity in property association and operation. Protégé and its plug-ins have been used to construct an ontology-based reasoning system. Based on ontologies and rules that have developed a reasoning system, IoT smart healthcare devices and biomedical data currently use several data types, protocols, and technology to function [29, 30]. IoMT is developing a framework for smart health patient monitoring related to IoT smart health devices based on interoperability principles. The IoT framework allows for interoperability with heterogeneous devices, and data collected from various smart health devices attached to patients and entire hospitals are annotated using a semantic annotation model. RDF is used to deliver semantic information and functionality data and to aid in the provision of heterogeneous health data collected from IoT health devices.

The main challenges addressed in this chapter are the development of semantically interoperable systems that are unified ontology-based knowledge representations of required data produced by IoT test beds. The purpose of the ontology reasoning mechanism is to improve required data, ensure processing, and share the method of interpreting IoT health data. The complexity of SWT lies on end-users who are provided with unified health sector services, enriched IoT secured-access data, real-time processing, and the scalability of health sector challenges. The proposed architecture is a collaborative approach to SWT interoperability for IoT in healthcare applications. To function effectively in the healthcare domain, physicians and patients may need to communicate with one another. This scenario focuses on two different use cases of the SWT and IoT in the healthcare system. Experimental analysis of semantically interoperable IoT on healthcare data is carried out using the

Protégé tool. Furthermore, the supporting semantic tools and methodologies in IoMT for a variety of healthcare applications are summarized.

3.2 Related Works

Emerging technologies that have been applied in various aspects of the healthcare sector, such as IoMT technology, have been demonstrated in medical research [31, 32], such as remote-based treatment cure, medical diagnosis, and tracking of patients in the different waves of the COVID-19 pandemic [7]. The main goal of various applications in IoT healthcare services is to provide users with low-cost experiences and access to wearable IoT smart devices and real-time positioning systems in radio frequency identification (RFID) bracelets for personal healthcare services as well as automatic monitoring of daily-wise patient activity. IoT technologies innovated in future perspectives include RPM clinical practice and major research challenges, which have been highlighted as innovative technologies for significant implications in the field of IoMT remote medical services. To establish standard ontologies, remote healthcare applications have been employed to offer healthcare services in rural and distant places via remote monitoring [5].

According to the IoMT, stakeholder security problems should automatically develop a security control for each threat. Ontology tools employ semantic security approaches. Ray(2014) [8] proposed an IoT-based model called home health hub IoT (H3-IoT) for tracing remote monitoring patient records and installing various biosensors and hardware platforms that allow physicians to interact with information on patients' remote monitoring health status [12]. Eero [9] presented an e-health portal based on a shared metadata sensor schema and different ontologies that have been enhanced with updated health information across the search technique for medical data or to maintain records using the SWT. Ali and Samina [10] created an ontology-based healthcare system for automating clinical pathways to unify the post-cancer management system [10]. Verma *et al.* [11] introduced smart wearable sensors and electrocardiogram sensors to collect and monitor health data conditions [11]. Furthermore, Ranganathan *et al.* (2003) described a cosmology implementation that is used as middleware, DAML+OIL, which is used to establish mindfulness and semantic interoperability [33, 34]. In healthcare services, IoMRT robotic surgery based on Li-Fi technology relies on arm movements and positioning and reduces major risks to patients [6, 36, 37].

A remote patient health monitoring system analyzes medical data to enhance decision making, processes, and early diagnosis [5]. Pathak *et al.* [12, 16] investigated the use of semantic web networks of things to maintain patients' electronic health records (EHRs) from multiple sources through resource description framework (RDF) graphs. The RDF graphs with SPARQL query end point assess research with the high penetration of other smart devices, including mobile technologies, for raw health data collection among hospitals, health practitioners, and patients via connected smart health devices [17, 18].

3.3 Network architecture of SWT for healthcare

Many e-health applications have adopted IoT-based industry technology to allow domain-connected IoT smart medical gadgets to interact more easily with each other through connected information. In e-health systems that communicate with WPAN, wireless LAN is widely used. The treatment of patients without consulting a physician clinic, as well as the preservation of all critical patient data even when the individual consults with more than one clinic and intends to identify LAN vulnerabilities in IoT medical smart gadgets that provide information to the e-health system architecture. This architecture comprises three domains, namely IoT health domain, multi-cloud domain, and user action domain. These domains are focused on communication vulnerabilities. Figure 3.2 shows the SWoT-layered architecture, which includes the design of a data management system for the graph of things, as well as a data processing flow linked to the middleware stream. The data acquisition layer, which comprises a variety of plug-in wrappers, handles data consumption. These wrappers analyze and stream data from a range of formats, protocols, and device platforms to link stream triples to the graph of things layer.

It indexes and stores RDF-based data in distributed permanent partitions, as well as the processing cluster and distributed in-memory storage. Data access interfaces for two query processing engines, namely modified SPARQL and CQELS, are implemented in the graph of things layer using a SPARQL endpoint or a stream subscribing to the channel. These two engines allow application developers to query data in the application layer. The SPARQL endpoint provides services by extending the SPARQL query language with built-in geographic, temporal, and free text features. The stream subscribing channel uses a web socket channel to serve continuous questions in the CQELS-QL query language. CQELS is a stream processing

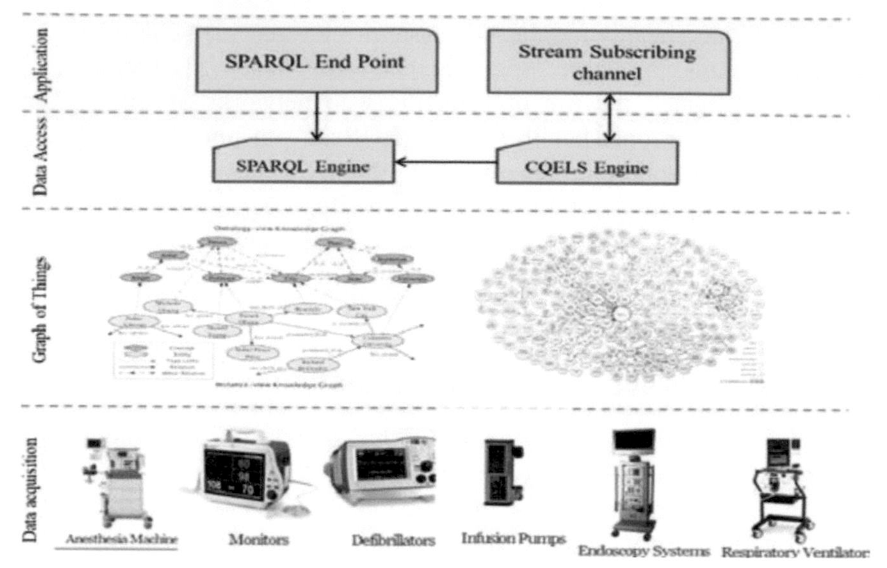

Figure 3.2 SWT-layered architecture.

engine that allows users to query RDF stream data in real time. A continuous query continues to execute indefinitely and is triggered anytime fresh relevant data are received.

Several protocols at the data acquisition layer are used to collect or push data into the system, including HTTP, FTP, TCP/IP, web socket, and MQTT. On dispersed processing nodes, it is processed asynchronously. These operations are organized into groups by utilizing a wrapper to acquire data from an XML-based patient monitoring service or an RDF-based wrapper to extract streams. Given that the connections to external data sources and the data processing wrappers are unsecured, the processes used to feed data to the stream graph data bus must be fault tolerant. The data management architecture requires robustness because the incoming stream throughput varies.

3.4 Methodology

3.4.1 Proposed semantic web technologies of interoperability using IoT

SW technologies provide knowledge sharing and reuse mechanisms that facilitate interoperability approaches, which adhere to recommendations from the SW community to avoid the creation of heterogeneous models that impede interoperability to reuse and enhance existing knowledge, refer to other levels, and perform semantic mapping tasks.

The data extraction process, storage approaches, and tools for the SWT methodologies for IoT e-health ontologies are presented in Figure 3.3. Moreover, the SWoT for the healthcare interoperability of IoT devices is shown in Figure 3.3. The sensor data are communicated through IoT-based SWT with ontology. Sensors allow IoT devices to communicate with one another. A sensor network API is built into every device and is used to filter data by domain. The filtered data are subsequently provided to the web service of the sensors, which allows them to communicate with the rest of the world. The sensor web enablement framework is used for web services to deliver heterogeneous IoT devices. SWoT is used for locating, accessing, and using IoT devices. These keywords are identifiers for medical information on human disorders. The dataset is then sent to the semantic interoperability portion, where each token is assigned to one of several illness domains. Each token is directed to the illness classification section, along with its description. This section classifies human diseases to illustrate which patients have certain diseases.

3.4.2 Ontology validation tools

To ensure the ontologies, tools that validate syntax, detect available incorrect formats, and measure performance are used to evaluate e-health smart ontologies for IoT and smart cities that are available in the incorrect format.

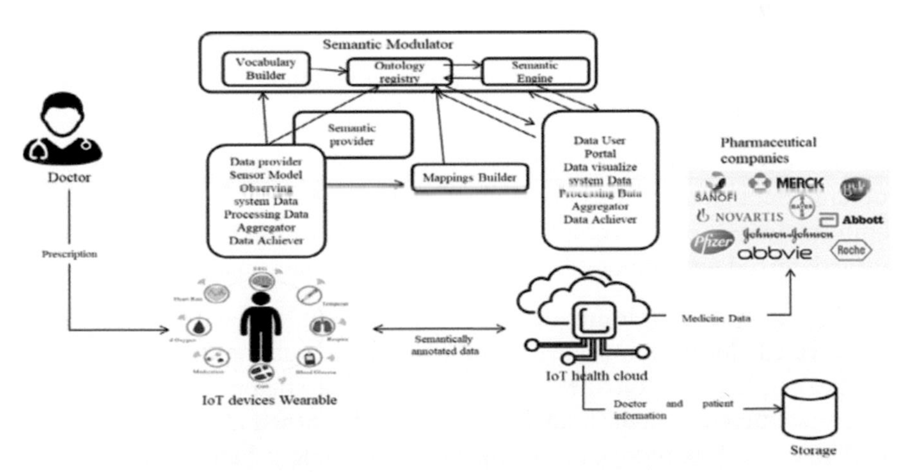

Figure 3.3 SWoT for healthcare interoperability.

Compared with automatic documentation ontologies, the evaluation has concluded that ontologies and revealing errors are superior. The semantic web community employs practices that are uncommon in the IoT smart health context. Amid the lack of semantic web experts, modeling ontologies have been made available in the context of e-healthcare, logistics, and transportation ontologies represented by SW languages, such as RDFs or OWL, and information on the ontologies' capacity to transmit data across agents and platforms has been shared. Several semantic sensor network ontologies that are more expressive in terms of normalization than others in the IoT health domain are required, as are formal implicit ontologies.

3.4.3 Biomedical ontology domain

Biomedical ontologies have grown in popularity in the computational biology field due to their crucial role in providing formal descriptions of biomedical knowledge, categorization, and domain-wide concepts. Biomedicine ontologies are used to provide a distinct representation of concepts derived from unstructured or structured datasets. In particular, the case study's knowledge extraction methods rely on annotating concepts extracted from clinical and scientific methods.

Knowledge-driven framework for big data: A knowledge-driven framework takes enormous amounts of data in many formats, such as clinical images and scientific patterned structured data, and turns it into a knowledge graph from which unknown patterns and relationships can be identified. The framework is divided into four major sections: i) knowledge extraction, ii) knowledge graph development, iii) knowledge management and discovery, and iv) data access control and privacy. A knowledge graph can combine and describe various data sources.

Knowledge extraction: This component applies data mining methods to unstructured data sources, such as clinical photographs and scientific organized information. Processes employ ontologies to express extracted concepts using standardized terminology from a wide range of data sources.

Knowledge graph development: This component receives annotated datasets created during knowledge representation extraction and creates a knowledge graph based on the assessment of RML mapping rules, thus allowing annotated data in the knowledge graph to be turned into RDF triples. A knowledge graph is produced by employing a single schema to describe the entities semantically. The semantic goal of annotations is to determine the relationships between entities and knowledge representation graphs, in addition to a duplicate for inconsistency detection. Entity approaches are used to connect analogous knowledge networks by linking RDF entities to knowledge graph policies, ontological axioms, and annotation datasets.

Knowledge management and discovery: The exploration of the knowledge graph for new relationships or patterns between entities is possible with the knowledge management component. Once created, the knowledge graph can be explored, and the federated queries can be used as input for data analytics or knowledge discovery tasks, as well as patterns among entities on a knowledge graph that can be discovered.

Data access control and data privacy enforcement: This component provides the definition of access policies, which specify the actions that can be performed on the data knowledge graph.

3.4.4 Security and privacy concerns of semantic web of IoMT

IoT has several security risks that have harmed authentication systems for responding to emergencies for monitoring and improving IoT security properties.

Confidentiality: Services for data protection in the SWoT and data from IoT smart health industry connected gadgets should be kept confidential. Several existing asymmetric and symmetric encryption techniques are guaranteed to work. A healthcare setting in which patient activity data are stored at multiple hospitals should be considered.

Integrity: Critical data are exchanged between IoT services and other services and third parties, and these services strictly enforce the requirement that data be stored, sensed, and transferred without being intentionally harmed. The integrity of IoT sensor data must be protected to build a dependable system.

Availability: Sensor node-hosted services may be included in a SWT ecosystem. To provide semantically based data, services must be accessible from anywhere at any time. No security procedure meets this requirement. Instead, a variety of pragmatic approaches is used to ensure availability.

Authentication: Identity verification must be applied in terms of required authentication in actuating processes, and the customer service provider that needs to ensure authenticated service by the user is to provide any authentication process that must be registered to identify user resource constraints. The purpose of SWoT object restriction is to enable authentication techniques.

Authorization: SWoT environment necessitates the employment of policies outlining mechanisms that are re-usable, fine-grained, dynamic, and easy to update. Externalizing the policy and enforcement process of SWoT services is critical.

Access control: It is a mechanism for ensuring that only authorized users have access, and such access is frequently enforced based on the consequences of access control. It is extremely sensitive information that should only be shared with those who have been permitted to view it.

3.5 Implementation of Knowledge-driven Framework in TIMER

3.5.1 Temporal information modeling, extraction, and reasoning (TIMER)

The semantic data routing flow for TIMER for healthcare interoperability of IoT devices is shown in Figure 3.4. According to the researchers, splitting a

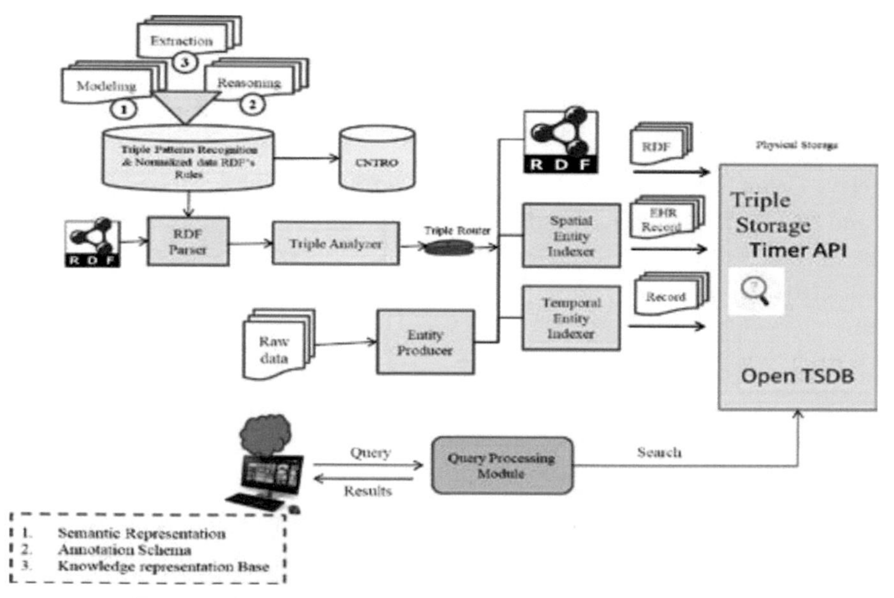

Figure 3.4 Semantic data routing flow for TIMER application.

large RDF graph into smaller subgraphs to store many processing nodes could be efficiently parallelized. The physical design of RDF storage triples may minimize the incoming data from the streaming graph data bus in the router to data routing flow after fetching. The activities are designed as asynchronous jobs that are scheduled in clusters.

The triple analyzer receives streaming RDF data and processes them using the client's medical items to define the triple-pattern recognition rules. Each triple-pattern recognition rule is composed of many triple patterns that specify how to extract values from streaming data. The retrieved values' attributes and graph patterns will be used to index them. Stream subgraphs with geographic context go to the clustered index, whereas time series with numeric values go to regular triple storage.

3.5.2 Clinical narrative temporal relation ontology (CNTRO)

CNTRO is a real-time search use case to encourage the building of a medical graph of things. For example, Patient 1 is enrolled online at a given hospital on a specified date and time. His or her medical information is sent to a specific physician automatically. Patient 1's wristband and implanted IoMT devices send the information automatically through SMS.

The patient's waiting time is greatly shortened, and the GPS wristwatch tracks his or her whereabouts. When an emergency is recognized, an alarm is sent automatically to the physician. Self-diagnosis can be detected and tracked automatically by smart U-IoMT devices. SSN clinical narrative temporal relationship ontology-based stream data subgraph is shown in Figure 3.5.

3.5.3 Semantic ontology-driven translator

The semantic process based on an ontology-driven translator is shown in Figure 3.6. A clinically focused organized ontology drives the behavior of the ontology-driven modeling system. This ontology's main purpose is to develop a model that encompasses (1) illness associations and (2) the relationship between the diagnosis and relevant observations. (1) The ontology is enhanced with links from concepts included in it to the data recorded in the necessary information to provide value to researchers and physicians.

The data comprise the knowledge needed to identify patients with EHR records, the knowledge representation needed to select the clinical data items that are most relevant, and the knowledge needed to locate these data-gathering patients. The taxonomic explosion can be used by the system to aid in the creation of queries for finding patient groups. The system will traverse the hierarchy and bind all the relevant codes into a query that will be executed against a target. The results will be compiled into a list of patients with non-EHR records.

3.5.4 Semantic knowledge representation ontology

The healthcare entity relationship serves as the proposed framework's high-level graph representation of entity knowledge. Each instance can be described using the annotation and a set of quantitative attributes. Description logics are a group of logistical languages that have been selected as the framework's reference formalism, which assumes that the reader is familiar with the fundamentals of ontology web language (OWL) domain things whose semantics map the attribute language with constraints proposed by the formal concepts of medical domain, namely, patient medical records and pharmaceutical annotation records.

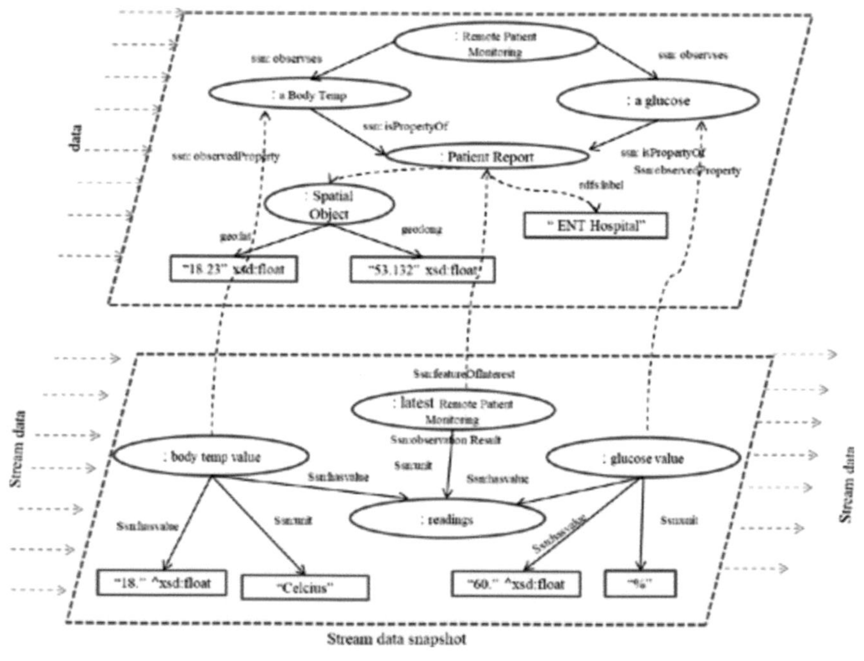

Figure 3.5 Stream data subgraph based on SSN clinical narrative temporal relation ontology.

- Medical record ontology (T).
- Anatomy: The study of bodily structures and systems; the classification of healthcare therapy classes, including side effects and interactions with other treatment classes.
- Disease classification, including which body structures are affected, specific measures, and usual therapy classes.

3.5.5 Connectivity management semantics ontology (CMTS)

The block diagram of CMTS is shown in Figure 3.7. The semantic ontology's connectivity management tool aims to provide tools for monitoring physical infrastructures using three common types of IoT devices: gateways, nodes, and sensor smart devices. The ontology is based on two-way concepts, specifically the monitored physical area and the devices connected to the gateway, sensor nodes, and sensor devices. A sensor can be both hardware and software in nature, and the ontology can also specify the structure of gateways and nodes. The proposed health ontology is represented by a UML class diagram

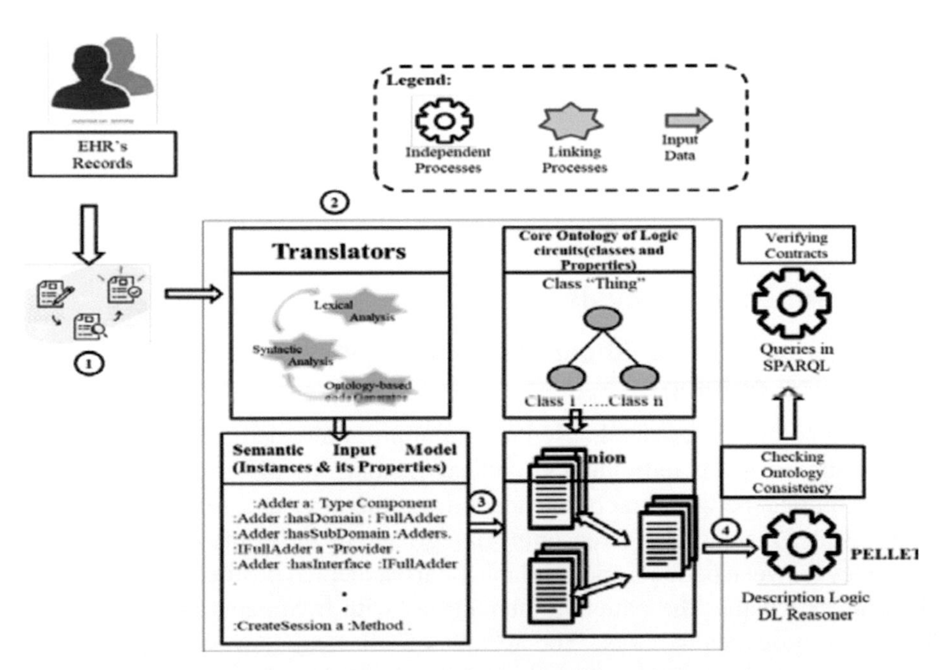

Figure 3.6 Semantic process using an ontology-driven translator.

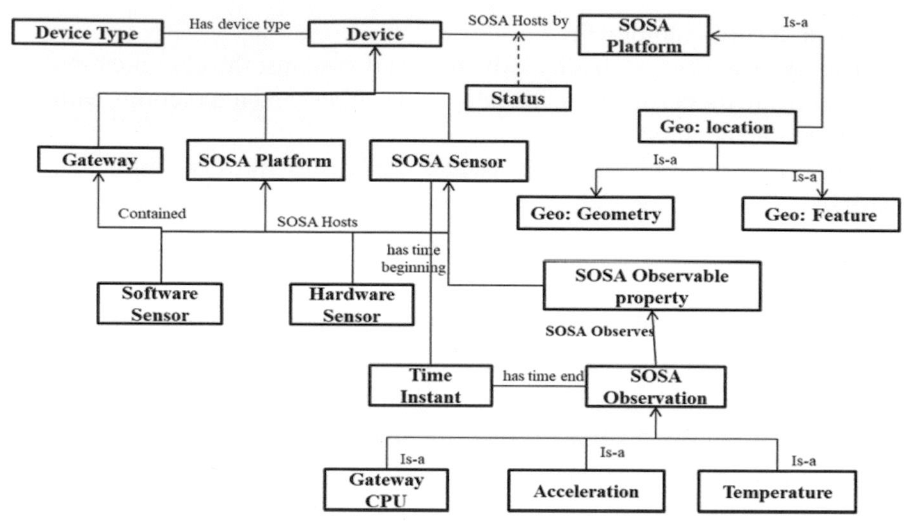

Figure 3.7 CMTS healthcare ontology.

stating the concept of the health domain ontology that is provided to offer IoT health sensing connection management across numerous devices and settings to be monitored. An ontology can describe the topology of gateways and nodes, and sensors can be both hardware and software. A UML class diagram represents the suggested health ontology, which specifies the notion of the health domain ontology that is described to offer the aim of connectivity management in IoT health sensing across multiple devices and settings to be monitored.Fig 3.8 represents the classes of healthcare management ontology.

3.6 Experimental Analysis

3.6.1 Reasoning for healthcare context-rule-based decision support ontology

The non-monotonic inference is used to find appropriate medical treatments that are tailored to an individual's medical history. To discover potential incompatibilities between recommended drug therapy and to assess the patient's state, matchmaking reasoning integrated in the semantic web of consistency between patients and available pharmaceutical profiles is applied. It determines whether the patient's interactions with treatments are already effective or pose health risks. The results of all cases are subsequently described.

The ultimate goal of the proposed approach is to determine how appropriate a specific medicine is for the patient's ailment. Doing so can provide practitioners with therapeutic decision support. To determine this aspect, pharmaceutical classes relevant to the condition's treatment are added to the disease annotation. Medication treatments are used to describe disease concepts by referring to bodily organs and systems. As a result, the given pharmaceutical exhibits negative effects of the concept abduction. Semantic inconsistency is checked. Then, the concept contraction technique is utilized to find interactions between the new therapies.

Rule-1: Cortisone's semantic annotation contains the following potential side effects on the eyes, bones, and cardiovascular system.

Cortisone: Corticosteroid affects the eye, bone, and cardiovascular system
S. P has a blood disorder, which, according to the ontology, can impact the circulatory and skeletal systems. Cortisone treatment of paraneoplastic dermatomycosis is requiring caution. When the doctor continues to check the patient's status in the system, this inference is made automatically by the system, which then sends a warning along with a semantic-based affinity level between the cortisone and the patient's profile.

Give up: Affects only cardiovascular system and bone

Keep: Corticosteroid, corticosteroid for therapy, and affects eye
Embedded reasoning calculates the penalty CA, including the patient's description based on the outcome of the inference. If a conflict exists, concept contraction is used to extract the conflicting request and analyze the context-specific parameters of the combination function:

- The patient's age, the frequency of pharmaceutical side effects, and the WHO ICF degree of disease-related impairment in the patient.

Rule-2: The second rule for context specifies how the proposed system resolves conflicts between therapy and a patient who has previously been profiled for patient details and has records that provide information about the medicine.
C. C: Mild SLE and musculoskeletal system disease and (serious effect on only beta blocker) and (reduced by only beta blocker) and (increases only beta blocker) and (increased by only beta blocker).

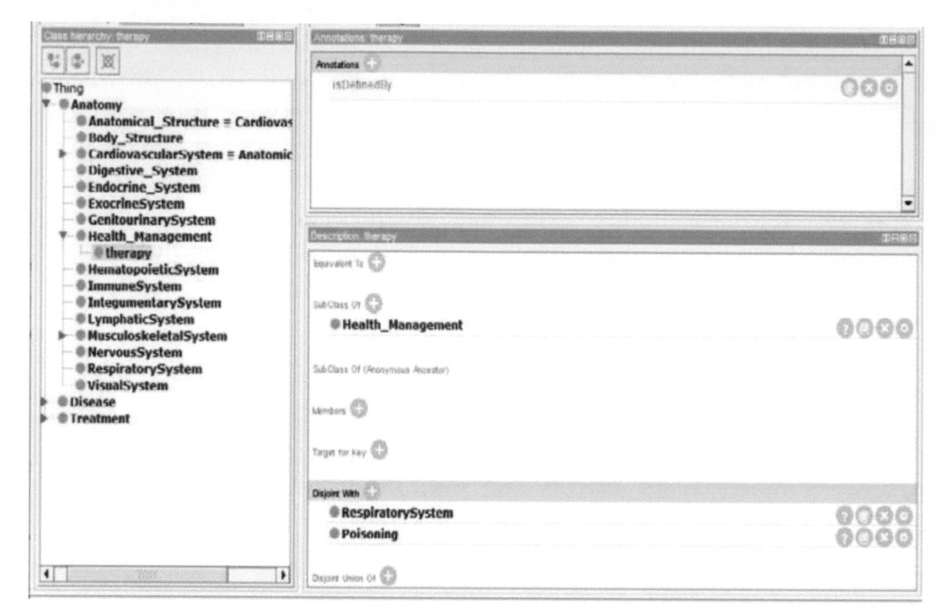

Figure 3.8 Inferred classes with healthcare management ontology.

3.6.2 Evaluation of ontology modeling for IoMT services

Data that have been detected and processed are referred to as low-level context or preliminary context. The low-level context solely pertains to sensed data in this instance. Using the context model and inference rules, this low-level context should be transformed into a high-level context before being used in healthcare services. As a result, the purpose of context reasoning is to use a context ontology model and rules to transform low-level reasoning into high-level logic. To deliver context-aware healthcare services, the ontology's axiomatic semantics-based inference and domain-specific rules are applied. The healthcare service rule component contains the domain-specific rule. According to the healthcare service regulation, personalized healthcare is defined as recognizing the user's physiological situation.

The healthcare service rule, which is dynamically established, deleted, and updated when the user's status changes, also describes the general rule. The OWL ontology and healthcare instance are processed so that they can be loaded into working memory. The instance of ontology is parsed and translated to a triple format, which is a first-order logic form. It is the changing knowledge base in the rule component of the healthcare service. The healthcare service rule is made up of high-level context rules

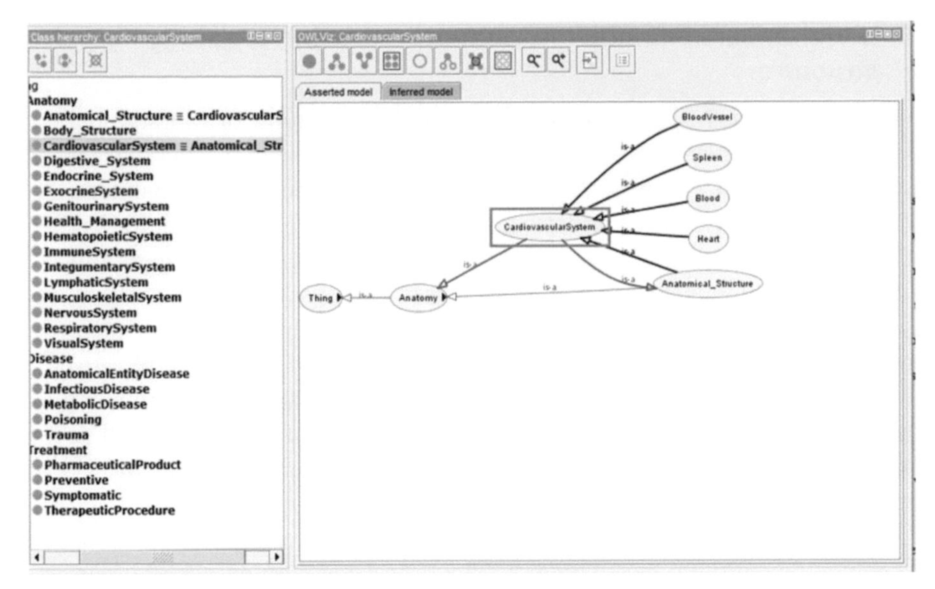

Figure 3.9 An ontology structure of semantic information and reasoning for healthcare services.

that cause context-aware service rules to be triggered. Figure 3.9 depicts the hierarchy and relationship of health data in the cardiovascular system setting. The cardiovascular system context, such as blood vessels, spleen, heart, and anatomical structure, is shown in relation to the subclass of health data context. The same relationship can interpret any context providers who can observe the cardiovascular system. Patient context is a subclass of person context in our context model, and patient context has health data context.

Similarly, Figure 3.10 describes the ontology representation of the semantic information and reasoning of diseases and their relations. The disease system context, such as respiratory, urinary, digestive, and cardiovascular systems, with the subclass of the health data context is presented in relation to one another. The same relationship can interpret any context that provides the observation of various systems. Low-level context is converted into high-level context by the high-level context generating rule. To capture the links between methodological and topological aspects, high-level context rule generation is required. Relationships between ontology properties, as well as relationships to other domain properties, must be captured by rules. Some of these rules can be viewed and represented in SWRL.

3.6.3 Hierarchical semantic information modeling ontology structure

Figure 3.11 depicts the ontology structure of the modeling of hierarchical semantic information with Protégé for medical objects. The two super classes depicted in Figure 3.11 are anatomical structure and health management. The cardiovascular system is a data superclass that includes physiological health data. Physiological data are shown in the raw data format in location-based services. The physiological context offers the raw data context. When an application requests physiological raw data, the context-aware system retrieves the raw data as well as temporal information from the physiological raw data context. The identification of physiological data is part of the physiological data context and various types of health data. The rule engine infers the status of the physical data and assigns it as safe or dangerous.

Data that have been detected and processed are referred to as low-level context or preliminary context. The low-level context solely pertains to sensed data in this instance. This low-level context is converted into a high-level context for use in location-based identification services using the context model and interference rules. Context reasoning turns low-level context into high-level context using an ontology model

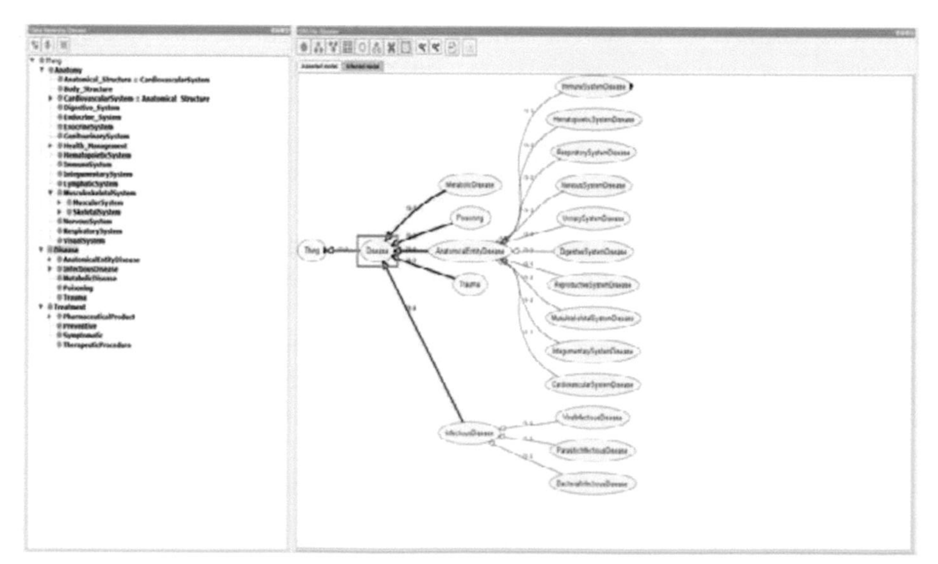

Figure 3.10 An ontology representation of semantic information and reasoning for diseases.

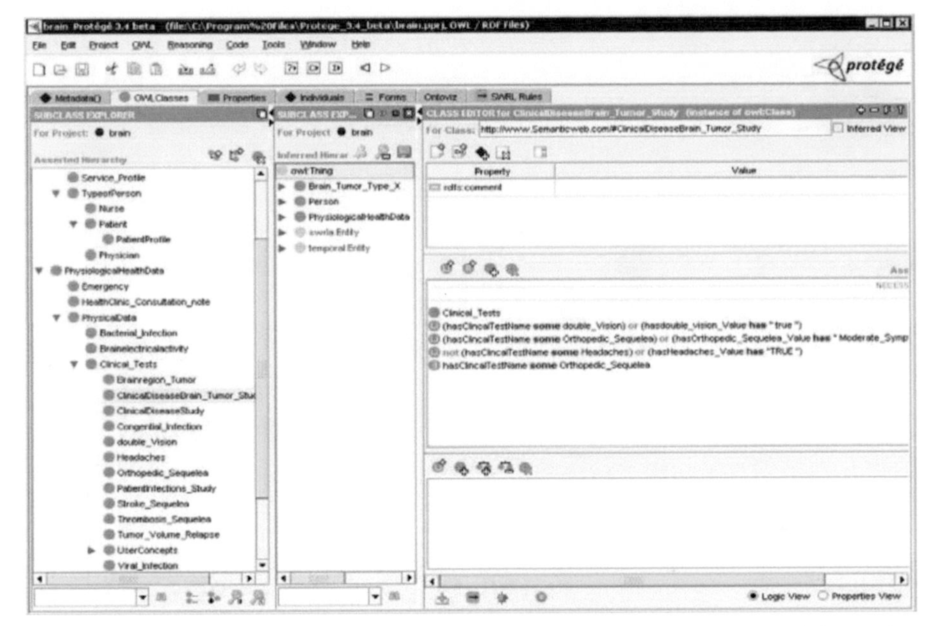

Figure 3.11 Ontology structure of hierarchical semantic information modeling using Protégé.

and applied rules. Ontology-based context-aware healthcare services are delivered via axiomatic semantics-based inference and domain-specific rules.

3.7 Semantic Industry for Applications

3.7.1 Applications of smart health semantic industry

The IoT is a network connection of sensors, software, and other technology embedded in physical things that enable them to share resources with other devices over the internet. IoT is a concept that uses the power of the internet to connect several things, including computing devices, to achieve data communication among people, processes, and things. Sensors and other smart devices are emerging as a result of e-health content on other smart devices. The architecture of smart semantic IoT-based health risks used in a hospital setting is depicted in Figure 3.12. When a patient's ID card is scanned, a secure cloud is automatically connected to store their electronic health record vitals and medical lab results from their prescription histories.

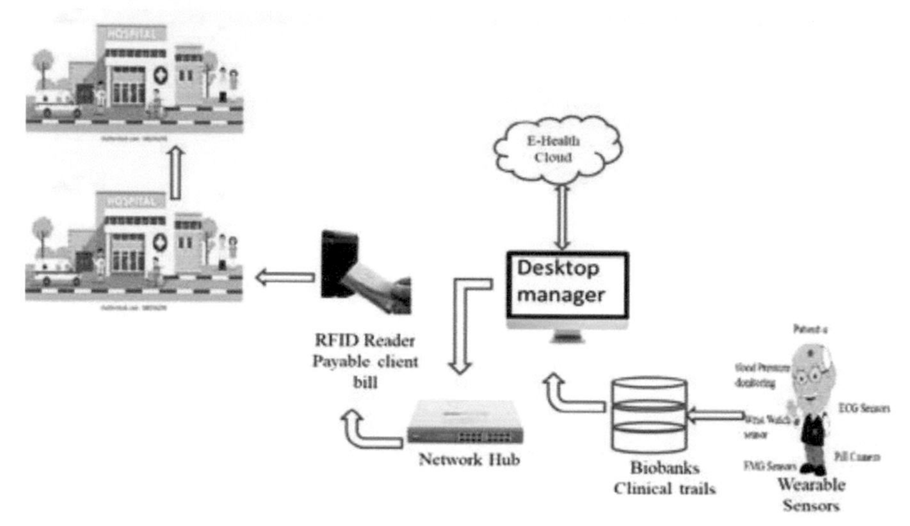

Figure 3.12 Architecture of smart semantic IoT-based health industry.

Cost reduction: IoT enables real-time patient monitoring. It significantly reduces unnecessary human–human interactions and hospital queues.

Drugs and equipment management: Drug and medical equipment are used in the healthcare business, where linked devices are controlled efficiently and at a minimal cost.

Error reduction: Data created by IoT health devices not only aids in better decision making but also assures seamless healthcare operations with deceased errors, waste, and system expenses.

Improved treatment: It allows doctors to make educated judgments based on evidence, improves treatment outcomes, and ensures complete transparency. Continuous patient tracking and real-time data assist in the early diagnosis of diseases caused by symptoms associated with the COVID-19 pandemic.

Proactive treatment: Continuous remote patient health monitoring for proactive medical diagnosis and increased hospital patient engagement. Many smart health platforms are being developed to provide smart innovative health services to various patient categories based on the concept of smart IoT healthcare and AI. The use of smart-IoT-based mobile health devices is also improving telemedicine. Some emerging healthcare IoT-based technology devices, such as Bluetooth-enabled devices with smartphone apps, can record patient data for statistical analysis and sharing. Health workers and physically challenged patients benefit from IoT-based devices that improve the provision of data that can be used in patient surveillance to reduce the frequency of hospital visits.

3.7.2 Semantic web technologies in e-healthcare

SWTs aid in the annotation of data using common healthcare and supply chain management ontologies, which then interlink the data on the web and enable effective and easy query of a web of things and knowledge representation of facilities with easy access to medical data on the web of things. In recent years, the medical industry has focused on the SWT to provide efficient complex management and distributed health data to achieve standard interoperability and the implementation of web-based knowledge representation of SPARQL queries.

SWTs have the potential to improve the storage and sharing of medical graphical representations of knowledge and skills, thereby promoting health-related interoperability in the fundamental building blocks in universal resource identification (URI), extensible markup language (XML), resource description framework (RDF), RDFS-RDFS schema, SPARQL protocol, RDF query language, and ontology web language (OWL). The architecture of semantic sensor e-health applications is depicted in Figure 3.13. SWTs are altering medical research and are being used in more general situations. Ontologies serve as both a conceptual foundation for information exchange and standards for exchanging data between different systems. Ontologies can help with the discovery of context-based medical research information for future research, as well as the development of context-based rules for appointments, procedures, and tests (RDF). It is an object-oriented XML-based standard for describing concepts and creating documents in semantic web sensors. IoT health-based sensor services and remote monitoring will be provided to end-users through collaboration among entities, thus making them more comfortable and efficient. With semantic techniques, clinical data modeling and integration can be accomplished by extracting and querying from various sources of ontologies and data mapping to build a semantic health data model. Moreover, IoMT developed a web-based prototype that enabled interoperability between various formats.

SWTs have pioneered a novel method for creating and managing data-based semantic metadata, which frequently provides semantic descriptions of health data. Semantic encoders could be used to improve efficiency in healthcare workflows, such as complex event processing, in which a sequence

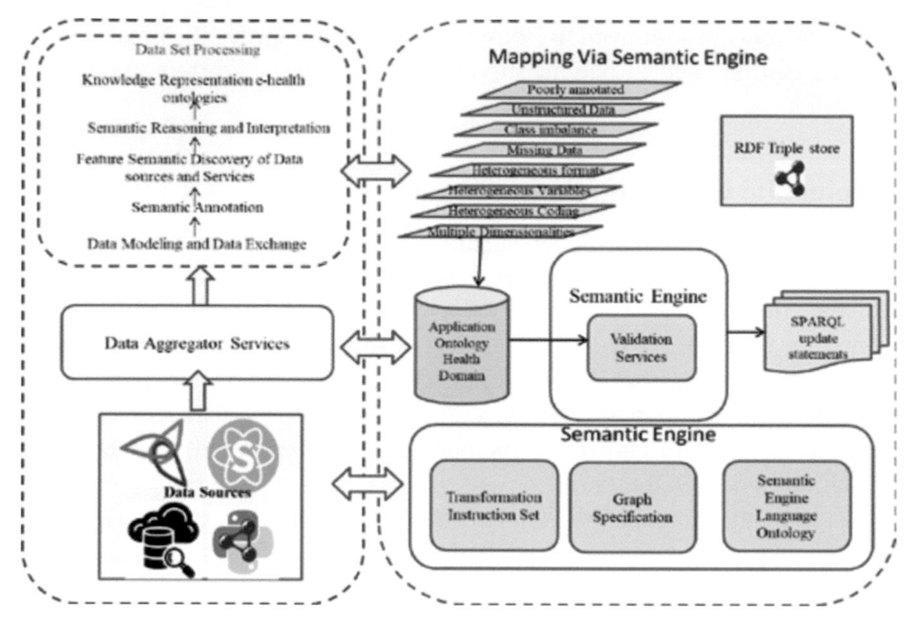

Figure 3.13 Architecture of semantic sensor e-health applications.

of data is analyzed in real time in search of events using complex event processors, common data models, and machine learning algorithms. Many communities use semantic web technologies in healthcare and smart medical devices, and they have access to personalized healthcare. Both IoT and SWT enable secure connectivity between patients and healthcare providers through seamless and secure connectivity between patients, clinics, and healthcare. Semantic web functions will continue to play an increasingly important role in enabling humans and machines to collaborate in the health sector, as well as in improving the semantics and quality of health information on the web.

Data modeling and data exchange: One of the major issues in semantic interoperability is the standardized data modeling and representation in smart e-health IoT industry environments. A semantic model describing the meaning of occurrences can be used for the data model. Interoperability between various systems and applications in IoT medical care devices is enabled by semantic modeling.

Semantic annotation: Another challenge is providing an exchange of heterogeneous data from IoT health semantic sensor descriptions. One way to support the integration is to improve the interoperability and the data integration in the knowledge representation of the health domain graphically.

Semantic discovery sources of data and services: This information is annotated to the IoT environment of health supply chain management to aid in the discovery and support of resources, entities, and services. RDF schema ontology resource-constrained IoT smart health devices are used for data generation and services.

Ontologies for knowledge representation and health: In IoMT, ontologies are used in health apps to express knowledge and share health data. When different systems communicating with each other use different knowledge representations, ontologies have been utilized to address important interoperability health challenges. Semantic web annotations are necessary to represent knowledge-structured information models. Supply chain management principles and components are described using IoT frameworks.

3.7.3 IoT e-health ontologies framework

The most important component of the semantic web is ontology, which serves as a repository for information and knowledge about things. Simple semantic annotation of raw IoT health data is aided by ontologies. Adding an abstraction layer and building an OWL ontology with class, property, and individual descriptions also ensure compatibility. Several ontological models have been proposed within the framework of the semantic web of things.

Semantic sensor network ontology: It was created by the W3C Semantic Sensor Network Incubator and describes sensors and their observations. SSN can support a wide range of technologies and use cases, including several ontologies that have been extended to create a semantic sensor ontology.

Wireless semantic sensor network (WSSN): It was proposed as an addition to the SSN ontology. The wireless sensor network concept provides communication and data transformation for sensing streams.

Sensor core ontology (SCO): It introduces fundamental classes in the context space, time, and theme by extending the SSN ontology, which includes modules such as the component module, service module, and context module.

M3 ontology: It defines the concept of the SSN ontology. Metadata such as observation, measurement units, and application domain are examples of metadata. Ontologies and rules are used to describe sensor data and the main feature of the M3 ontology on its ability to link domain e-health to derive knowledge from sensor data.

Onto sensor ontology: It expands on the SensorML concept by identifying sensors, their behaviors, and relationships, as well as new concept characteristics. It enables data discovery and processing when sensor observations cannot be properly described.

Commonwealth scientific and industrial research and organization: It is a general ontology that describes sensor capabilities, properties, functionality, physical qualities, and metrological aspects.

One M2M ontology: The term semantic interoperability refers to a generic ontology that represents the capacities of various IoT e-healthcare technologies.

IoT health ontology: It describes the representation of device knowledge. Intelligent, physical, control, electronic gadgets, and networks are all examples of entities.

IoT-O ontology: It is designed to be expanded to satisfy unique application requirements as well as real-world devices and services. It is intended to represent information about IoT systems and applications.

Smart onto sensor: It is an ontology designed for smartphone applications. The goal of the SOS developing ontological model is to formalize the conceptualization of smartphone resources and sensors, including their categories and applications.

3.7.4 Semantic interoperability in IoT applications

Ontology web language for services (OWL-S): OWL-S is an abbreviation for ontology web language for services for health services and its framework, which provides a language for specifying a semantic web operation, task, or service in detailed process specifications by expressing semantic definitions and terms associated with sub-ontologies and their profile process model.

Onto sensor: It includes a sensor descriptive specification model as well as a semantic sensor describing model, an extensive knowledge model, enhanced data representation for sensor observations, and a descriptive model for measurement data.

SEEK extensible observation ontology (OBOE): It specifies the measurement kinds, the context, and the type of thing being measured. The ontology's drawback is that it focuses more on sensor results than IoT feature items. Extensions are necessary to incorporate sensor descriptions from various ontologies.

Sensor data ontology: Sensor data ontology is an ontology that, in addition to SensorML standards, focuses on observations and measurements. It employs a gateway component to collect sensor observation data and analyze it in the gateway node using ontology-based descriptions.

SemSOS O&M-OWL ontology: Sensor observations and measurements are described using the OWL ontology model. It visualizes and annotates a stream of sensor data linkages with semantic annotations. The e-health ontology encompasses both observations and e-health IoT concepts.

CSIRO sensor ontology: Sensors are semantically described using the CSIRO sensor ontology in terms of platform and specialized actions. Sensor measurement ideas, as well as CSIRO concepts for domain knowledge, units of measurement, and location, are missing from the ontology.

IoT-Lite: IoT-lite is a straightforward paradigm for explaining essential Internet of Things (IoT) e-healthcare ideas. It is based on the creation of ontology for a semantic sensor network (SSN). In heterogeneous IoT e-health platforms, the lightweight semantics ontology is employed to facilitate sensor data interchange and discovery. IoT-Lite, on the other hand, merely covers the sensory data, location, and type, which are examples of data analytics foundations in IoT applications. It also outlines medical device processes and storage sites.

Open IoT middleware platform: The open IoT middleware platform is a free and open-source IoT health platform that allows IoT services in the cloud to communicate semantically. Open IoT connects sensor data to the cloud environment.

Methodology of systematic mapping study: A systematic mapping study identifies, evaluates, and interprets relevant research on a specific research phenomenon. The outcomes of controlled processes and formal research are assumed to benefit the community of interest, gaps, and challenges. We also performed procedures on IoT supply chain management to identify the portions considered in our research for a systematic mapping study to perform the planning, conducting, and reporting of activities. The primary goal is to outline the procedures for planning, organizing, and carrying out research. In our work, the conducting activity is carried out in primary studies using digital e-health content.

3.8 Limitations and Challenges

Privacy critics of IoT have expressed concerns about the privacy of patients or hospitals, as the technology frequently shares extreme details of data without the user's knowledge.

Complexity: IoT health systems and devices often appear highly difficult to implement and maintain due to the utilization of many technologies and a huge number of new enabling technology applications and deployment.

Compliance: IoT in healthcare is being deployed in the absence of medical ethics and regulations. The complexity of the healthcare sector, as well as its linkage with other enabling supply chain management technologies, causes more difficulty for users to meet both professional and software compliance standards.

Security: Security is still a major concern in many technologies, and IoT creates an ecosystem in which connected health smart devices communicate over a network to control any security measures to protect medical data from cyber threats and malware clinical attacks.

Flexibility: Stakeholders in the health sector are concerned about an IoT system's ability to integrate easily with other ecosystems.

Network issues: IoT health domain sectors are internet based. They require systems or users to connect to reliable networks to improve their operation and performance. The adoption of IoT healthcare devices may be difficult in rural areas or developing countries.

Power supply issues: Most of the time, the use of smart IoT health devices in healthcare systems necessitates developing countries or rural areas.

3.9 Conclusions and Future Enhancements

In this chapter, we present an SWT interoperability system that is being built, as well as knowledge IoT for medical devices that are linked to patients. It offers a unified graph-based view of data generated by connected medical devices, as well as the comprehension of the processing data from sensors and the world around physical sensor things, such as sensor reading, sensor observation, sensor context, and real-time relationships among the semantic web of medical things. As such, we implemented secured algorithms on the context-structure-based recommendation features. This analysis was critical in developing a federated and unified semantic engine for the medical IoTs and smart e-healthcare applications. The main research challenges were highlighted, and three use cases were discussed: machine-to-machine (M2M), TIMER application, and clinical narrative temporal relation ontology. IoMT-O techniques benefit from the heterogeneity and interoperability provided by semantic web technologies. SSN links to a range of sensors automatically to access a variety of applications, including medicare, electronic health records, fast healthcare interoperability resources, and security issues. Semantic web technologies integrate, process, and manage a variety of sensors, thus enabling the resolution of technological challenges while focusing on the concept. Health ontology requires the specification and scope of the implementation, as well as the demands of the applications. The hierarchical ontology structure was first used in the Protégé ontology-building environment.

References

[1] Qu, C., Tao., M., Liu., F and Yuan R. F., (2016) Design and application of topontologies for the transactions on IoT, International Journal of Embedded Systems. 8(2-3), 155-163.

[2] Kuziemsky, C. E., Archer, N., and Peyton, L., (2009) Towards E-Health interoperability: challenges, perspectives, and solutions, Journal of Emerging Technologies in web Intelligence. 1(2), 107-109.

[3] Henson, C., Sheth, A., and Thirunarayan, K., (2012) Semantic perception: Converting sensory observations to abstractions. 007AInternet Computing, IEEE, 16(2); 26–34.

[4] Compton, M., Barnaghi, P., Bermudez, L., Garcia-Castro, R., Corcho, O., Cox, S., Graybeal, J., Hauswirth M., Henson C., Herzog A. (2012) The SSN ontology of the w3c semantic sensor network incubator group, Web Semantics: Science, Services, and agents on the World Wide Web.

[5] Guntur, S. R., Gorrepati, R. R., Dirisala, V. R., (2018) Internet of medical things remote healthcare and health monitoring perspective. Medical

Big Data and Internet of Medical Things: Advances, Challenges, and Applications, Chap. 11. CRC Press Taylor & Francis Group, Boca Raton.

[6] Guntur, SR., Gorrepati, R. R., Dirisala, V. R., (2019) Robotics in healthcare: An Internet of Medical Robotic Things (IoMRT) perspective. Machine Learning in biosignal analysis and Diagnosis Imaging, Chap. 12. Elsevier, Amsterdam.

[7] Onyema, E. M., Shukla, P. K., Dalal, S., Mathur, M., Zakariah, M., Tiwari, B. (2021) Enhancement of patient facial recognition through Deep learning algorithm: ConvNet. Hindawi Journal of Healthcare and Engineering, 38; 154-166, 2021.

[8] Ray P. P. ,(2014) Home Health Hub Internet of Things (H3IoT): An architectural framework for monitoring health of elderly people. IEEE- 32331 International Conference on Science, Engineering and Management Research 3–5.

[9] Eero, H., Kim, V., Osma, S., (2008) Health Finland Finnish Health Information on the Semantic Web, Semantic computing research group.

[10] Ali, D., and Samina, RA., (2009) Computerizing clinical pathways ontology-based modelling and execution, Medical informatics in a united and healthy Europe, Ios press. Uk.

[11] Verma, P., Sood, S. K., and Kalra, S., (2017) Cloud-centric IoT based student healthcare monitoring framework. Journal of Ambient Intelligence and Humanized Computing, 1–17.

[12] Pathak, J., Kiefer, R., Chute, C., (2012) Using Semantic Web Technologies for Cohort Identification from Electronic Health Records for Clinical Research.

[13] Nguyen, D. C., Ding, M., Pathirana, P. N., Seneviratne, A., Li, J., and Poot, H. V., Federated Learning for Internet of Things: A Comprehensive Survey. IEEE.

[14] Eid, M., Liscano, R., et al.: (2007) A universal ontology for sensor networks data. In: IEEE International Conference on Computational Intelligence for Measurement Systems and Applications, 59–62.

[15] Anowicz, K., Haller, A., et al.: (2019) Sosa: A lightweight ontology for sensors, observations, samples, and actuators. Journal of Web Semantics 56, 1–10.

[16] Gyrard, A., (2013) A machine-to-machine architecture to merge semantic sensor measurements, in: Proceedings of the 22nd International Conference on World Wide Web, 371–376.

[17] Lee, K., Lee, J., Kwan, M. P., (2017) Location-based service using ontology-based semantic queries: A study with a focus on indoor activities in a university context. Computers, Environment and Urban Systems 62, 41–52.

[18] Sudhana, K. M., Raj, V. C., Suresh, R., (2013) An ontology-based framework for context-aware adaptive e-learning system, in: International Conference on Computer Communication and Informatics, IEEE. 1–6.

[19] W3C OWL Working Group. OWL 2 Web Ontology Language Document Overview (Second Edition). W3C recommendation, W3C. 2012.

[20] Antunes, R. S, Seewald, L. A, Rodrigues, V. F, Costa, C. A. D, Righi, R. R, Maier, A., Eskofier B, Ollenschläger, M., Naderi, F., Fahrig, R., et al. (2018) A survey of sensors in healthcare workflow monitoring. ACM Comput. Surv. (CSUR) 51.

[21] Skillen, K. L., Chen, L., Nugent, C. D., Donnelly, M. P., Burns, W., Solheim, I., (2014) Ontological user modelling and semantic rule-based reasoning for personalization of Help-On-Demand services in pervasive environments. Future Gener. Comput. Syst, 34: 97–109.

[22] W3C SPARQL Working Group. SPARQL 1.1 Overview. W3C Recommendation, W3C. 2013. Available online: https://www.w3.org/TR/sparql11-overview.

[23] Ruta, M., Scioscia, F., Noia, T., Sciascio, E., (2009) RFID-based Semantic-enhanced Ubiquitous Decision Support System for Healthcare. In Proceedings of the 3rd International Workshop on RFID Technology. Concepts, Applications, Challenges, Milan, Italy,79–88.

[24] Horrocks, I., Patel-Schneider, P., Boley, H., Tabet, S., Grosof, B., Dean, M., (2004) A Semantic Web Rule Language Combining OWL and Rule ML. W3C Member Submission, W3C. Available online: https://www.w3.org/Submission/SWRL.

[25] Parsia, B., Motik, B., Patel-Schneider, P., (2012) OWL 2 Web Ontology Language Structural Specification and Functional-Style Syntax (Second Edition). W3C Recommendation, W3C. Available online: http://www.w3.org/TR/owl2-syntax.

[26] Horridge, M., Patel-Schneider, P., (2012) OWL 2 Web Ontology Language Manchester Syntax (Second Edition). W3C note, W3C. http://www.w3.org/TR/owl2-manchester-syntax.

[27] Motik, B., Patel-Schneider, P., (2012) OWL 2 Web Ontology Language Mapping to RDF Graphs (Second Edition). W3C Recommendation, W3C.

[28] El-Sappagh, S., Franda, F., Ali, F., Kwak, K. S., (2018) SNOMED CT standard ontology based on the ontology for general medical science. BMC Med. Inform. Decis. Mak. 18, 76.

[29] Scheuermann, R. H., Ceusters, W., Smith, B., (2009) Toward an ontological treatment of disease and diagnosis. Summit Transl. Bioinform, 116–120.

[30] Gyrard, A., Serrano, M., (2015) A unified semantic engine for internet of things and smart cities: From sensor data to Ebd-users applications. Conference Paper.

[31] Gyrard, A., and Bonnet, C., (2014) Semantic web best practices: Semantic web guidelines for domain knowledge interoperability to build the semantic web of things, 04.

[32] Compton, M, Barnaghi, P., Bermudez, L., Garcia-Castro, R., Corcho, O., Cox, S., Graybeal, J., Hauswirth, M., Henson, C., Herzog A *et al.*, (2012) The SSN ontology of the w3c semantic sensor network incubator group, Web Semantics: Science, Services and Agents on the World Wide Web, http://www.w3.org/2005/Incubator/ssn/ssnx/ssn.

[33] Ranganathan A. , AL-Muhtadi J., Chetan S., Campbell., (2004) MiddleWhere: A Middleware for Location Awareness in Ubiquitous Computing Applications.Lecture Notes in Computer Science, Springer.

[34] Horrocks I.,(2002) DAML+OIL: a reason-able web ontology language. Lecture Notes in Computer Science 2287, Springer.

[35] Henson, C., Sheth, A., and Thirunarayan, K., (2012) Semantic perception: Converting sensory observations to abstractions, Internet Computing, IEEE, 16(2); 26–34.

[36] Nguyen, D. C., Ding, M., Pathirana, P. N., Seneviratnem A., Li, J., Poor, H. V., (2021) Federated learning for internet of things: A comprehensive survey IEEE, 16.

[37] Lim, W. Y. B., Luong, N. C., Hoang, D. T., Jiao, Y., Liang, Y. C., Yang, Q., Niyato, D., Miao, C., (2020) Federated Learning in Mobile Edge Networks: A Comprehensive Survey, IEEE Communications Surveys & Tutorials, 1–1.

4

AI Compatible Key Hardware Design for Smart Warehouse: A Practical Implementation

Ngoc-Bich Le[1,2], Ngoc-Huan Le[3], Manh-Kha Kieu[4,5], Xuan-Hung Nguyen[3], Vu-Anh-Tram Nguyen[4], Tran-Thuy-Duong Ninh[4], Duc-Canh Nguyen[3], and Narayan C. Debnath[6]

[1]School of Biomedical Engineering, International University, Vietnam
[2]Vietnam National University Ho Chi Minh City, Vietnam
[3]Mechanical and Mechatronics Department, Eastern International University, Vietnam
[4]Becamex Business School, Eastern International University, Vietnam
[5]School of Business and Management, RMIT University, Vietnam
[6]School of Computing and Information Technology, Eastern International University, Vietnam
E-mail: lnbich@hcmiu.edu.vn; huan.le@eiu.edu.vn; hung.nguyenxuan@eiu.edu.vn; canh.nguyen@eiu.edu.vn; kha.kieu@eiu.edu.vn; tram.nguyen@eiu.edu.vn; duong.ninh@eiu.edu.vn; narayan.debnath@eiu.edu.vn

Abstract

This study conducted a hardware design approach of critical smart warehouse key components with the desire to best support digital transformation and AI implementation. Specifically, three important constituents related to the three main areas of a smart warehouse (i.e., mechanical, controller, and data collection algorithm), including telescopic fork mechanisms for AS/RS stacker cranes, multilayer industrial controller, and an effective algorithmic data collection solution was created and demonstrated. The strict requirements of pallet storage smart warehouse comprise (1) mechanical mechanism: high payload, high positioning accuracy, low taping deformation, high acceleration transmission, and low noise; (2) controller: industrial standard, cloud database communication, and storage, accurate inventory, stock keeping unit planning, inventory management, circulating conveyor system, and control of incoming and exiting items; (3) data collection: high capacity with 1000 pallets storage, pallet circulation, and management code reassignment, and continuous operation; and the desire for the compatibility and possibility of complementing IoT and AI add-ins in the future has been considered and guaranteed. The results show that the input design parameters and constraints are well captured and assured.

Keywords: Artificial intelligence, digital transformation, Industry 4.0, testbed, telescopic fork, smart warehouse, simulation modeling.

4.1 Introduction

Emerging markets in many countries have been experiencing fast changes in the last two decades [1, 2], in which warehouses play an important part in the food and agricultural product supply chain. In the world market, warehouse service management determines the success or failure of e-commerce companies. Through a survey in the Vietnamese market, most warehouses are built and managed using the traditional approach. As a result, typical warehouses have a number of drawbacks, including poor capacity planning, perishable products, operational inefficiencies, item overhandling, and ineffective material handling apparatus [3, 4].

The logistics business has expanded at a dizzying speed in several countries in recent years. As a result, warehouses and storage services play an important role in boosting logistics performance [4, 5]. Warehouse services are limited by simple storing, handling, value-adding, and piece-picking [6]. One of the difficulties and barriers to the growth of warehousing services is the ability to incorporate modern technologies such as artificial intelligence, machine learning, and others into warehouse operations.

Artificial intelligence (AI) techniques, such as machine learning or deep learning, have been successfully utilized in a variety of industries in recent years [7]. AI is being hailed as a game-changer in the manufacturing industry. Artificial intelligence (AI) has the ability to transform manufacturing operations in every way. At least one AI use application in industrial processes has been deployed by more than 50% of Europe's large corporations [8].

On the other hand, automated storage and retrieval systems (AS/RS) have been widely used to distribute and produce goods in the smart warehouse.

AS/RS typically includes racks and cranes that run through the middle rail between the racking structures to collect and stock goods. An AS/RS is capable of automatically handling pallets without operator intervention. AS/RSs are used to store products (e.g., raw materials or (semi-)finished products) and to retrieve those products from storage when required [9, 10].

The telescopic fork is the most important aspect of the mechanical framework. A static body and sliding parts make up the telescopic fork. The latter moves, thanks to racks and pinion mechanism or chains. Rollers, suspensions, actuators, transmitting mechanism, and position sensors are the essential components of the telescopic fork. Telescopic forks are generally classified as follows: (1) "single deep fork" includes two movable parts and one fixed body. These forks can handle loads stored in a single rack; (2) "double deep fork" comprises three movable parts and one fixed body. These forks can handle loads stored on two shelves; (3) "triple deep fork" includes four movable parts and one fixed body. These forks can handle loads stored on three line shelves.

ICS comprises a range of control systems that are widely used in industrial and critical infrastructure applications, such as supervisory control and data acquisition (SCADA) systems, distributed control systems (DCS), and various control system configurations, such as PLC [11]. An industrial control system (ICS) comprises a cluster of controllers (electrical, mechanical, hydraulic, and pneumatic) which work hard to meet a given industrial goal. The expected outcome or performance specification is included in the system's control. Control could be completely automated or have a human component. A typical ICS is made up of a variety of control loops, human–machine interfaces (HMIs), and remote diagnosis and maintenance tools, all of which are built utilizing a variety of network protocols. ICS is frequently used to manage industrial processes.

Figure 4.1 depicts the possible benefits of using AI in smart warehouses, such as reduced operating energy and logistical costs, as well as benefits from a system automation standpoint. Specifically, (1) reduce operating energy consumption: The energy used in warehouse operations has been a hot topic in green warehousing, attracting a slew of studies aimed at lowering the carbon footprint of the supply chain. Energy usage in warehouses is a complicated topic with numerous stages. Cost reduction, profit gain or marketing initiatives, and regulatory compliance difficulties all have a role in lowering warehouse operating energy use [12]. Furthermore, the optimization of energy consumption in warehouse operation is to reduce logistical costs and increase efficiency. Consequently, the application of AI to the operation

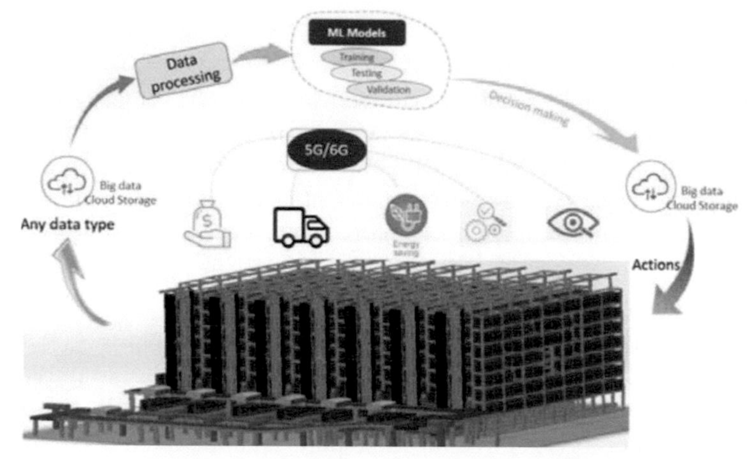

Figure 4.1 Benefits of AI implementation in smart warehouse system [4].

of the warehouse system will help save operating energy, thanks to optimal management and operation algorithms. Based on the previous operating parameter results, these algorithms will find the best parameters to save energy on their own. According to Amazon's statistics, thanks to Amazon's intelligent goods handling and operation system, which effectively combines machines and people, warehouse productivity has increased by 40%, and operating costs have been reduced to only less than 10% of revenue. This is a clear demonstration of the efficiency that smart warehouses bring to businesses. (2) Logistic cost reduction: According to the Vietnam Logistics Report 2019, outsourcing transportation services of enterprises in Vietnam is quite diverse, from 0% to 100%. The highest choice of warehouse leasing is up to 20% and the lowest of 3.6% for businesses, with the outsourcing proportion from 51%–75% and 75%–100%, respectively. On the other hand, warehousing and warehouse management costs range from 10% to 41% of a business's total logistics costs [13]. It is evident that smart warehouses, especially AI-powered ones to cut logistics costs, are very potent. Typically, traditional warehouses manage storage manually. A few other warehouses use semi-manual operation and management software. In both circumstances, the placement of commodities is based on some basic guidelines to keep the operation and management as simple as possible. The use of AI models to determine the best sorting solution, together with the ability to evaluate more inputs and boundary conditions, will considerably improve the optimization of the commodities arrangement process. (3) Advantages in terms of system automation: It is self-evident that automation is a prerequisite for AI implementation. As a result, the benefits of automation will be passed on to AI applications. Automation in the smart warehouse, in particular, provides various benefits to investors, including lower labor costs, faster pick-up and delivery times, fewer delivery failures, and improved overall management efficiency. Many technologies and techniques are currently applied in smart warehouse operation and management toward a smarter warehouse.

If one stays ahead of the data with AI, it may be easy to become the winner. This shows the massive importance of data in AI applications. It is observed that the areas that attract significant development of AI have excellent data platforms, such as healthcare, medicine, education, finance, natural language processing, etc. Consequently, Vietnam and other nations have established AI networks and communities in a variety of industries to boost community strength in data creation. The Vietnam–Australia Artificial Intelligence Cooperation Network (Vietnam–Australia AI) is the most current one, having been started in August 2021. These are very good starters. However, it is necessary to build networks and communities systematically and synchronously in specialized fields to be effective. The AI network and community for smart warehouses is one of them.

This study proposed and described in detail how a smart warehouse testbed works in practice. Specifically, a hardware design approach of critical smart warehouse key components with the desire to best support digital transformation and AI implementation were conducted.

4.2 System Description

This study developed a smart warehouse testbed, including a pallet circulation for the data collecting system (see Figure 4.2 (a)). The recommended system handles both storing and retrieval activities (i.e., transport of pallets containing objects from input stations to storage racks, and transport of pallets with items from storage racks to output stations). This system's automated guide vehicles (AGVs) may traverse between storage racks. As illustrated in Figure 4.2 (b), AGVs are stacker gantries with retractable prongs that perform storage and retrieval operations.

Pallets are circulated around the system deploying horizontal conveying structure to feed packages into the system, save packages as they exit, and circulate packages out of the system back to the entrance. Pallets are collected from the horizontal conveyor before being picked up by an AGV and sent to

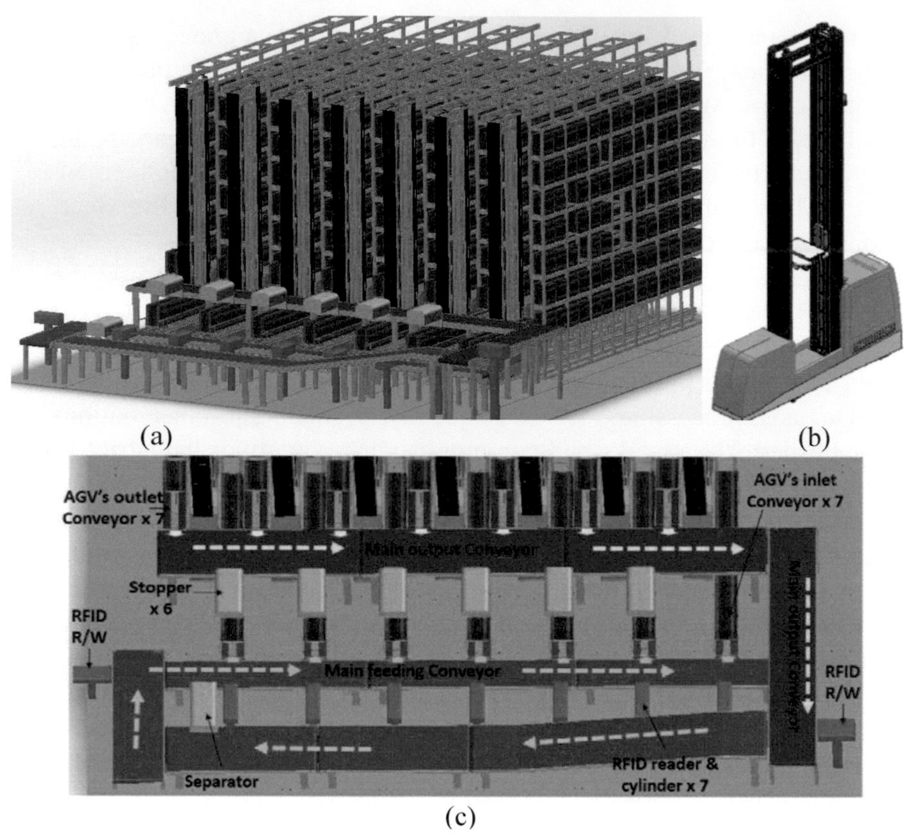

Figure 4.2 (a) Warehouse system. (b) AS/RS AGV with a telescopic fork. (c) Conveyor system.

storage. Two vertical conveying structure in front of each racking structure carry packages from the AGV to the horizontal conveying structure.

A sensing and levering station on the horizontal conveying structure regulate the smallest interval to stock packages into the storing area just before they reach the vertical conveyor position. In Figure 4.2 (c), an RFID station is installed on the horizontal conveying structure directly in opposite of individually vertical conveying structure, scanning the package's RFID information to define where it should be deposited.

The operation of the testbed can be summarized as follows: (1) The WMS software utilizes an optimizer to determine the best pallet placement when an order is received. After that, the RFID writer will assign management codes to the appropriate number of pallets (each pallet has one RFID tag) and put items on the horizontal conveying structure. (2) Packages are transported to the separating equipment and then to the storage area at regular intervals. After that, the RFID reader will identify the RFID tag on the pallet in front of each rack to decide which racking structure the package will be stocked in. The step is performed until all pallets in the warehouse have been stowed. (3) When the customer requests that the pallets be removed, the WMS software will perform an algorithm to determine the pallet removal sequence, and the AGVs will carry the packages to the vertical conveying structures. The pallets will next be transported from the vertical conveyors to the horizontal conveyor. The sensing and levering stations hold the traveling packages in place till the packages from the vertical conveying structure have entered the horizontal conveying structure, preventing a crash amongst them. The output package will then be stowed on the horizontal conveying structure till a subsequent order is received.

4.3 Key Hardware Design

4.3.1 Telescopic fork

4.3.1.1 First version approach

The first option approach, represented in Figure 4.3, uses a timing belt transmission coupled with a pulley cable transmission with one degree of freedom to meet the following technical requirements. The linear backward and forward motion of the intermediate link is actuated by a stepper motor through a timing belt transmission. The pulley cable transmission aids in moving coupling between the middle and upper links, allowing the middle and top links to reciprocate in the same direction. Only one drive source is required to achieve the requisite elongation in both sides of the two-stage stretching mechanism with this method.

Figure 4.4 presents the 3D design of the structure according to the first approach. The top, middle, and bottom links are interconnected by a linear single ball array.

Figure 4.5 shows the implementation result of the first approach. Operation results meet the following requirements: (1) long distance; (2) payload; (3) noise level; and (4) operation speed satisfaction. However, the experimental results revealed some disadvantages of the mechanism, including (1) backlash due to unsecured cable tension; (2) the taping deformation when elongated is quite large. To overcome the existing problems, an improved

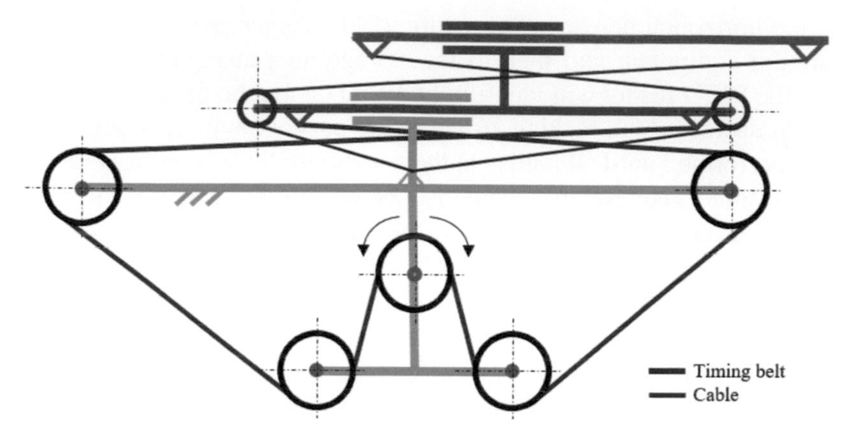

Figure 4.3 Schematic diagram.

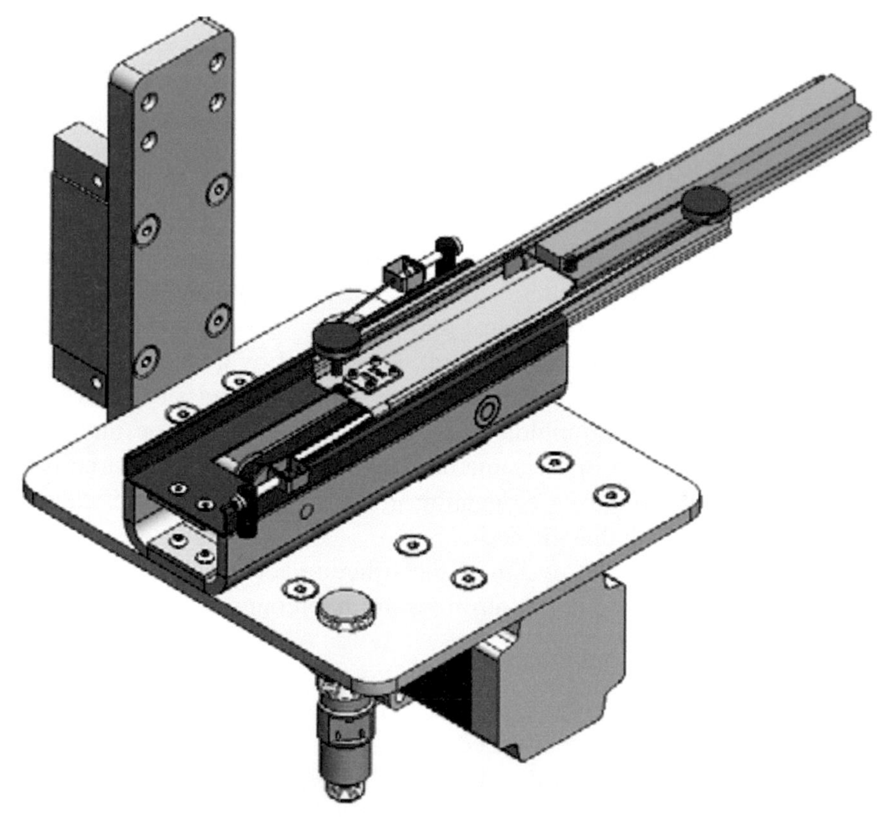

Figure 4.4 Timing belt and cable mechanism 3D design approach.

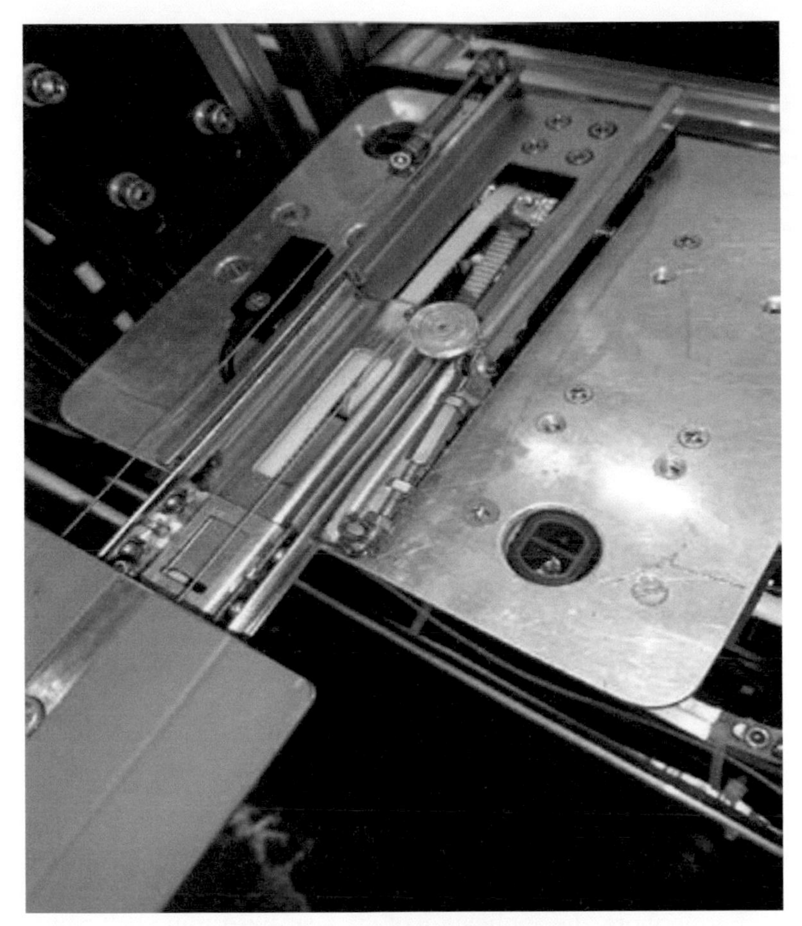

Figure 4.5 First approach prototype.

approach was proposed adopting timing belt transmission as described in Figure 4.6. Timing belt and cable mechanism 3D design approach.

In the revised version, the movement of the middle link is driven by a gear-like connection using a timing belt transmission. The use of this gear-like transmission simplifies the structure. There is more space for reinforcing the connection between the three links, thereby helping to enhance thc phenomenon of taping deformation. Specifically, the relationship of the three links was substituted by a roller system with ball bearings to increase the rigidity compared to the single ball array linear sliding bearing in the first option. This is expected to reduce the taping deflection when extending

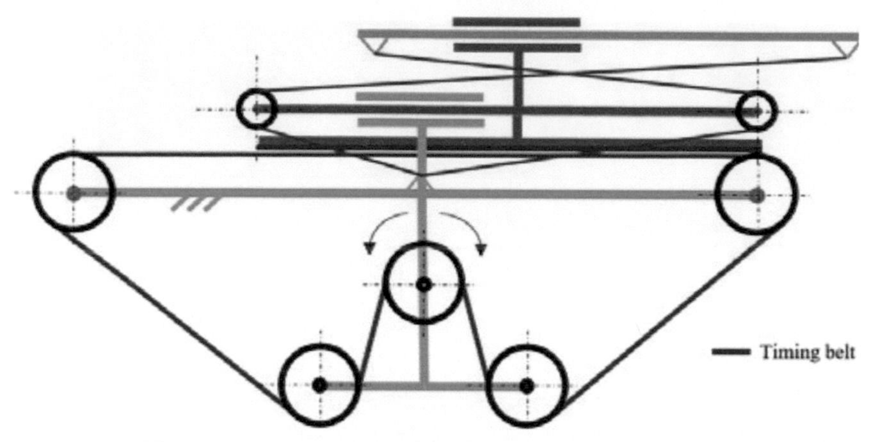

Figure 4.6 Telescopic fork mechanism schematic diagram.

significantly. Furthermore, in the new version, the timing belt transmission was deployed to generate the bounding connection between the middle and top links, which is the bully cable transmission in the primary approach.

4.3.1.2 Improved version of the design

The telescopic fork is designed to operate at high acceleration and braking levels. It is also required to have precise positioning at high speed. Consequently, timing belts are a primary interest. Furthermore, the timing belt is nominated in the system because of the following advantages: positive drive; no slip between the belt and pulleys and no variation in output speed; high torque transmission; high strength-to-weight ratio; high-speed operation; thin and flexible. Furthermore, the application of the gear-like timing belt connection of the bottom and middle links generates the same motion and advantages as the rack pinion gear transmission. This enables further extended movement, which is limited to a half-link length distance in the first approach. The deployment of gear-like timing belts is even more advantageous than the rack pinion mechanism because there is no phenomenon of joint in and joint out when the middle link moves. Thus, it enhances the complications of noise and shaking.

Figure 4.7 demonstrates the revised version's 3D design (Figure 4.7 (a)) and prototype (Figure 4.7 (b)). In addition to changing the transmission method, thanks to the simplicity of the transmission structure, the new version allows the use of a roller system with ball bearings to increase rigidity compared to the single ball array linear sliding option in the first approach.

The enhanced design also demonstrates the excellent ability of the machine to assemble.

4.3.2 Controller design

Figure 4.8 depicts the functional diagram of the proposed controller. In terms of multi-level management, the system requires the following layers: (1) Field layer manages 7 optical sensors, 7 RFID readers, 2 RFID read/write stations, 15 pneumatic cylinders, 20 DC motors, 7 stepper motors, and 14 AC servo motors; (2) the control layer manages 8 S71200 PLCs, 14 AC servo motor drives, 7 stepper motor drives, power drives for 20 DC motor, and 15 air cylinders; (3) the HMI layer manages and monitors the operating parameters of the system; and (3) the management layer is the warehouse management software (WMS). Regarding the required control feature, the system needs the following functions. (1) Upload data to the cloud to serve the WMS as well as for other analysis purposes, including AI. (2) Three-axis movement of seven AGVs. Each AGV needs two servo motors and corresponding drivers to control horizontal and vertical motion, one stepper and driver to execute the movement of the telescopic fork, nine optical sensors for homing purposes, and limiting travel of three movements. These devices are managed by one PLC S7-1200. (3) Conveyor operation to ensure pallets circulation. With this requirement, the system needs seven optical sensors and seven pneumatic cylinders to recognize and block the packages on the major output conveying structure to prevent impacts with moving out pallets from AGV's outlet conveyors. A separator station consisting of an optical sensor and a pneumatic cylinder is installed at the inlet of the main feeding conveyor to help regulate the number of pallets supplied to the inlet conveyor of the AGVs. (4) Code reassignment: To fulfill this task, the system needs 1500 RFID tags permanently attached to the respective pallets, and seven sets of RFID readers and pneumatic cylinders are necessary to read the code and redirect pallets to the corresponding AGV's inlet conveyors. Two RFID read/write stations are necessary at the main conveyor's outlet and inlet to clear the old code and assign a new code, respectively. (5) Supervisory, management, and connection requirement of AGV's controllers and conveyor system's controller. To fulfill this requirement, an Ethernet communication network configuration was utilized.

Modbus was the most widely used serial communication protocol for connecting industrial electronic devices in the past. MQTT is a newer protocol than Modbus that was created for Internet of Things (IoT) applications. As

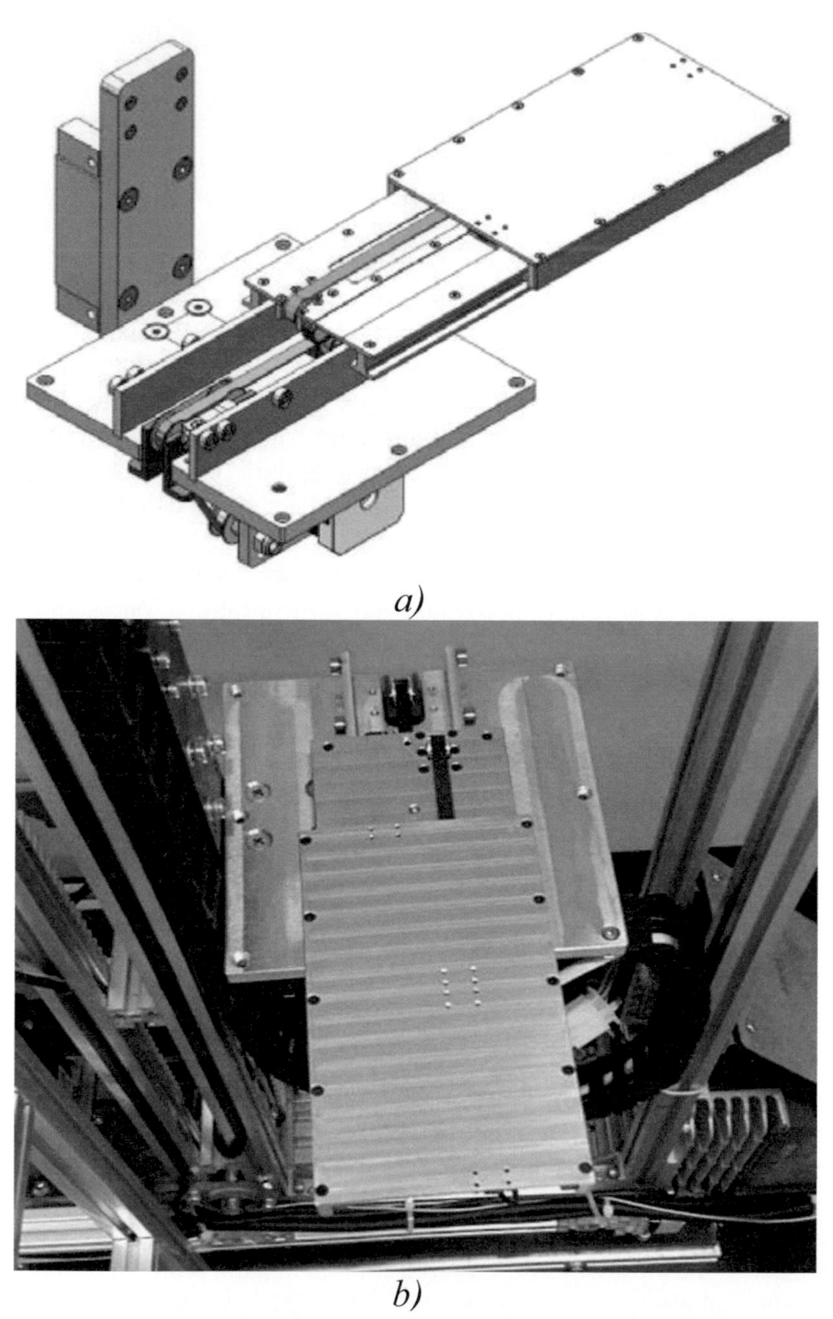

Figure 4.7 Improved version. (a) 3D design. (b) Prototype.

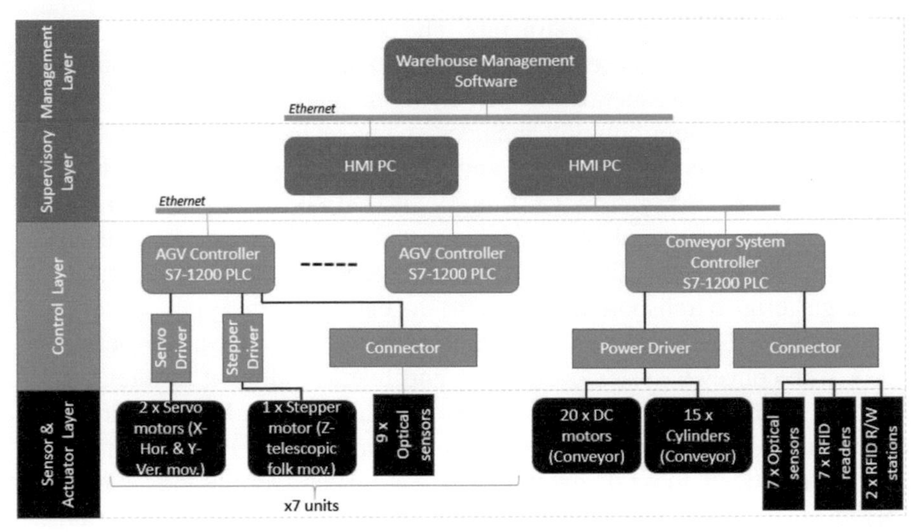

Figure 4.8 Controller schematic diagram.

a result, older equipment only support Modbus TCP and not MQTT. Using MQTT and a message broker, we can deliver data to numerous recipients. We just send data to the Modbus TCP server. Modbus TCP is less secure than MQTT. MQTT is currently supported by a wide range of software languages and cloud service providers. The authors tested and verified the performance of MQTT implementations in [14]–[16]. In this project, we used MQTT as the connection protocol between the WMS application and the PLCs.

4.3.3 Data collection

The smart warehouse system was created with two key goals in mind. First, the system is used as a regional demand testbed. Second, it is employed for research, particularly in the areas of logistic and AI algorithms. These requirements necessitate continuous procedure and a large amount of data collection. As a result, pallets must be recycled and circulated. This circulation necessitates the removal of two major roadblocks: the mechanical hardware system for transferring pallets from the exit to the entrance, and the deletion of the old administration code and reassignment of the fresh managing code to the pallets.

The following options were examined to handle the problem of pallet circulation: (1) manual collecting and depositing; (2) AGV robotic system; (3)

gantry system; and (4) conveying structure. The conveying system option was evaluated better utilizing the evaluation matrix and the assessment criteria stated in Table 4.1. Consequently, the conveying method was employed.

The conveyor system solution includes many features: (1) Pallets are continuously circulated between exit and entrance; (2) the entrance (blue conveying structure in Figure 4.2 (a)) and exit (red conveying structure in Figure 4.2 (a)) system implementations were designed in two heights to increase the system's versatility; (3) pneumatic cylinder pushers were located at right angle turns to transport the pallet; (4) the entrance conveying structure of each AS/RS AGV serves as a buffer for line-up packages; (5) before the AGV. The conveyor system's structure and features allow it to integrate automation solutions such as reading, writing, and erasing administration codes for pallets, and ensuring continuous functioning. If AGV autonomous systems are deployed, these demands can be easily met. Despite this, a large number of AGV mobile robots is required to achieve the continuous operation objectives. As a consequence, the price would rise dramatically. Charging up AGV robotic system after each working cycle also actually increases the number of robots.

The following options – provided in Table 4.2 – have been examined to tackle the pallets' allocated administration code writing and removing issues: (1) manually operating, (2) barcode scanners, and (3) RFID. The highest rating score of the RFID solution is obtained by using a decision matrix with analogous evaluation criteria.

It is evident that the RFID approach offers a simple method for writing and deleting management codes. For coding purposes, each pallet was allocated an RFID tag. Utilizing manual or barcode methods, however, this work becomes rather difficult. Both procedures necessitate removing the old written label from the pallet and replacing it with the new one. Furthermore,

Table 4.1 Evaluation matrix for pallet rotation.

Approaches	**Measures**			**Total**
	Budget	**Auto. companionable**	**Cont. process**	
Manual collecting and depositing	4	1	1	6
AGV robotic system	1	4	3	8
Gantry system	2	2	2	6
Conveyor structure	**3**	**3**	**4**	**10**

Table 4.2 Evaluation matrix for the pallets' allocated administration code writing and removing.

Solutions	**Measures**			**Total**
	Budget	**Auto. companionable**	**Cont. process**	
Manually operating	3	1	1	5
Barcode scanner	2	2	2	6
RFID	**1**	**3**	**3**	**7**

by configuring the correct RFID scanning stations, obtaining codes to reroute packages into the corresponding AS/RS AGV's conveying structures is performed. Using the barcode method, you may tackle this problem in a similar way. The manual operating solution is still difficult since separate sensors must be placed along the conveyor in order to locate and identify all of the pallets on the system. As a result, RFID is thought to be the ideal option in this situation.

It is critical to compute the values (T-sepa: separator time, V-conv: conveyor velocity, and T-stor: storing time) in the suggested system in order to run continuously without becoming blocked while transporting packages from horizontal to vertical conveying structures. T-sepa is the lowest time-span for a new package on the horizontal conveying structure to access the storing zone; V-conv is the velocity of the horizontal and vertical conveying structures (these two velocities are supposed to be identical); and T-stor is the AGV's travel time from the time the pallet is received, taken to storing, and returned to the entrance location. The simulation model for the system was created using the Simio simulation software. This section describes the simulation model and the conditions that were considered.

T-sepa, V-conv, and T-stor definition simulation model:

In the instances considered, T-stor is fixed (averaged handling time at one rack). The problem of defining the appropriate sorting time is continuously being researched. Similarly, conveyor speeds in real-world systems range from [0.1−3] m/s. It was also the metric that is used to calculate the simulation results.

T-sepa determination based on dataset modification (V-conv and T-stor) is a multifactorial challenge. To avoid package obstruction on the vertical conveyor, the experiment was performed by holding the value couple (T-stor and V-conv) and progressively altering the T-sepa rate. Five racking structures with a total capacity of 1000 pallets were used in this experiment.

The smallest interval desired to manipulate 1000 packages into the full stock was observed at each acceptable T-sepa value.

The sets of parameters (T-stor and V-conv) used in the model were trialed to determine T-sepa and the operating times are shown in Figure 4.9.

Simulation model:

In this section, the simulation model for the suggested concepts is presented. Simulated models of a vertical conveying structure, a horizontal conveying structure, and a separating station are shown in Figure 4.10.

The simulated model has the boundary conditions of the following. (1) The largest quantity of packages that may be lined up on the vertical conveying structures is 3. Vertical conveying structure length and width are 2.5 and 2 m, respectively. (2) The horizontal conveying structure length and width are

T-stor (s)/V-conv (m/s)	7	10	13	18	25	35	50	70
0.1								
0.2								
0.3								
0.5					T-sepa(s)?			
0.8								
1.0					Operating time?			
1.2								
1.5								
2.0								
3.0								

Figure 4.9 Dataset for simulation (V-conv and T-stor).

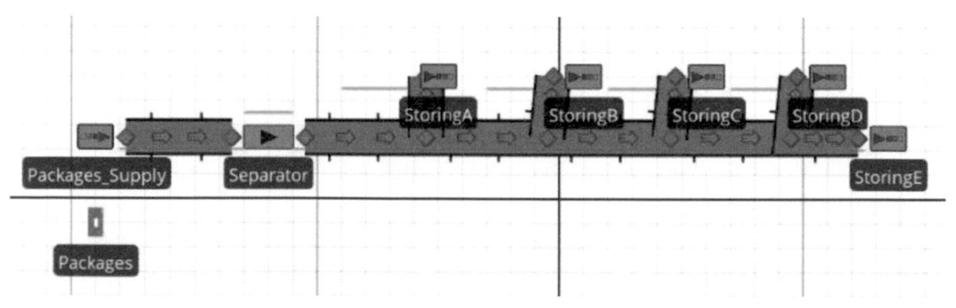

Figure 4.10 Simulated model of a vertical conveying structure, a horizontal conveying structure, and a separating station.

20 and 2 m, respectively. (3) Each pallet package length is 0.63 m, width is 1.2 m, and height is 0.37 m.

4.4 Results and Discussion

Figure 4.11 presents the prototype of one module, including AGV, double side racking, and conveyor. The whole system, including seven modules and a complete conveyor system, will be developed right after the termination of the COVID-19 pandemic. The current achievements are as follows.

4.4.1 Telescopic fork

Practical operating results yielded the following results: (1) Payload – up to 3 kg; (2) manipulating time – 3 seconds for each side; (3) the mechanism and rigidity ensure the stable operation of the telescopic forks. In addition, Figure 4.12 presents the key performance indicators. The results demonstrate that the taping deformation of fully loaded and fully extended has the mean value of about 0.6 mm (standard deviation = 0.04 mm); the positioning accuracy varies within ±0.2 mm; the backlash of the mechanism is recorded with an average value of 0.037 mm (SD = 0.005 mm); the noise level of the structure reached an average value of 43.6 dB (SD = 3.5) and 45.6 dB (SD = 2.5) for the measurement results in the room with the door closed and the

Figure 4.11 One module prototype including AGV, double side racking, and conveyor.

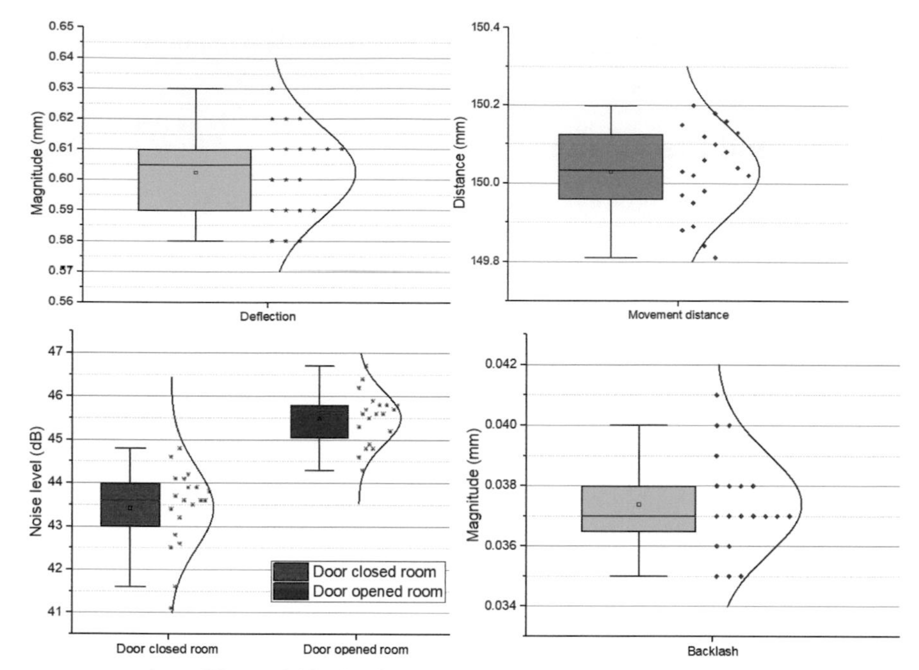

Figure 4.12 Performance index experimental results.

room with the door open, respectively. These key performance indicators all meet the requirements. The application of a timing belt transmission greatly improves the taping deflection, elongation, and other performance indicators. Comparing the experimental results of the two versions in Table 4.3 shows that the improved version is primarily superior, only the noise level is worse than the first version. However, the noise value is still within the initial design limit. The lower noise level of the first version is attributed to the use of a roller cable transmission and less number of balls for connecting the links.

4.4.2 Controller

With the aforementioned hardware configuration required, the electrical connection diagram of AGV and the conveyor system were proposed and shown in Figure 4.13. To ensure the requirements, stability, and reliability of the entire control system, the essential devices were selected from reputable manufacturers. Specifically, the PLC controllers were selected from the Siemens manufacturer and are among the latest configuration (i.e., S7-1200). Similarly, the AC server motor and driver were selected compatibly

Table 4.3 Experimental result comparison.

Performance indicators	First version	Improved version
Payload	Up to 3 kg	Up to 3 kg
Manipulating time	3 seconds	3 seconds
Mechanism and rigidity	Good	Excellent
Taping deformation of fully loaded and fully extended	2 mm	0.6 mm
Positioning accuracy	±0.8 mm	±0.2 mm
Backlash	0.9 mm	0.037 mm
Noise level	41.2 dB (closed-door room)	43.6 dB (closed-door room)

from the Mitsubishi manufacturer. The installation process showed that the implementation of these circuit diagrams is relatively straightforward except for one consideration of servo driver exciting signals. Specifically, an NPN to PNP converter should be applied between the PLC's high-speed pulse outputs and the servo driver's pulse inputs to mismatch the incompatible voltage level.

Figure 4.14 presents the control flowchart of the AGV and conveyor system. As illustrated in Figure 4.14 (a), the AGV executes the main commands, including "Storing," "Retrivaling," "Home X," "Home Y," and "Home Z," in which to execute the "Storing" instruction, the AGV needs to execute the subcommands including "Taking from conveyor," "Moving to selected position," and "Putting to selected position." Similarly, to execute the "Retrivaling" instruction, the AGV needs to execute the subcommands, including "Moving to selected position," "Taking from selected position," "Move to conveyor," and "Putting to conveyor." In a simpler routine, the "Home X" and "Home Y" commands are executed through two subcommands, "Moving to home X" and "Moving to Home Y," respectively. These commands were organized into subroutines. Due to the complexity of the program structure and a large number of subroutines, details are not presented here. Simultaneously, as depicted in Figure 4.14 (b), the conveyor system has three main groups of activities that operate independently and in parallel.

The first is to control the delivery of pallets to the AGV inlet conveyor (AIC). The controller checks the pallet's RFID code at each AIC; if the state of the corresponding RFID bit is ON, then the corresponding cylinder will be activated for 1 second to redirect the pallet to the corresponding AIC. The second group of control operations is to block pallets on the main output conveyor if there is a risk of collision with the pallet coming out of the

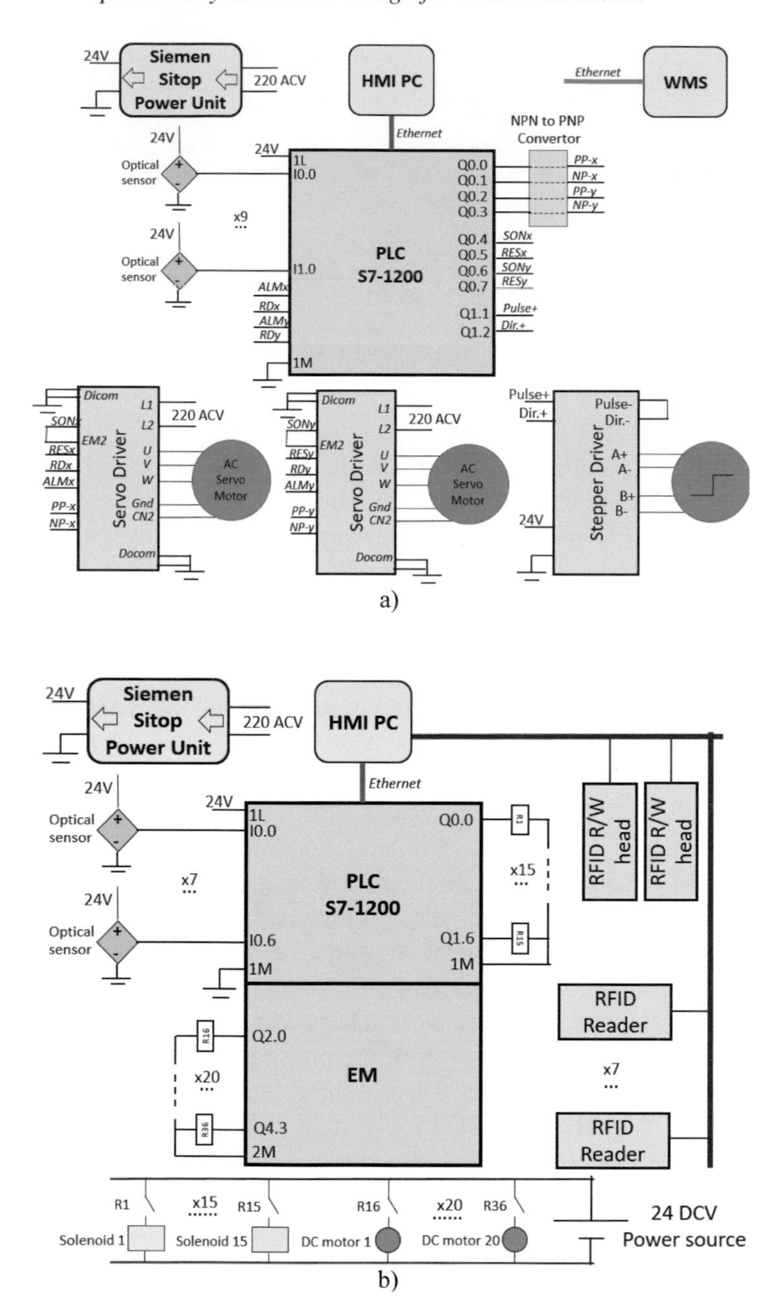

Figure 4.13 (a) AGV and (b) conveyor system electrical connection diagram.

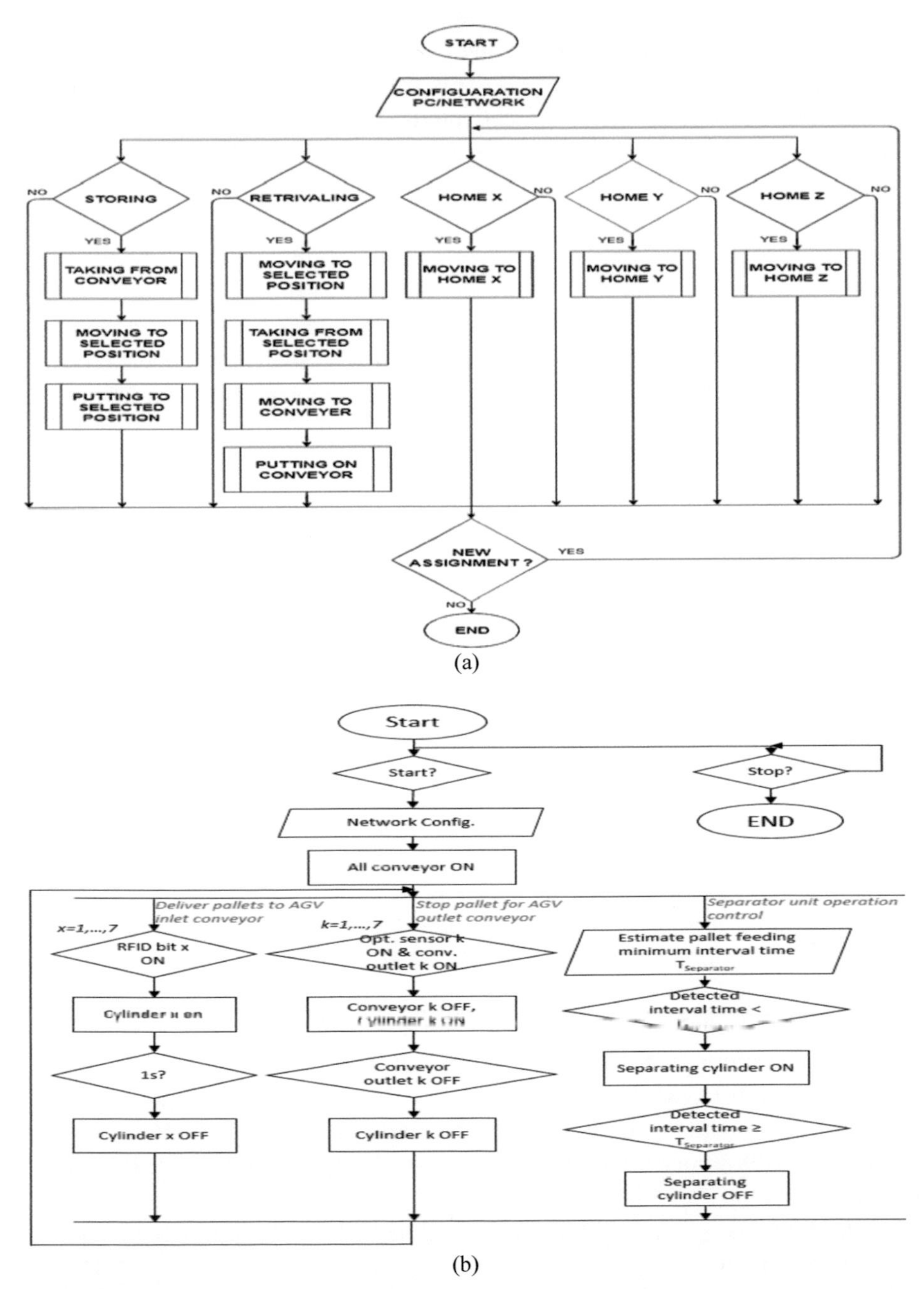

(a)

(b)

Figure 4.14 (a) AGV control flowchart. (b) Conveyor system control flowchart.

AGV outlet conveyor (AOC). The controller checks the status of seven optical sensors mounted on the main output conveyor before the AOCs perform this task. Suppose the *k*th AOC (i.e., $k = 1-7$) is transporting the pallet, and the corresponding *k*th optical sensor is activated. The corresponding cylinder is activated to block the pallet until the *k*th AOC stops working. Finally, the third control operation is the control of the separator unit at the input of the main feeding conveyor to moderate the number of pallets available on the main feeding conveyor. To realize this task, the minimum feeding interval time (i.e., T-sepa) needs to be estimated based on the setting value of conveyor speed (i.e., V-conv) and AGV operation time (i.e., T-stor) to ensure the suitable number of pallets on the main feeding conveyor. Accordingly, the controller detects the next pallet using an optical sensor after a pallet has been passed. If the interval time between the two pallets is less than T-sepa, the cylinder will be activated to block the upcoming pallet. After ensuring that the time interval between the two pallets is greater than or equal to the T-sepa, the cylinder is deactivated to allow the pallet to pass through. The success of the corresponding codes demonstrated the feasibility of these flowchart algorithms.

4.4.3 Data collection

In this section, the simulation results are demonstrated. When the optimum value of T-sepa is attained for each couple (T-stor and V-conv), the quantity of packages dispersed in each racking structure (A, B, C, D, and E) is preserved. Pallet distribution is arbitrary and varies around the predicted value at each racking system, as presented in Figure 4.15 (200 pallets).

In multisource feedback studies, response surface polynomial regression investigation is a complicated arithmetical procedure that has gained prominence. This method can be used to understand how two predictor variables interact to affect a consequence variable, which is particularly valuable once the discrepancy amongst the two regressors is a significant component. The T-sepa distribution based on the last two variables is shown in Figure 4.16. T-sepa changes linearly without any singularities, as shown in Figure 4.16. This aids the polynomial regression function's ability to levitate prediction outcome precision.

For finding T-sepa, a polynomial surface regression function (i.e., eqn (4.1)) was deployed:

The coefficients obtained from eqn (4.1) are listed in Table 4.4.

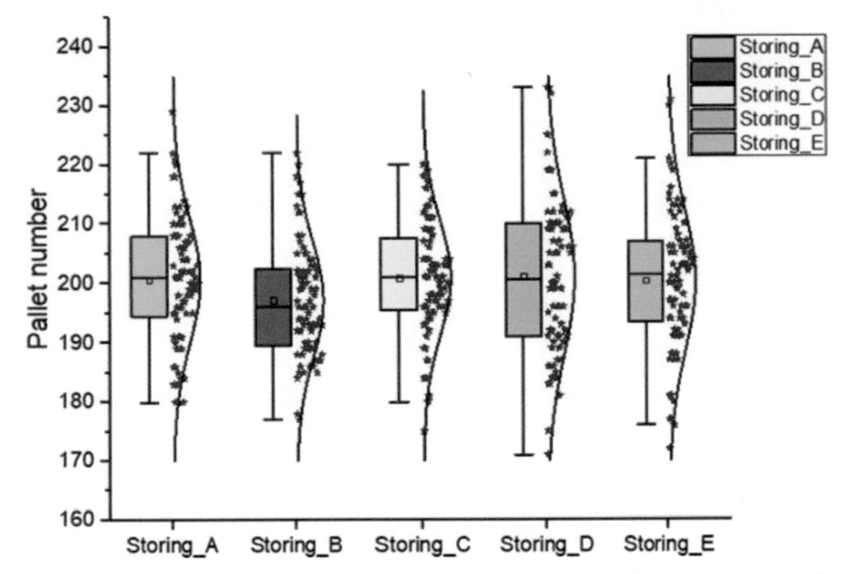

Figure 4.15 Pallets were divided onto five racks in 80 different simulated configurations.

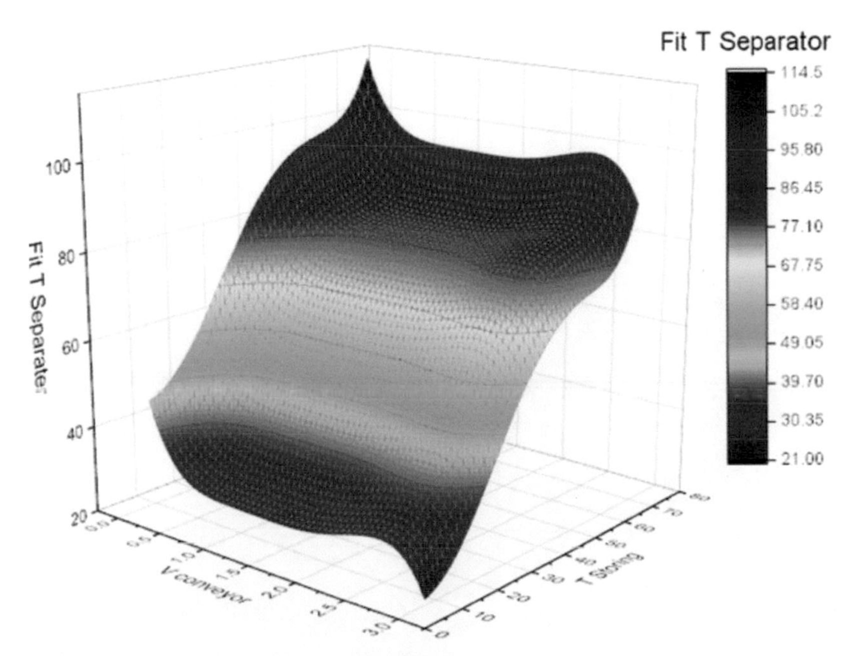

Figure 4.16 T-sepa regression findings with respect to T-stor and V-conv graph.

The time required to position whole stock regarding to the set of three fundamental factors is shown in Figure 4.17 (T-sepa, T-stor, and V-conv).

Table 4.4 Regression results.

	Value	Standard error	*t*-Value	Prob > (*t*)	Dependency
z0	21.63063	26.02391	0.83118	0.40874	0.99852
A1	−43.15849	53.93387	−0.80021	0.42633	0.99982
A2	12.46406	129.25855	0.09643	0.92346	0.99999
A3	16.51231	126.05197	0.131	0.89616	1
A4	10.15704	52.26997	−0.19432	0.8465	1
A5	1.51109	7.57091	0.19959	0.84239	1
B1	0.35862	5.99022	0.05987	0.95243	0.99998
B2	−0.00331	0.48842	−0.00678	0.99461	1
B3	0.00243	0.01735	0.14027	0.88886	1
B4	−6.5260E-5	2.73477E-4	−0.23863	0.8121	1
B5	4.74016E-7	1.55581E-6	0.30468	0.76153	1

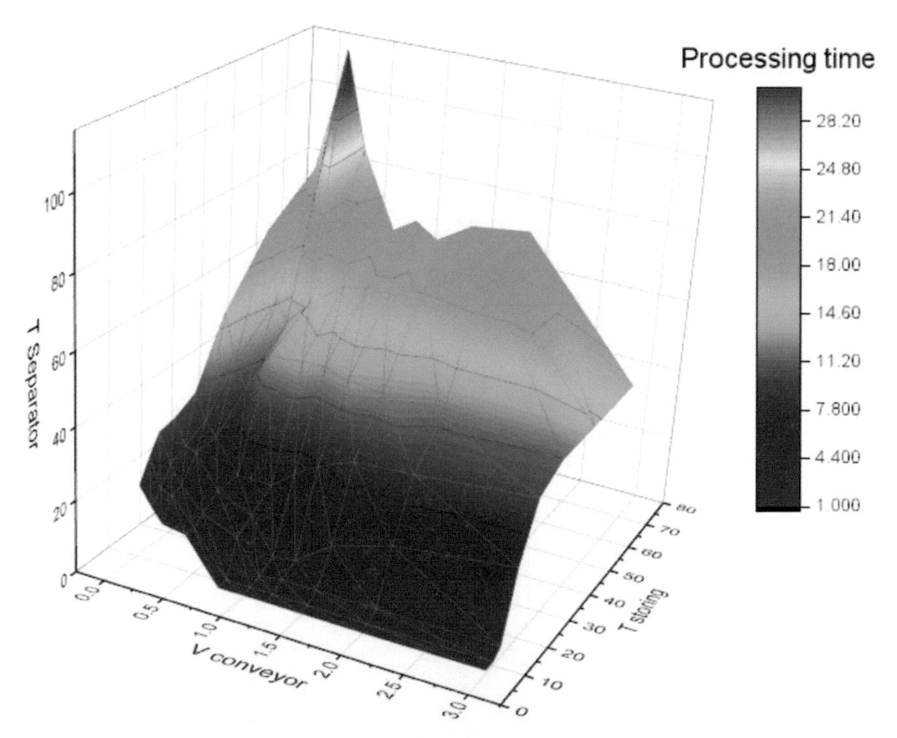

Figure 4.17 Working time regression graph with respect to T-sepa, T-stor, and V-conv.

4.5 Conclusion

This chapter proposed and described in detail how a smart warehouse testbed works in practice. The system consists of a warehouse with seven racks, a pallet circulation system, AGVs that move pallets, and an RFID unit (reader/writer/tag) that manages the necessary information of the pallets. The design of an ICS, including motors, RFID, PLCs, sensors, cylinders, etc., as well as the flowchart of the control algorithm of the main objects (AGVs, conveyors, and WMS), is clearly described.

Regarding the first key hardware-telescopic fork, the experimental results demonstrate the superior performance of the improved version using timing belt transmission compared to the first version applying pulley cable combined with the timing belt transmission. The obtained results for the enhanced version include: the maximum taping deformation is 0.6 mm; the result of controlling the mechanism to the 150-mm extension position ranges from ± 0.2 mm; the backlash of the mechanism is 0.037 mm; the noise level is 43.6 dB for closed-door room operation. These performance parameters all meet the initial requirements. Moreover, the performance indicators of the improved version are mostly superior except for the noise level compared with the first version.

Initial results show that ICS configuration (multilayer controller) is optimal, leading to accurate control and data collection as expected. The system will be completed with the hardware system of a full testbed as well as the ability to optimize the operational management of WMS using AI applications as soon as the epidemic situation is under control.

Finally, the findings revealed that one of the testbed's strong point is pallet circulation. It can process transactions quickly and with minimum human engagement, permitting it to gather large quantities of data for AI and deep learning optimization complications.

Acknowledgment

This research is financially supported by Eastern International University, Binh Duong Province, Vietnam.

References

[1] Le, N. H., Kieu, M. K., Nguyen, V. A. T., Ninh, T. T. D., Nguyen, X. H., Nguyen, D. C. et al. (2022). Implementation of a Multi-Disciplinary

Smart Warehouse Project with Applications. *3rd International Conference on Emerging Technologies in Data Mining and Information Security (IEMIS 2022).*

[2] Tran, V. L., Kieu, M. K., Nguyen, X. H., Nguyen, V. A. T., Ninh, T. T. D., & Nguyen, D. C. et al. (2022). Reinforcement Learning for Developing an Intelligent Warehouse Environment. *The 8Th International Conference On Advanced Machine Learning And Technologies And Applications (AMLTA2022)*, 11-20. https://doi.org/10.1007/978-3-031-03918-8_2

[3] Kamali, A. (2019). Smart warehouse vs. traditional warehouse-review. *Automation and Autonomous Systems*, *11*(1), 9-16.

[4] Nguyen, V. A. T., Le, N. B., Kieu, M. K., Nguyen, X. H., Nguyen, D. C., & Le, N. H. et al. (2022). Artificial Intelligence Based Solutions to Smart Warehouse Development: A Conceptual Framework. *The 8Th International Conference On Advanced Machine Learning And Technologies And Applications (AMLTA2022)*, 115-124. https://doi.org/10.1007/978-3-031-03918-8_11

[5] Blancas, L. (2014). *Rapid growth, limited connectivity: Challenges and opportunities in Vietnam's logistics* [PDF]. Retrieved 5 May 2022, from https://drive.google.com/file/d/1qdvAfnwfGcMYosUlmWIRg3Grr64wKHC0/view.

[6] Arnold, J., Arvis, J., Mustra, M., Horton, B., Carruthers, R., & Ojala, L. (2010). Trade and Transport Facilitation Assessment. https://doi.org/10.1596/978-0-8213-8412-1

[7] Copeland, B., & Proudfoot, D. (2007). Artificial intelligence. *Philosophy Of Psychology And Cognitive Science*, 429-482. https://doi.org/10.1016/b978-044451540-7/50032-3

[8] Wendehorst, C., & Hirtenlehner, J. (2022). Outlook on the Future Regulatory Requirements for AI in Europe. *Available at SSRN 4093016.*

[9] Roodbergen, K., & Vis, I. (2009). A survey of literature on automated storage and retrieval systems. *European Journal Of Operational Research*, *194*(2), 343-362. https://doi.org/10.1016/j.ejor.2008.01.038

[10] Le, N. H., Nguyen, D. C., Nguyen, X. H., Kieu, M. K., Nguyen, V. A. T., & Ninh, T. T. D. et al. (2022). An Intelligent Algorithmic Approach for Data Collection in a Smart Warehouse Testbed. *The 8Th International Conference On Advanced Machine Learning And Technologies And Applications (AMLTA2022)*, 557-566. https://doi.org/10.1007/978-3-031-03918-8_46

[11] Le, N. H., Phan, M. D. K., Nguyen, D. C., Nguyen, X. H., Kieu, M. K., Nguyen, V. A. T. et al. (2022). Multilayer Communication Based Controller Design for Smart Warehouse Testbed. *3rd International Conference on Emerging Technologies in Data Mining and Information Security (IEMIS 2022).*

[12] Official Journal of the European Union. (2019). *Directive 2009/125/Ec Of The European Parliament And Of The Council: Establishing a framework for the setting of ecodesign requirements for energy-related products.* Strasbourg.

[13] Ministry of Industry and Trade. (2019). *Vietnam logistics report 2019* [PDF]. Tran Gia Trading And Printing Company Limited. Retrieved 5 May 2022, from https://gosmartlog.com/wp-content/uploads/2019/12/Bao-cao-logistics-viet-nam-2019.pdf.

[14] Collina, M., Corazza, G., & Vanelli-Coralli, A. (2012). Introducing the QEST broker: Scaling the IoT by bridging MQTT and REST. *2012 IEEE 23Rd International Symposium On Personal, Indoor And Mobile Radio Communications - (PIMRC).* https://doi.org/10.1109/pimrc.2012.6362813

[15] Lee, S., Kim, H., Hong, D., & Ju, H. (2013). Correlation analysis of MQTT loss and delay according to QoS level. *The International Conference On Information Networking 2013 (ICOIN).* https://doi.org/10.1109/icoin.2013.6496715

[16] Thangavel, D., Ma, X., Valera, A., Tan, H., & Tan, C. (2014). Performance evaluation of MQTT and CoAP via a common middleware. *2014 IEEE Ninth International Conference On Intelligent Sensors, Sensor Networks And Information Processing (ISSNIP).* https://doi.org/10.1109/issnip.2014.6827678

5

A Knowledge Graph-based Integration Approach for Research Digital Artifacts

Zaenal Akbar, Arya Adhyaksa Waskita, Dadan Ridwan Saleh, Ariani Indrawati, and Yulia Aris Kartika

Research Center for Computing, Research Organization for Electronics and Informatics, National Research and Innovation Agency, Indonesia
E-mail: zaenal.akbar@brin.go.id; arya.adhyaksa.waskita@brin.go.id; dadan.ridwan.saleh@brin.go.id; ariani.indrawati@brin.go.id; yulia.aris.kartika@brin.go.id

Abstract

We continuously generate digital artifacts in today's digital era. Similarly, digital artifacts are created purposefully or accidentally in research and development activities. For example, research data (raw or intermediate findings), access logs, computer program codes, calculation processes or pipelines, pre-printed articles, and so on are all produced due to a research and development activity. Preserving such artifacts for future use and, most crucially, gaining meaningful insights as they are created requires a methodical strategy. Unfortunately, most of the existing research data management solutions only manages them partially. Furthermore, the distributed nature of artifacts has contributed to the under-utilization of several types of artifacts. To the best of our knowledge, very few works have looked into this case. This chapter presents a solution, based on a knowledge graph. First, we identified and represented numerous types of research items using a common schema. Then we built our knowledge graph by aligning various digital assets mentioned in the scientific literature with the schema. Thus, such artifacts may be combined, integrated, and their relationships studied further for essential insights.

Keywords: Data integration, data management, knowledge graph, ontology, research artifact..

5.1 Introduction

We intensively consume and produce digital artifacts daily in today's digital era. Digital artifacts include digital documents, emails, comments on discussion forums, instant messaging, social media posts, and other digital objects that have been processed by using digital devices. Likewise, during research activities, we may consume data from various sources and in a variety of formats, process it using specialized data analytic software, sometimes using pre-defined data models, and generate multiple digital outputs as results. Scientific articles, charts, tables, pictures, workflows, computer program code, research data, and access logs are only a few examples of digital artifacts generated during research. Processing these artifacts are a novel concept in this era where their interactions will result in a better and more meaningful understanding of knowledge, mainly when a research activity is computation, and data-enabled [1].

Digital artifacts as a source for rich scientific data analysis may be linked to other digital artifacts created during other research projects. In order to have a consistent relationship, digital artifacts are supposed to be immutable so that they can be appropriately verified and reproduced correctly [2, 3]. While persistent linkages are necessary for verifiability and reliability, semantic information about the artifacts is also necessary. This information makes it easier for humans and algorithms to find vast and complicated digital artifacts and link them to other pieces of data [4]. In the last few years, the interaction between research activities can be seen only through scientific articles. However, information regarding the data, the performed computational processes, and the code from data to inference is rarely or only briefly supplied. As a result, the articles alone are insufficient to comprehend the entire study process. There are many more digital artifacts related to research activities that currently remain neglected by such analytics.

A few attempts have been made to provide more advanced analytics on digital artifacts. For example, the OpenAIRE[1] project collects metadata descriptions about organizations, data sources, projects, funding programs, people, publications, and datasets in order to populate, curate, and enrich an information space [5]. Another example is through the use of the research graph[2] data model and the research data switchboard[3] to connect data repositories to publications and grants across numerous research data infrastructures [6]. In this case, it will be able to discover and connect the related research datasets based on publication co-authorship or jointly funded grants. Another example, the RO-Crate[4] solution, provides an approach to packaging all items that are contributed to a research outcome, including datasets, software, and methods with their metadata [7]. In this case, the RO-Crate simplifies making research outputs FAIR (findable, accessible, interoperable, reusable) while also enhancing research reproducibility.

Despite those efforts, multiple challenges remain. There is currently no unified representation for all digital artifacts. Most current solutions were created to address specific needs. Therefore, their representations differ from one solution to another. For example, the granularity of objects to represent research on digital artifacts is not equal, leading to ambiguity. Moreover, a better representation should be available for those who focus on actors as well as on activities. In this work, we would like to have a generic representation that brides those schema discrepancies using knowledge graph technology. A knowledge graph is a group of data that is meant to acquire and disseminate information about the actual world, with nodes representing items of interest and edges representing relationships between these entities [8]. Using a knowledge graph opens up a range of techniques that can be brought to bear for integrating and extracting value from diverse data sources on a large scale. For example, to integrate information on various touristic services for marketing and sales purposes [9] and to integrate biodiversity information such as taxa, publications, people, places, specimens, sequences, and institutions as a single, shared knowledge space [10], including information generated from scientific literature [11]. The knowledge graph approach also has been used for intelligent survey systems to provide the type of on-the-fly content planning and to guarantee that selection

[1]https://www.openaire.eu/, Accessed: 2022-05-11

[2]https://researchgraph.org/, Accessed: 2022-05-11

[3]https://www.rd-switchboard.org/, Accessed: 2022-05-11

[4]https://www.researchobject.org/, Accessed: 2022-05-11

of subsequent questions is based on response to previous questions in a survey [12].

A knowledge graph has an ontology as its schema defining the vocabulary used in the knowledge graph [13]. It is the main component in the knowledge creation phase as the core data for a knowledge graph [14]. The novelty of this work lies in the generation of the schema for our knowledge graph. We use a bottom-up (data-driven) approach. Instead of acquiring knowledge from domain experts, we rely on our schema from relations of digital artifacts extracted from datasets of published scholarly articles.

5.1.1 Motivation

Our work was motivated by the utilization of various digital artifacts consumed and/or produced during research activities. Those digital artifacts contribute equally to the activity and should also be considered in data analytics. Since data analytics rely on data integration, a solution to integrate those artifacts is highly desired. Knowledge-graph-based data integration is necessary due to its dataset's wide variety of representations.

Unfortunately, using the existing ontology, such as the OpenAIRE data model [5], is not sufficient. As shown in Table 5.6 in the Appendix, there are so many entities and attributes that cannot be mapped into the data model. Second, generic terms such as "Funding" and "Software" provide no detailed information. Therefore, it is necessary to have a better representation. However, instead of defining schema from domain experts, which might differ from one project to another, we use a data-driven approach to realize our ontology.

5.1.2 Contribution

Our main contribution lies in integrating various types of research on digital artifacts through a data-driven knowledge graph technology. Our main contribution can be listed as follows:

1. A knowledge graph technology for research on digital artifacts. The technology was selected to accommodate multiple types of representations of related schema.
2. A data-driven development of schema for knowledge graph. As the core component, an ontology defines how data will be stored, managed, and queried.

The rest of the chapter is organized as follows. Section 5.2 discusses a few related works, introduces a gap, and explains how our work filled the gap. After that, we present our research methodology in Section 5.3, followed by our results as well as a discussion on our findings in Section 5.4. Finally, we list a few conclusions and some future works in Section 5.5.

5.2 Related Work

This section describes a few related works from two research areas: knowledge graph utilization in various domains and construction from the scientific literature. After that, we outline our contributions at the end of this section.

Knowledge graph solutions have been implemented in various applications across various domains.

An intelligent survey system can benefit from the usage of a knowledge graph solution, which can create a dynamic and informative solution [12]. In this case, the questions will be arranged on their level of acceptance, allowing for the collection of more informative data in a shorter amount of time than would otherwise be possible. The solution also allows for the classification of participants based on their responses, allowing for the delivery of relevant follow-up questions according to participants' responses. A knowledge network with shared global IDs can be created by connecting isolated databases of biodiversity data that have been assigned particular identifiers [10]. The integration allows for a better understanding of changes in taxonomic publication patterns over time due to data integration. Furthermore, a knowledge graph solution can also be used to open biodiversity knowledge that has been buried in scientific publications, allowing for the establishment of open science practice in the biodiversity domain to be established [11]. Incorporating heterogeneous data from the education domain might extract concepts from subjects or courses and identify educational relationships between concepts in a knowledge graph. [15]. The automatic construction of a knowledge graph from electronic medical information, which is accomplished through the learning of high-quality knowledge bases that link diseases and symptoms, is also possible [16].

A combination of knowledge graphs, natural language processing (NLP), and machine learning (ML) can assist journalists in the live harvesting of possibly news-relevant information from different data sources [17]. On the other hand, a combination of fuzzy logic and a knowledge graph on chemical substances and their impact on a human was proposed as a building block of an AI-based analysis system to diagnose and mitigate dangerous chemical accidents [18], while Martinez *et al.* proposed the use of knowledge graph for power transformer root cause analysis [19]. The use of knowledge graph in financial fraud analysis [20] and detection [21] was also proposed. A computational framework based on knowledge graphs allows for the exploration of the network of links between essential materials properties found in the literature on material science [22]. Moreover, the creation of connections between biological entities, published scientific articles, authors of scientific articles, affiliations of scientific articles, and funding is made possible through the use of multiple data sources such as PubMed[5], ExPORTER[6], ORCID[7], and MapAffil [23].

Like other generic knowledge-based solutions, reasoning, which refers to the ability to infer unknown connections from existing ones, has also emerged as a strength of the knowledge graph solution. The ability will also ensure the consistency and integrity of the knowledge graph, allowing them to be fully comprehended [24]. Many applications could benefit from the ability to reason. These include applications inside the computer, such as knowledge completion and entity classification, and external applications, such as intelligent question–answering systems and recommendation systems.

In particular, for the construction of knowledge graphs, several existing works have taken advantage of scientific articles as the primary source of information for their work. Aside from the rich information provided by such articles, scientific articles are also required to be FAIR [25] in order to increase the overall value of the research conducted [26, 27]. As a result, it is widespread for them to be used as primary sources. For example, the construction of knowledge graph from biodiversity scientific literature [10, 11], biomedical literature [28, 29, 30, 31], cyber-security literature [32] or geo-science literature [33, 34], and the other seven domains literature [35]. Another method of constructing a knowledge graph from the scientific literature is to combine NLP and ontology [36] or combine the scientific literature with other data sources [23]. It is also possible to build models from selective or more specific scientific literature, like when related literature was used to build the models as in the case of COVID-19 models [37, 38]. And RO-Crate is an open and community-driven approach in packaging research artifact from various scientific domains based on Schema.org[8] [39].

---

[5]https://pubmed.ncbi.nlm.nih.gov/, Accessed: 2022-05-11
[6]https://exporter.nih.gov/, Accessed: 2022-05-11
[7]https://orcid.org/, Accessed: 2022-05-11
[8]https://schema.org/, Accessed: 2022-05-11

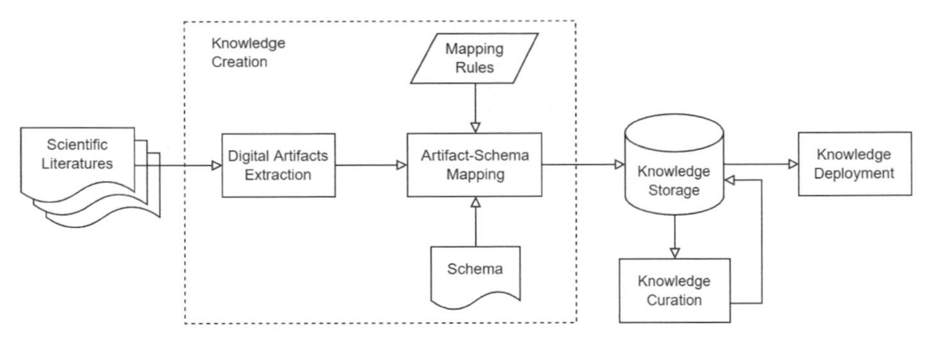

Figure 5.1 Research methodology.

In line with related works from two broad research areas explained above, we outline our contributions as follows:

1. In the knowledge creation step of our knowledge graph, we constructed knowledge using a data-driven approach instead of elicitation from domain experts. We argue that this approach is highly suitable for dynamic conditions of scientific publication where the number and topics are rapidly increasing.
2. Instead of focusing on one specific scientific field/domain, our knowledge graph was constructed from 240 selected papers from three domains: life, engineering, and social science.

5.3 Method

In this section, we explain our research method. After introducing our research methodology, we explain our approach to constructing our research on digital artifact knowledge graph step by step.

5.3.1 Methodology

In general, a knowledge graph entails several technologies, including knowledge representation and reasoning (languages, schema, and standard vocabularies), knowledge storage (graph databases and repositories), knowledge engineering (methodologies, editors, and design patterns), and knowledge learning, such as schema learning and population [13]. Figure 5.1 shows our methodology, which was adopted in [14]. It consists of a few steps as follows:

1. Knowledge creation, which refers to the process of knowledge acquisition from specified knowledge sources. The vital element in this step

is the definition of the ontology to describe and represent all required information in the modeled domain.
2. Knowledge hosting, which refers to the process of storing the acquired knowledge for further use. The knowledge can be stored in a graph-based repository, but any storage should facilitate knowledge retrieval through a standardized method, for example, through SPARQL.
3. Knowledge curation, which refers to the process of assessing the stored knowledge and performing required actions such as cleaning, enriching the knowledge to ensure their qualities.
4. Knowledge deployment, which refers to the process of deploying the curated knowledge in different use cases. As final results, in our work, we should be able to integrate data of digital artifacts from multiple domains.

5.3.2 Research Digital Artifact Knowledge Graph

Figure 5.1 depicts our research methodology for constructing our knowledge graph. As we can see, we use scientific articles as the input for the knowledge creation step in our method. It is indicated that our knowledge creation is based on data, not elicited from the knowledge of domain experts.

Knowledge creation:

In this step, we performed digital artifact extraction through manual identification. To be precise, we performed the following procedure:

1. Identify all related artifacts that are mentioned in the article. They can be inputs, processes, as well as outputs.
2. From each artifact, identify essential properties such as names of organizations or people, titles of publications, names and versions of software, licenses of produced source code, and so on.
3. Identify an identification for every artifact. For example, the URL of the artifact.
4. It is also essential to ensure the completeness of the properties. Therefore, in some cases, a further investigation can be conducted by visiting the URL of the artifact and extracting information from the website.

As a result, a list of digital artifacts will be obtained from every article. It is worth noting that the richness of properties varies from one artifact to the next. Therefore, it is necessary to align each artifact to a pre-defined schema. We adopted the existing schema (the OpenAIRE), where a few extensions are required to represent the entities better. A detailed explanation of the schema generation will be described further in Section 5.4.2. The extracted list of artifacts will be mapped to the schema, where a few mapping rules will be constructed. Further explanation of the mapping rules will be explained in Section 5.4.3. As a final result of this step, the extracted digital artifacts will be uniformly represented.

Knowledge hosting:
In this step, the extracted digital artifacts that have been represented uniformly will be stored in an RDF triplestore. We use the GraphDB[9], a widely used triplestore for data integration as well as relationship exploration and data publishing and consumption. The triplestore has provided multiple advantages to managing our knowledge, including a reasoner and a workbench to interact with the knowledge conveniently.

Knowledge curation:
This process is essential to ensure the quality of the stored knowledge. As mentioned in the previous section, further investigation can be performed during knowledge creation by visiting relevant websites to enrich the information. Information in scientific articles can be very limited. For example, when mentioning an organization, the article could mention only the organization's name.

Therefore, it is necessary to seek more information by visiting the organization's website where a richer organization can be obtained, for example, the type of organization, the location of the organization, etc. In the case of source code, the article sometimes only mentioned the name and a link to its website. Richer information, such as people who contributed to the code, can be identified by following the link.

Knowledge deployment:
Assuming the quality of integrated knowledge has been curated, the last step will be to deploy the integrated knowledge for multiple objectives. In this work, our main objective is to enable data interoperability and reusability through FAIR principles [40]. For example, to make data or workflow that was utilized during research activities findable by machines [41], we argue that exposing those multiple types of digital artifacts would increase science's reproducibility.

[9]https://graphdb.ontotext.com/, Accessed: 2022-05-11

5.4 Result

In this section, we explain our results. First, we introduce our dataset generated based on a well-known bibliographic database. Furthermore, we describe our schema, which was used as the core of our knowledge graph, followed by mapping rules that dictate how to align our data to the schema. Finally, we discuss our analysis and findings at the end of this section.

5.4.1 Dataset

We generated our dataset from Scopus[10], a well-known bibliographic database and one of the largest curated abstract and citation databases, with global and regional coverage of scientific journals, conference proceedings, and books while guaranteeing only the highest quality data [42, 43]. Furthermore, the growth of publications and collaboration in this database has been rapid over the last five years and is expected to continue [44]. Therefore, we selected this database as the source of our dataset because it would capture the most recent developments in various scientific domains.

To generate our dataset, we performed multiple searches on the database specifically for three scientific fields/domains, namely life sciences, engineering sciences, and social sciences. We collected only the most recent published articles (the last 10 years) and divided them into two time periods: 2013-2017 and 2018-2022, intended to acquire the article's time representation.

Table 5.1 shows the query parameters of six queries to collect our dataset. The selection of scientific fields is determined by the definition of subject areas as defined in the Scopus search criteria[11]. For example, the life sciences scientific field is represented by the subject areas of agricultural and biological sciences (*agri*), biochemistry, genetics, and molecular biology (*bioc*), immunology and microbiology (*immu*), neuroscience (*neur*), and pharmacology, toxicology, and pharmaceutics (*phar*). We also filtered our search results based on the two types of documents: conference proceedings and journal articles.

As shown in Table 5.1, each query produced a different number of articles, including across two time periods within a scientific domain. From each query result, we collected the 50 highest citations under the assumption that those articles would be sufficient enough to represent each domain.

[10]https://www.scopus.com/, Accessed: 2022-05-11

[11]http://schema.elsevier.com/dtds/document/bkapi/search/SCOPUSSearchTips.htm, Accessed: 2022-05-11

Table 5.1 Query parameters and results of database search.

No.	Scientific field	Subject area	Period	Results
1	Life sciences	agri, bioc, immu, neur, phar	2013–2017	1.182.668
2	Life sciences	agri, bioc, immu, neur, phar	2018–2022	1.308.792
3	Engineering sciences	ceng, chem, comp, eart, ener, engi, envi, mate, math, phys	2013–2017	1.678.770
4	Engineering sciences	ceng, chem, comp, eart, ener, engi, envi, mate, math, phys	2018–2022	2.367.488
5	Social sciences	busi, econ, psyc	2013–2017	184.025
6	Social sciences	busi, econ, psyc	2018–2022	25.158

Ultimately, we obtained a dataset consisting of 240 articles, comprising 80 articles from each domain, spanning two time periods.

5.4.2 Schema

Figure 5.2 shows our schema, which was adopted from the OpenAIRE data model [5]. It comprises a few entities and properties to extend the existing model. We regularly update our schema, the so-called research digital artifacts ontology (RDAO), and the latest version is available online[12]. As mentioned in Section 5.1.1, our direct mapping to the existing model reveals some limitations, especially the limitation in the representation of a few entities such as "Results" and "Software." Therefore, an extension is necessary to ensure better representation for several entities, including their properties. More detailed explanations of the extended entities can be found in Section 5.4.3.

It is also worthy of mention that we encountered several entities with no individuals in our direct mapping. Therefore, it is also urgent to consider if we should include them in our model. We acknowledge that our approach to constructing the schema driven by data is different from the existing schema. While most existing schemas were designed to answer specific questions, we argue that our approach is more natural and can be implemented in broader applications.

5.4.3 Mapping rules

As mentioned in the previous section, we encounter the necessity to extend the existing schema. As a consequence, it is also required to specify how to

[12]http://ontology.pubnesia.com/rdao/, Accessed: 2022-05-11

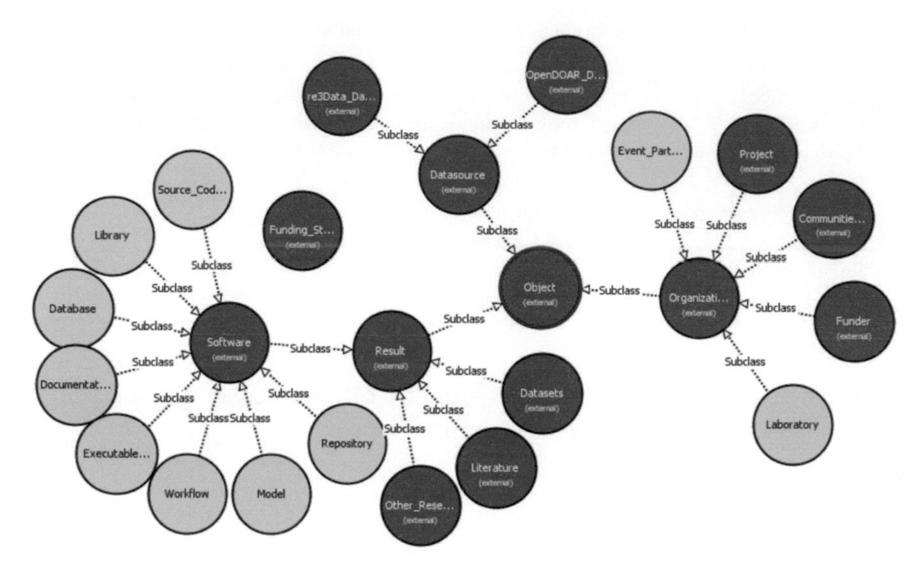

Figure 5.2 The proposed research digital artifacts ontology (RDAO), adopted from the OpenAIRE v1.3 data model.

map the data to the new schema. In the two subsequent sections, we describe how to define the mapping rules for two expanded entities, namely "Funding" and "Software" of OpenAIRE.

Expand the funding entities:

It is important to connect publications with their supporting systems in the research knowledge graph. This connection will give us some valuable information, such as funding recommendations for similar research topics.

OpenAIRE v1.3 has entities that link publications to the funding system. However, this funding entity focuses solely on financial support, even though research support involves more than just financial support. Other types of support, such as laboratories that provide laboratory equipment and/or materials for research, computing infrastructures that facilitate research with high-performance computing, teams of programmers who help develop some of the tools for their research, communities or people who provide feedback through their discussions, and so on, are equally important. These non-funding-supporting objects cannot be mapped into the initial funding entity, and critical information may be lost as a result.

Table 5.2 Expanded research support entities and mapping rules.

No.	Entity	(1)	(2)	(3)	(4)	(5)	(6)	(7)
1	Funder	Yes						
2	Communities					Yes		
3	Organization	Yes	Yes	Yes	Yes	Yes	Yes	
4	Project	Yes	Yes	Yes	Yes	Yes	Yes	
5	Event participant					Yes		Yes
6	Laboratory		Yes			Yes	Yes	

Note: (1) Grant, (2) infrastructure, (3) publication, (4) scholarship, (5) subject Matter, (6) material, (7) other

This study proposes adding numerous entities to the original funding entity to cover non-funding-supporting objects. Table 5.2 shows extended support entities and mapping rules. In the table, there is an x-axis (column) and a y-axis (row), and also "Yes" statement. The table column represents the support forms for research, divided into several types of support, including grant, infrastructure, publication, scholarship, subject matter, material, and others. At the same time, the row section represents the extended support entities, including funder, communities, organization, project, event participant, and laboratory. The word "Yes" in the column indicates that the form of support in the row section has the form of the type of support in the column section. The following is an example of how to read the table: A laboratory entity has the word "yes" in columns 2 (infrastructure), 5 (subject matter), and 6 (material). It means that the laboratory entity has the type of support related to infrastructure, subject matter, and material.

The support provided by the funder is support in the form of financial research support. Support in providing equipment for research is included in the infrastructure. Support in the form of funding support for journals or proceedings that require publication costs is included in the publication entity. The scholarship includes financial assistance in the form of a scholarship or fellowship. Support in the form of research support for the substance of the research is contained in the subject matter. Material aid for research, such as chemical assistance, is a type of support that includes material. Another support is in the form of non-material facilities.

Expand the software entities:

In OpenAIRE v1.3, the resulting entity has four sub-entities: literature, datasets, software, and other research products. When mapping research outcomes to those original result entities, we think that these entities are excessively generic, which can lead to ambiguity. For example, if the research develops an algorithm that is then implemented into a Python library, there will be uncertainty as to whether the research is classified as a software entity because there is a computational process and the output can be used as a tool or classified as "other research product entity" as stated in the guidelines. To tackle this ambiguity problem, we attempt to specify the result entity, particularly in the software entity section, by breaking it down into numerous new entities. In this section, we examine the research outcomes that will be mapped into new entities using the mapping rules shown in Table 5.3. In the table, there is an x-axis (column) and a y-axis (row), and also "Yes" statement. In the table column, the forms of software are divided into several parts of software, including source code, documentation, binary code, algorithm, and dataset. While in the row section (extended software entities), the forms of software are classified into library, source code, executable file, documentation, model, workflow, repository, and database. The word "Yes" in the column indicates that the form of software in the row section has the form of the type of software in the column section. The following is an example of how to read the table: A library entity has the word "Yes" in columns 1 (source code), 2 (documentation), and 3 (binary code). It means that the library entity has the type of software related to source code, documentation, and binary code.

Source code is a component of software that contains a collection of computer programs written in a specific language that comprises an application or software. The documentation is a component of the software that serves as both a method of communicating information about the software and a means of communicating information about the software. The binary code is the compiled software's source code file. An algorithm is a piece of software with a collection of instructions or steps written logically to solve a problem. A dataset is a set of data that is ready to be manipulated to generate new information.

5.4.4 Analysis

We compared the results of mapping our dataset to the original schema and the extended schema with mapping rules as explained in the previous section. The results of mapping individuals that are provided support during research activities can be seen in Figures 5.3 and 5.4, respectively, for the original and the extended schema. We have only four relevant entities in the original

Table 5.3 Expanded software entities and mapping rules.

No.	Entities	(1)	(2)	(3)	(4)	(5)
1	Library	Yes	Yes	Yes		
2	Source code	Yes				
3	Executable file			Yes		
4	Documentation		Yes			
5	Model	Yes		Yes		
6	Workflow				Yes	
7	Repository	Yes	Yes	Yes	Yes	
8	Database					Yes

Note: (1) source Code, (2) documentation, (3) binary Code,
(4) algorithm, (5) dataset

schema and six entities in the extended schema. Several individuals that were mapped to the "Organization" entity in the original schema have been distributed to multiple entities in the extended schema. It indicates that the extended schema has enriched the available information. Furthermore, the number of individuals that can be mapped has increased with the extended schema. It is another indication that the ambiguity when mapping individuals to the available entities has been reduced.

Similar results are also obtained from the second analysis. Figures 5.5 and 5.6 depict the results of mapping individuals that are related to produced software. There is only one relevant entity in the original schema, namely

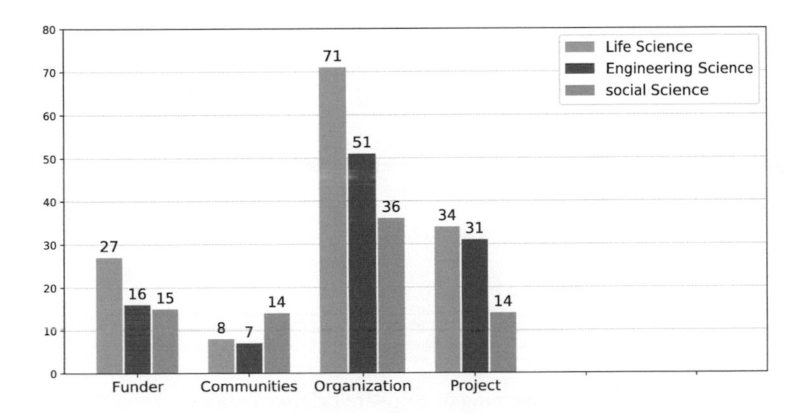

Figure 5.3 The mapping of research support original entities.

"Software," compared to the eight entities in the extended schema. As a result, richer representation can be obtained, reducing the ambiguity, where more individuals can be mapped to the available entities in the extended schema.

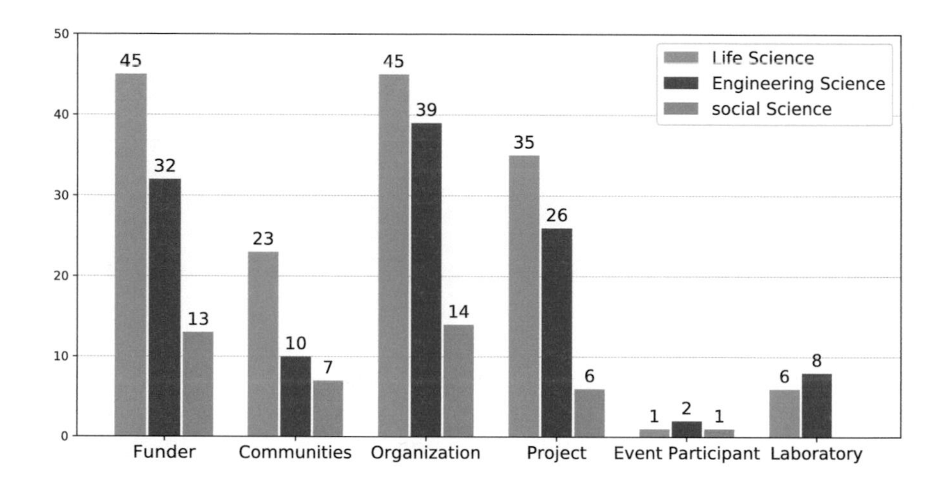

Figure 5.4 The mapping of research support expanded entities.

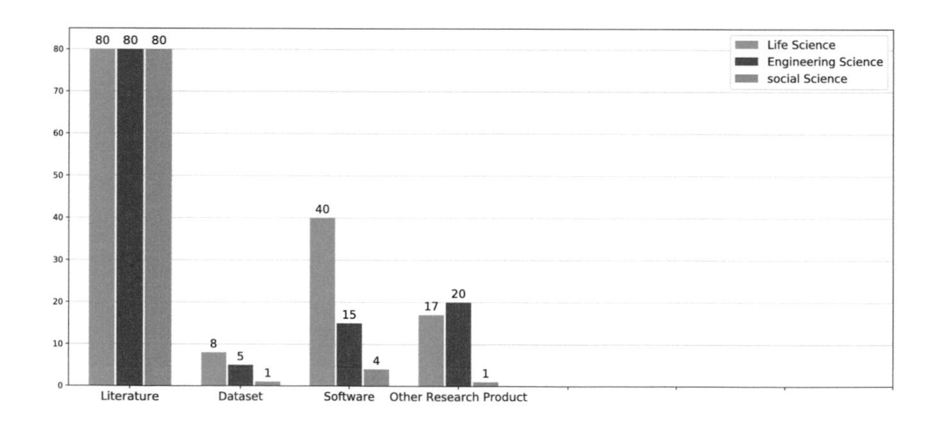

Figure 5.5 The mapping of software original entities.

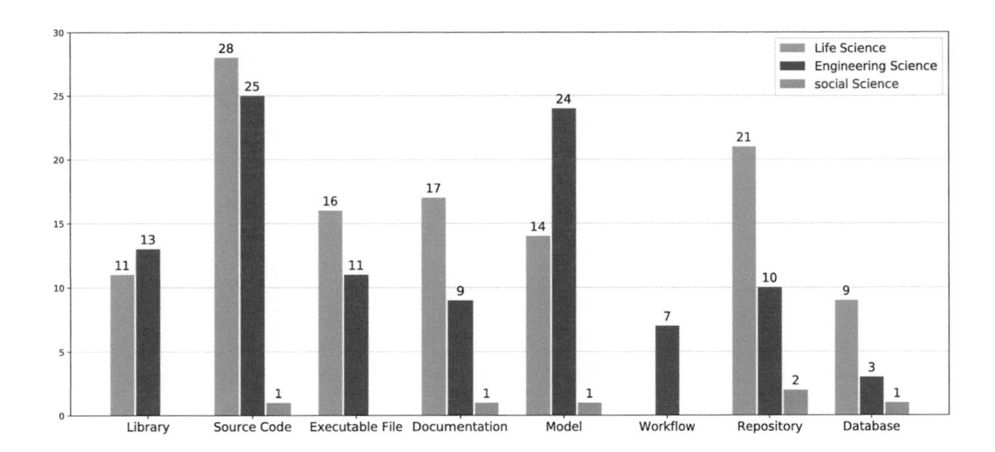

Figure 5.6 The mapping of software expanded entities.

5.4.5 Discussion

As depicted in previous sections, the knowledge creation step of our knowledge graph has taken most of our efforts. The efforts are necessary to ensure the reliability of the constructed knowledge graph. First, a better representation of digital artifacts is required, and, therefore, an extension to the existing schema was proposed. Second, mapping rules for the new schema must be defined as clearly as possible to minimize knowledge discrepancy. From the analysis of the results, we outline several important things as follows:

1. The extension and the mapping rules have minimized the ambiguity in performing a mapping. As shown in the comparison between Figures 5.3 and 5.4 or between Figures 5.5 and 5.6, the number of mapped individuals was increased. A few individuals cannot be mapped to any entity available in the original schema.
2. A few individuals from the entities of the original schema have been distributed to the newly introduced entities. In this case, our solution presents richer information.
3. The distribution of individuals across domains can be different when using the original and the extension. Once again, it demonstrates that the ambiguity in the mapping has been significantly reduced.

An ontology is essential as it shares a common understanding of entities and their relationships within a domain. In a knowledge graph solution, an ontology dictates how to store and retrieve facts. In this work, we have demonstrated that even though a knowledge graph solution is promising, it relies heavily on an ontology in its knowledge creation step.

5.5 Conclusion

While scientific articles can be seen as the ultimate results of research activities, it is hard to understand the whole process of the activities based on the articles alone. Information about the data, the performed computational steps, and the code leading from the data to inference is described minimally in the articles. This situation leads to the limitation of reproducibility of most results obtained from research activities.

In this work, we introduce an approach to overcome this limitation by integrating multiple types of digital research artifacts through a knowledge graph solution. The solution started with data-driven knowledge creation, where a collection of digital artifacts related to research activities will be extracted from scientific articles. After that, the extracted digital artifacts will be aligned to a pre-defined schema to represent them uniformly. As a result, the integrated and linked artifacts can be deployed for various objectives, such as enabling data interoperability and reusability.

Our work was mainly focused on the knowledge creation step as described throughout this chapter. This step was essential to ensure the quality of the produced knowledge. Our solution requires an extension of the existing schema to represent various types of digital artifacts better. Furthermore, the extension requires proper mapping rules to align a digital artifact with the correct entity. We evaluated our solution by extracting digital artifacts from articles in three prominent domains: life science, engineering science, and social science. Our findings indicate that a better distribution of artifacts can be achieved with richer representation. Finally, the combination of newly introduced entities and mapping rules has minimized the ambiguity of mapping artifacts to correct entities.

In the future, we would like to extend our solution to identify the level of contribution of each artifact in research activity. In the current solution, each artifact contributes equally to the activity. However, one artifact could contribute more or less than the other in reality. As a result, the importance of relationships between artifacts can be measured to understand better. We also consider establishing multiple intelligent applications on top of the knowledge graph. For example, to identify similar research projects among individuals or organizations and identify possible collaboration among them.

Acknowledgments

This research was supported by the Research Center for Computing, Research Organization for Electronics and Informatics, and National Research and Innovation Agency, Indonesia. We would like to thank all the members of the Knowledge Computing Research Group and the High Performance Computing Research Group for their valuable suggestions and feedback.

References

[1] V. Stodden, "Beyond open data: A model for linking digital artifacts to enable reproducibility of scientific claims," in *Proceedings of the 3rd International Workshop on Practical Reproducible Evaluation of Computer Systems*, P-RECS '20, (New York, NY, USA), p. 9âĂŞ14, Association for Computing Machinery, 2020.

[2] T. Kuhn and M. Dumontier, "Making digital artifacts on the web verifiable and reliable," *IEEE Transactions on Knowledge and Data Engineering*, vol. 27, no. 9, pp. 2390–2400, 2015.

[3] N. Juty, S. M. Wimalaratne, S. Soiland-Reyes, J. Kunze, C. A. Goble, and T. Clark, "Unique, Persistent, Resolvable: Identifiers as the Foundation of FAIR," *Data Intelligence*, vol. 2, pp. 30–39, 01 2020.

[4] M. Stuermer, G. Abu-Tayeh, and T. Myrach, "Digital sustainability: basic conditions for sustainable digital artifacts and their ecosystems," *Sustainability Science*, vol. 12, pp. 247–262, Mar 2017.

[5] P. Manghi, N. Houssos, M. Mikulicic, and B. Jörg, "The data model of the openaire scientific communication e-infrastructure," in *Metadata and Semantics Research* (J. M. Dodero, M. Palomo-Duarte, and P. Karampiperis, eds.), (Berlin, Heidelberg), pp. 168–180, Springer Berlin Heidelberg, 2012.

[6] A. Aryani, M. Poblet, K. Unsworth, J. Wang, B. Evans, A. Devaraju, B. Hausstein, C.-P. Klas, B. Zapilko, and S. Kaplun, "A research graph dataset for connecting research data repositories using rd-switchboard," *Scientific Data*, vol. 5, p. 180099, May 2018.

[7] S. Soiland-Reyes, P. Sefton, M. Crosas, L. J. Castro, F. Coppens, J. M. Fernández, D. Garijo, B. Grüning, M. La Rosa, S. Leo, E. Ó Carragáin, M. Portier, A. Trisovic, R.-C. Community, P. Groth, and C. Goble, "Packaging research artefacts with ro-crate," *Data Science*, vol. Preprint, pp. 1–42, 2022. Preprint.

[8] A. Hogan, E. Blomqvist, M. Cochez, C. d'Amato, G. de Melo, C. Gutiérrez, S. Kirrane, J. E. Labra Gayo, R. Navigli, S. Neumaier, A.-C. Ngonga Ngomo, A. Polleres, S. M. Rashid, A. Rula, L. Schmelzeisen, J. F. Sequeda, S. Staab, and A. Zimmermann, *Knowledge Graphs*. No. 22 in Synthesis Lectures on Data, Semantics, and Knowledge, Morgan & Claypool, 2021.

[9] A. Fensel, Z. Akbar, E. Kärle, C. Blank, P. Pixner, and A. Gruber, "Knowledge Graphs for Online Marketing and Sales of Touristic Services," *Information*, vol. 11, p. 253, May 2020.

[10] R. D. Page, "Ozymandias: a biodiversity knowledge graph," *PeerJ*, vol. 7, p. e6739, Apr. 2019.

[11] L. Penev, M. Dimitrova, V. Senderov, G. Zhelezov, T. Georgiev, P. Stoev, and K. Simov, "OpenBiodiv: A Knowledge Graph for Literature-Extracted Linked Open Data in Biodiversity Science," *Publications*, vol. 7, p. 38, May 2019.

[12] J. Z. Pan, E. Edelstein, P. Bansky, and A. Wyner, "A Knowledge Graph Based Approach to Social Science Surveys," *Data Intelligence*, vol. 3, pp. 477–506, Oct. 2021.

[13] J. M. Gomez-Perez, J. Z. Pan, G. Vetere, and H. Wu, *Enterprise Knowledge Graph: An Introduction*, pp. 1–14. Cham: Springer International Publishing, 2017.

[14] D. Fensel, U. Şimşek, K. Angele, E. Huaman, E. Kärle, O. Panasiuk, I. Toma, J. Umbrich, and A. Wahler, *How to Build a Knowledge Graph*, pp. 11–68. Cham: Springer International Publishing, 2020.

[15] P. Chen, Y. Lu, V. W. Zheng, X. Chen, and B. Yang, "KnowEdu: A System to Construct Knowledge Graph for Education," *IEEE Access*, vol. 6, pp. 31553–31563, 2018.

[16] M. Rotmensch, Y. Halpern, A. Tlimat, S. Horng, and D. Sontag, "Learning a Health Knowledge Graph from Electronic Medical Records," *Scientific Reports*, vol. 7, p. 5994, Dec. 2017.

[17] A. Berven, O. A. Christensen, S. Moldeklev, A. L. Opdahl, and K. J. Villanger, "A knowledge-graph platform for newsrooms," *Computers in Industry*, vol. 123, p. 103321, Dec. 2020.

[18] E. Shin, S. Yoo, Y. Ju, and D. Shin, "Knowledge graph embedding and reasoning for real-time analytics support of chemical diagnosis from exposure symptoms," *Process Safety and Environmental Protection*, vol. 157, pp. 92–105, 2022.

[19] J. Martinez-Gil, G. Buchgeher, D. Gabauer, B. Freudenthaler, D. Filipiak, and A. Fensel, "Root cause analysis in the industrial domain using knowledge graphs: A case study on power transformers," *Procedia Computer Science*, vol. 200, pp. 944–953, 2022. 3rd International Conference on Industry 4.0 and Smart Manufacturing.

[20] X. Mao, H. Sun, X. Zhu, and J. Li, "Financial fraud detection using the related-party transaction knowledge graph," *Procedia Computer Science*, vol. 199, pp. 733–740, 2022. The 8th International Conference on Information Technology and Quantitative Management (ITQM 2020 & 2021): Developing Global Digital Economy after COVID-19.

[21] H. Wu, Y. Chang, J. Li, and X. Zhu, "Financial fraud risk analysis based on audit information knowledge graph," *Procedia Computer Science*, vol. 199, pp. 780–787, 2022. The 8th International Conference on Information Technology and Quantitative Management (ITQM 2020 & 2021): Developing Global Digital Economy after COVID-19.

[22] D. Mrdjenovich, M. K. Horton, J. H. Montoya, C. M. Legaspi, S. Dwaraknath, V. Tshitoyan, A. Jain, and K. A. Persson, "propnet: A Knowledge Graph for Materials Science," *Matter*, vol. 2, pp. 464–480, Feb. 2020.

[23] J. Xu, S. Kim, M. Song, M. Jeong, D. Kim, J. Kang, J. F. Rousseau, X. Li, W. Xu, V. I. Torvik, Y. Bu, C. Chen, I. A. Ebeid, D. Li, and Y. Ding, "Building a PubMed knowledge graph," *Scientific Data*, vol. 7, p. 205, Dec. 2020.

[24] X. Chen, S. Jia, and Y. Xiang, "A review: Knowledge reasoning over knowledge graph," *Expert Systems with Applications*, vol. 141, p. 112948, Mar. 2020.

[25] H. Cousijn, R. Braukmann, M. Fenner, C. Ferguson, R. van Horik, R. Lammey, A. Meadows, and S. Lambert, "Connected research: The potential of the pid graph," *Patterns*, vol. 2, Jan 2021.

[26] S. Kanza, C. L. Bird, M. Niranjan, W. McNeill, and J. G. Frey, "The ai for scientific discovery network$^+$," *Patterns*, vol. 2, Jan 2021.

[27] U. Zeigermann, "Knowledge integration in sustainability governance through science-based actor networks," *Global Environmental Change*, vol. 69, p. 102314, 2021.

[28] S. Sang, Z. Yang, L. Wang, X. Liu, H. Lin, and J. Wang, "SemaTyP: a knowledge graph based literature mining method for drug discovery," *BMC Bioinformatics*, vol. 19, p. 193, Dec. 2018.

[29] S. Sang, Z. Yang, X. Liu, L. Wang, H. Lin, J. Wang, and M. Dumontier, "Gredel: A knowledge graph embedding based method for drug discovery from biomedical literatures," *IEEE Access*, vol. 7, pp. 8404–8415, 2019.

[30] A. Rossanez, J. C. dos Reis, R. d. S. Torres, and H. de Ribaupierre, "KGen: a knowledge graph generator from biomedical scientific literature," *BMC Medical Informatics and Decision Making*, vol. 20, p. 314, Dec. 2020.

[31] Z. Li, Q. Zhong, J. Yang, Y. Duan, W. Wang, C. Wu, and K. He, "DeepKG: an end-to-end deep learning-based workflow for biomedical knowledge graph extraction, optimization and applications," *Bioinformatics*, vol. 38, pp. 1477–1479, 11 2021.

[32] S. Katsikeas, P. Johnson, M. Ekstedt, and R. Lagerström, "Research communities in cyber security: A comprehensive literature review," *Computer Science Review*, vol. 42, p. 100431, 2021.

[33] C. Wang, X. Ma, J. Chen, and J. Chen, "Information extraction and knowledge graph construction from geoscience literature," *Computers & Geosciences*, vol. 112, pp. 112–120, 2018.

[34] X. Ma, "Knowledge graph construction and application in geosciences: A review," *Computers & Geosciences*, vol. 161, p. 105082, 2022.

[35] B. Abu-Salih, "Domain-specific knowledge graphs: A survey," *Journal of Network and Computer Applications*, vol. 185, p. 103076, 2021.

[36] H. Chen and X. Luo, "An automatic literature knowledge graph and reasoning network modeling framework based on ontology and natural language processing," *Advanced Engineering Informatics*, vol. 42, p. 100959, 2019.

[37] Q. Wang, M. Li, X. Wang, N. Parulian, G. Han, J. Ma, J. Tu, Y. Lin, H. Zhang, W. Liu, A. Chauhan, Y. Guan, B. Li, R. Li, X. Song, Y. R. Fung, H. Ji, J. Han, S.-F. Chang, J. Pustejovsky, J. Rah, D. Liem, A. Elsayed, M. Palmer, C. Voss, C. Schneider, and B. Onyshkevych, "Covid-19 literature knowledge graph construction and drug repurposing report generation," 2020.

[38] Y. Yang, Z. Cao, P. Zhao, D. D. Zeng, Q. Zhang, and Y. Luo, "Constructing public health evidence knowledge graph for decision-making support from covid-19 literature of modelling study," *Journal of Safety Science and Resilience*, vol. 2, no. 3, pp. 146–156, 2021.

[39] S. Soiland-Reyes, P. Sefton, M. Crosas, L. J. Castro, F. Coppens, J. M. Fernández, D. Garijo, B. A. Grüning, M. L. Rosa, S. Leo, E. Ó. Carragáin, M. Portier, A. Trisovic, R. Community, P. Groth, and C. A. Goble, "Packaging research artefacts with ro-crate," *CoRR*, vol. abs/2108.06503, 2021.

[40] M. D. Wilkinson, M. Dumontier, I. J. Aalbersberg, G. Appleton, M. Axton, A. Baak, N. Blomberg, J.-W. Boiten, L. B. da Silva Santos, P. E. Bourne, J. Bouwman, A. J. Brookes, T. Clark, M. Crosas, I. Dillo, O. Dumon, S. Edmunds, C. T. Evelo, R. Finkers, A. Gonzalez-Beltran, A. J. Gray, P. Groth, C. Goble, J. S. Grethe, J. Heringa, P. A. `t Hoen, R. Hooft, T. Kuhn, R. Kok, J. Kok, S. J. Lusher, M. E. Martone, A. Mons, A. L. Packer, B. Persson, P. Rocca-Serra, M. Roos, R. van Schaik, S.-A. Sansone, E. Schultes, T. Sengstag, T. Slater, G. Strawn, M. A. Swertz, M. Thompson, J. van der Lei, E. van Mulligen, J. Velterop, A. Waagmeester, P. Wittenburg, K. Wolstencroft, J. Zhao, and B. Mons, "The FAIR Guiding Principles for scientific data management and stewardship," *Scientific Data*, vol. 3, p. 160018, Dec. 2016.

[41] T. Weigel, U. Schwardmann, J. Klump, S. Bendoukha, and R. Quick, "Making Data and Workflows Findable for Machines," *Data Intelligence*, vol. 2, pp. 40–46, Jan. 2020.

[42] A. Martín-Martín, M. Thelwall, E. Orduna-Malea, and E. Delgado López-Cózar, "Google scholar, microsoft academic, scopus, dimensions, web of science, and opencitations´ coci: a multidisciplinary comparison of coverage via citations," *Scientometrics*, vol. 126, no. 1, pp. 871–906, 2021.

[43] J. Baas, M. Schotten, A. Plume, G. Côté, and R. Karimi, "Scopus as a curated, high-quality bibliometric data source for academic research in quantitative science studies," *Quantitative Science Studies*, vol. 1, pp. 377–386, 02 2020.

[44] H. Y. Khudhair, D. Jusoh, A. Bin, A. F Abbas, A. Mardani, and K. M. Nor, "A review and bibliometric analysis of service quality and customer satisfaction by using scopus database," *International Journal of Management*, vol. 11, no. 8, 2020.

Appendix

Table 5.4 The mapping of extracted digital artifacts to OpenAIRE.

Entity	Property	Life science	Engineering science	Social science
Object	–			
Datasource	Type	45	40	26
	OpenAIRE compatibility	0	0	0
	Official name	52	49	32
	English name	44	38	27
	Web site url	41	38	24
	Logo url	4	10	3
	Email contact	17	5	3
	Namespace Prefix	0	0	0
	Latitude	40	0	0
	Longitude	0	0	0
	Date of validation	1	1	1
	Description	24	32	3
	Subject	1	11	1
OpenDOAR Datasource	Number of items	0	0	0
	Date of number of items	0	0	0
	Policies	0	0	0
	Languages	0	0	0
	Content types	0	0	0
	Access info package	0	0	0
re3Data Datasource	Release start date	0	0	0
	Release end date	0	0	0
	Mission statement URL	0	0	0
	Data provider	0	0	0
	Service provider	0	0	0
	Database access type	0	0	0
	Database access restrictions	0	0	0
	Data upload type	0	0	0
	Data upload restrictions	0	0	0
	Versioning	0	0	0

Table 5.4 Continued.

	Citation guideline url	0	0	0
	Quality management kind	0	0	0
	PID systems	0	0	0
	Certificates	0	0	0
	Policies	0	0	0
Organization	PersistentIdentifier	16	13	3

Table 5.5 The mapping of extracted digital artifacts to OpenAIRE.(Continued)

	Legal short name	53	44	9
	Legal name	71	51	36
	Web site URL	29	20	23
	Logo URL	21	17	21
	Equivalent shortnames	1	1	1
	Equivalent names	2	1	1
Funder	Jurisdiction	27	12	15
	PID	24	16	11
Project	Project code	29	22	10
	Title	34	31	14
	Acronym	14	18	6
	Call identifier	1	1	1
	Contract type	1	1	1
	Keywords	1	1	1
	Web site URL	4	2	1
	Start date	6	8	3
	End date	6	8	3
	Duration	1	1	1
	EC SC39	0	0	0
	OA mandate publications	1	1	1
	Subjects	1	1	1
Result	Title	80	77	80
	Creator	80	77	80
	Date of acceptance	48	24	32
	Publisher	74	77	59
	Description	61	58	59

Table 5.5 (Continued).

	PID	80	77	79
	Language	61	48	60
	Subject structured (string, qualifier, provenance)	21	21	21
	Instance	26	28	21
	External reference	10	1	1
	Source	1	1	1
	Context	25	16	9
	Country	41	46	47
	Best access rights	31	47	51
Literature	–	0	0	0
Datasets	–	0	2	0
Software	Contact person	26	10	3
	Contact group	12	3	1
	Software type	30	9	4
	Distribution location	40	13	3
	Documentation	38	13	2
	Programming language	19	12	1
	Version	19	5	1
	Tool	8	2	1

Table 5.6 The mapping of extracted digital artifacts to OpenAIRE.

	Distribution form	33	15	2
Other research products	Contact person	1	1	1
	Contact group	2	1	1
	Distribution location	17	20	1
	Documentation	2	19	1
	Version	2	5	1
	Tool	0	0	0

Table 5.6 (Continued.)

Funding stream	Identifier	0	0	0
	Name	0	0	0
	Description	0	0	0
Communities	Name	8	7	14
	Related Subjects	8	5	9
	Inference parameters	1	1	1
	Monitoring parameters	1	1	1
	Relevant Projects	6	4	9
	HasRelevant Datasource	6	4	9
	HasRelevant Zenodo Communities	6	4	9

6

A Review of Ontology Development Methodologies: The Way Forward for Robust Ontology Design

Enesi Femi Aminu[1], Ishaq Oyebisi Oyefolahan[2], and Muhammad Bashir Abdullahi[1]

[1]Department of Computer Science, Federal University of Technology Minna, Nigeria
[2]Department of Information Technology, Federal University of Technology Minna, Nigeria
E-mail: enesifa@futminna.edu.ng; o.ishaq@futminna.edu.ng; el.bashir02@futminna.edu.ng

Abstract

In this present age, the application of ontology as a data modeling technique across different fields of study, for example, knowledge management and information retrieval systems, is indispensable. This development is necessary to find viable solutions to the challenges of data heterogeneity and concept mismatch. Therefore, the end goal is geared toward achieving machine-represented data; in other words, the data are being modeled ontologically. There are existing ontology design methodologies; however, a single methodology is often not complete to design a robust ontology. Thus, this research aims to review the existing standard methodologies through concept-based analysis that suggests a way forward to design robust ontology. The analysis of the review is carried out by considering the goals of achieving robust ontology design, such as data integration, accessibility, reusability, and domain granularity. Based on the literature, this review shows that collaborative design with domain experts, application of standard evaluation techniques, modification of existing ontology development methodologies, types of ontology, and ontology-based machine learning models are determinant factors that define the robustness of ontology. Therefore, if an ontology developer pays attention to these criteria to design an implementable model, this would pave way for robust ontology to be designed.

Keywords: Data collaboration, data integration, robust ontology, ontology design, ontology methodologies.

6.1 Introduction

In this current age, while availability of data is no longer an issue, the astronomical growth of these data in heterogeneous forms calls for research attention. This is because the heterogeneous nature of data along with the unstructured state of its repositories poses difficulties to achieve the required collaborative operations on data, such as integration and reusability. Consequently, several data modeling techniques have been employed by researchers to advance data collaboration in the form of metadata. However, there are gaps to bridge in order to ascertain a more robust knowledge management system, such as information retrieval systems, question answering systems, and recommendation systems. In view of this research quest, a more robust data modeling technique called ontology is promising to this effect. Over some decades, researchers have been constantly employing the technique in the knowledge management fields of study as mentioned earlier in this section. An important strength of ontology technique lies on its ability to adapt to any series of modification either during the course of modeling or application. It is flexible toward any form of design approaches; either top down or bottom up [1]. Ontology, as semantic technology, has the potential to map the physical entities into computational entities that would make communication easier between human and machine [2]. This is because ontology has a standard mechanism to characterize domain knowledge.

Ontology is described as the prime stronghold of semantic technologies. As such, researchers in the field of ontology engineering are currently challenged to advance a solution on viable ontology development methodology that can assist to design robust knowledge-based systems [3]. There are required technologies to model ontology, which include but not limited to extensible markup language (XML), resource description framework (RDF), web ontology language (OWL), logic inference, and SPARQL as shown by the popular semantic technologies stack shown in Figure 6.1.

The semantic stack's architecture represented by Figure 6.1 shows the semantic technologies in different layers to realize knowledge systems. The

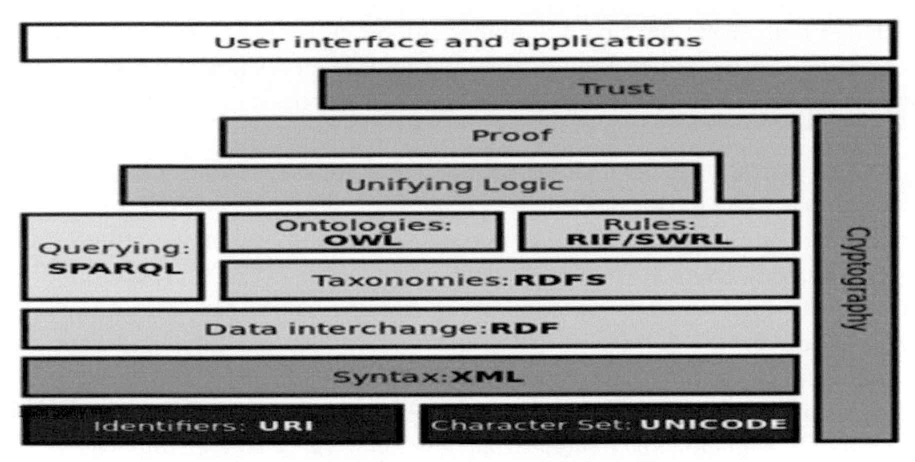

Figure 6.1 The semantic technologies stack [4].

bottom layer consists of basic techniques, which include URIs, Unicode, and XML. The machine-readable format is represented in the syntax of XML with URI as metadata and namespace identifier. The middle layer presents some techniques, which include RDF, RFDS, OWL, and SPARQL. The first three technologies are used to represent knowledge; however, their differences are premised on the degree of expressivity. Axioms and rules are encoded using SWRL or RIF; SPARQL serves as a medium to query data. The top layer, proof, trust, and cryptography, ensures that source documents for the web are from trusted sources. The last layer is the user interface that enables humans to use an application.

Based on the literature, ontology is a data modeling technique that leverages on gathering of entities by taking cognizance of their semantic relations, such as meronyms and holonyms. Also, constraints are enforced among the relations in order to achieve a structured and collaborative data repository [5]. In essence, this is to enhance the data communication between human and machine. The strength of a robust ontology hugely depends on its capacity to infer contextual (hidden) knowledge based on a given literal knowledge. The inference would occur based on the knowledge representation models such as RDF and OWL [6–8]. Similarly, a long-standing acceptable definition of ontology [9, 10] is that of Gruber [11], which defines ontology as an explicit and formal specification of a conceptualization. The term "*explicit*" in the definition suggests that entities of any knowledge must be defined in clear terms; also, the entities' integrity must be precisely defined as well. The term "*formal*" connotes and emphasizes that ontology is a structured knowledge representation model. Another important term in the definition is *conceptualization*, which implies that ontology is an abstract representation of a physical knowledge where the required entities are harvested for modeling.

The physical knowledge includes physical real objects; similarly, abstract objects are also part of the knowledge. Examples of objects include bioinformatics, biomedicine, sports, building construction and design, agriculture, and religion; they are described as domains of ontology [12, 13]. Therefore, following the established knowledge about ontology, it is also viewed as *shared* knowledge of domain to solve the underlying structural and data heterogeneous issues [14]. The term "shared" describes ontology as a mechanism that supports data collaboration and integration. Consequently, ontology is therefore a semantic data model that leverages on a well-defined data collaboration mechanism, thereby ameliorating the issue of data inconsistency [15]. Thus, the role of ontology in knowledge management, data sharing, integration, and reusability against other techniques is described as unmatched [16]. This is because ontology has been severally applied to model complex data of different domains such as biomedical data (gene ontology), informatics data, and agriculture data [17]. This is evident in the work of He *et al.* [18] who employed ontology to model the data of the world ravaging ailment, coronavirus (COVID-19). The next section of this chapter presents the common principles for ontology development and the technologies required.

6.2 Ontology Development

Researchers in knowledge management usually employ ontology to build structure and intelligent knowledge-based systems that assist in knowledge reuse [19]. More so, with the aid of ontology, a system capable to infer the contextual knowledge of a given domain is achieved [20]. There are several reasons behind ontology development; these are but not limited to creating a platform for machine or users to share knowledge, to allow knowledge to be reused, and to define and analyze domain knowledge unambiguously [21, 22]. These reasons are principles that define the robustness of a given ontology if they are successfully attained.

To develop such ontology, a standard methodology, which is an iterative engineering process like the standard software development principle, is required. However, part of the contending issues that surround ontology design is on the appropriate methodology to deploy [23, 24]. Noy and McGuiness [21], in their research, reported that among the existing methodologies, there is no one that can sufficiently claim superiority over others. In other words, each of them has their own design flaws, which makes it inadequate to be used effectively. Therefore, a developer has the prerogative to make a choice based on certain defined factors. Nevertheless, to design ontology with characteristics of shared knowledge and inference capability, the methodology employed must be robust. This implies that ontology development's activities must be spread across the three categories of design, which include predevelopment, development, and postdevelopment stages [15]. Considering the literature reviewed in this work, the most frequently used methodologies include Gruninger–Fox, Methontology, Noy–McGuiness methodology, and NEON [25].

The process of ontology development equally involves the choice of ontology knowledge representation technologies, which include the languages and editors [26]. However, the focus of this review is on the methodology for ontology design that provides structural framework. In addition, the developer determines the language in which the methodology is implemented based on certain criteria; for example, expressivity of language. In this research, ontology representation language is classified into two folds. They are World Wide Web Consortium (W3C) based standard and International Standardization Organization (ISO). Examples of the former include OWL [27–29], RDF [30–32], RDF Schema [33–35], DARPA Agent Markup Language, and Ontology Interchange Language (DAML+OIL) [36, 37]. A classic example of the latter is XML Topic Map (XTM) [38]. OWL is reported as the most commonly used language as a result of its semantic expressivity [30, 39, 40], regardless of the two standards. The ontology editors are Protégé [41–43], FAO AGROVOC Concept Server Workbench Tool [44], OBO-Edit [45], SWOOP [46], Apollo [47], IsaViz [48], and TopBraidComposer [49, 50].

6.3 The Existing Ontology Development Methodology: The Review

In this section, the existing ontology design methodologies are partly reviewed. Methodology specifies a set of procedures on how the identified activities in the process of ontology design are duly carried out. Therefore, the quest to employ an existing methodology or brainstorm on a total new approach is indispensable [51, 52]. This is because the developers are in most cases constrained on the existing methodologies to model a knowledge-based system. The constraint is owing to their shortfalls based on the standard of software development guidelines [53, 54]. The literature of Lenat and Guha [55] reported the first ontology development methodology dubbed as Cyc project. The activities of the methodology are encapsulated into three stages [21]. Dean *et al.* [56], in their work, reviewed some ontology development methodologies. They include TOVE (Toronto Virtual Enterprise) that consists of six approaches: enterprise model approach with four stages of activity and Methontology that consists of seven activities of the development process. Others are SENSUS, MENELAS, ONIONS, Ontolingua KBSI, and IDEF5 consisting of five activities. Similarly, the research of Fernandez-Lopez *et al.* [57] presents what is described as regularly used approaches for ontology design. They are Methontology, Gruninger–Fox, SENSUS, and Uschold–King. The functionalities of these approaches are analyzed against IEEE Standard 1074–1995, which is often described as a software development standard.

There are some relevant questions that normally arise when ontology design is required to address a particular problem. For instance, most often, the ontology developer and domain query the suitable methodology to design ontology. This question has to be diligently addressed because the robustness of ontology is a consequence of ontology methodology, which requires a lot of analysis. Similarly, a question such as "should ontology be developed from scratch or reuse an existing ontology?" normally arises. To address this question, it largely depends on the proposed approach to design ontology [58]. Therefore, the choice of methodology is a rigorous task that requires attention. This postulation is affirmed by the work of Dnyanesh and Rahul [15] who reported that the existing approaches for ontology design are deficient in terms of completeness in design; that is, inability of design activities' wholesomeness for predevelopment, development, and postdevelopment stages. In order to deal with this issue, some literature advocate for hybridized methodology for robust activities. Methontology and Gruninger–Fox [59–61] are some of the methodologies that have benefited from this approach.

Similar to standard software methodologies, the ontology design methodologies are described as iterative ontology engineering process. This is because the developer can effortlessly recall back to the previous activities of methodology whenever the need arises. Most of the domain-based ontologies developed are premised on the engineering process; for example, the soccer ontology in the research of [62]. Obrst *et al.* [63], in their work, equally harped for the significance of methodology to be an iterative process.

Some of the methodologies reported in the literature are Gruninger–Fox, Methontology, Noy–McGuiness (with example of wine ontology), and Uschold–King.

An important activity during the process of ontology development [64, 65] is the identification of ontology's terms or concepts. In order to achieve this goal, there are three strategies that can be employed [66]. They are bottom-up, top-down, and middle-out strategies. The bottom-up strategy works by first identifying the specific terms and generalizing into more abstract terms. On the other hand, the top-down strategy first identifies the most abstract terms and specializes the terms into more specific terms. Lastly and more importantly, the middle-out strategy first identifies the most significant terms, generalizes, and specializes into other terms [58]. Thus, the following subsections specifically review Gruninger and Fox's methodology, Methontology methodology, Noy–McGuiness methodology, and Uschold–Kings methodology.

6.3.1 Gruninger and fox's methodology

This is a methodology that is built based on the technique of first-order logic proposed by Gruninger and Fox [67] to develop a knowledge representation system. As stated earlier, the methodology leverages on the strength of logic because it has the capacity to transform informal scenarios into formal notations. It has five activities; they are identification of motivating scenarios and formalization of informal; specification of ontology's terms in formal language; formulation of competency questions; specification of axioms and rules for the ontology's terms; and creation of conditions for characterizing the completeness of ontology [58, 68, 69]. The methodology was initially conceived to design ontology for business-enterprise-related knowledge. However, the methodology has been constantly employed for different real-life scenarios. For example, Walisadeera *et al.* [23] adapted the methodology to design ontology for the agriculture domain to assist farmers' information needs in Sri Lanka. The implementation was achieved using OWL knowledge representation. Also, the middle-out strategy was employed to specify the core terms of the ontology designed. Similarly, the Gruninger–Fox approach partly constitutes the hybridized methodology for the development of the university ontology [70].

The methodology has an important activity called formulation of competency questions (CQs) that can serve as evaluation mechanism, to validate the correctness of ontology. It equally aids to set the scope of an ontology's domain. As a result of this development, the application of formal notations for machine-represented knowledge is indispensable [71, 72]. Therefore, the capacity of this methodology cannot be overestimated; this is why most often according to literature, it usually formed part of a hybridized approach [59]. However, the methodology is not without notable gaps as identified by the literature [15]; for instance, ontology mapping or versioning cannot be technically achieved. Similarly, the literature [59, 73] equally reported the deficiency of the methodology in terms of completeness of activities for the three stages of design. Another challenging issue is on its capacity to reuse existing ontology.

6.3.2 Methontology methodology

It is a research product of Artificial Intelligence Lab, which supports both ontology design approaches. That is, to design ontology based on reuse of existing ontologies or to entirely create a new ontology from scratch. WebODE and OntoEdit are the management tools that were primarily designed for it [20]. At the initial stage, chemical ontology was the first beneficiary of the methodology, which consists of seven activities. The activities that range from the starting step to the ending step are as follows. The first step is specification, where the objective of the ontology has to be defined. This is followed by knowledge acquisition; the required knowledge to model ontology has to be acquired from related sources. The third activity is conceptualization; the concepts of a domain are defined and classified as subjects, objects, and relations. Integration is another subsequent activity where the concepts are related together as hypernym, hyponym, holonymy, or meronym. The fifth activity is implementation; at this point, knowledge representation language and editor are required. The sixth is evaluation; that is, the consistency of ontology to develop has to be verified and validated. The last activity is documentation of ontology development process [74–76].

According to literature, this methodology has a wide application. The work of [77] designed graduation screen ontology based on this methodology. The ontology was implemented using OWL and protégé as representation language and editor, respectively. Similarly, the methodology was also adopted to develop an active waterfall protection based ontology [78], against the existing methodologies reviewed in the work of [79]. Furthermore, the research of Rizwan *et al.* [80] reaffirmed the standard of the methodology whose activities were claimed to be in total compliance with the procedure of software development life cycle. Even though the reviewed literature of [81] duly acknowledged that no methodology for ontology development is sufficiently robust, the work described Methontology as a promising approach. The work of Ibrahim *et al.* [82] also described the methodology as an outstanding and balanced approach for ontology design.

More so, Methontology was reported as one of the commonly used methodologies for ontology design [83, 84]. Similarly, the research of [85] equally reported the strengths of the methodology; among others is that the activities for the development process of the methodology are in full conformity with the IEEE standard 1074-1999.

Nonetheless, the methodology still has some areas in which an improvement can be made. For instance, a robust activity as that of competency questions stands to improve the validation process.

6.3.3 Noy–McGuiness methodology

Noy–McGuiness is another iterative ontology engineering principle that is designed based on certain rules. It equally works based on seven iterative activities. The first activity is to define the ontology's domain along with precise scope. The second activity is to ascertain the need to develop a new ontology; this is because the methodology encourages the reuse of existing ontology. The next activity is to specify the required and relevant concepts for ontology design. The fourth activity is to declare the classes along with hierarchy, especially when OWL is to be used for implementation. The fifth activity is to also declare the class's property; while the sixth activity is to determine the constraints of the properties. Lastly, individuals or instances (as in the case of OWL) are established [22, 86]. The functionality process of the methodology is explained based on wine ontology. Furthermore, Godspower and Esingbemi [66] adopted the methodology to design a cash crop based ontology for farmers in Nigeria's market. Top-down strategy was employed for terms identification of the ontology and was implemented using the protégé editor. Similarly, the work of Serna and Serna [87], whose aim was to develop software maintenance based ontology, equally employed the Noy–McGuiness methodology.

In a modified form, Tiffani *et al.* [88] adapted Noy–McGuiness and formulated the Arp methodology [89] as a six-activity ontology engineering process to design antimicrobial prescription based ontology. The OWL-Protégé platform was used for the implementation. Similarly, the research of Chen-Huei *et al.* [90] considered the first four activities of Noy and McGuiness approach and merged it with another methodology to design natural disaster management based ontology.

6.3.4 Uschold–King methodology

The idea of this methodology is conceived from an (enterprise) ontology, which comprises four activities [14]. They are as follows: to determine the sole objective of the proposed ontology; to develop the ontology; to perform evaluation measure; and, lastly, to conduct process documentation. However, as the application of the methodology progresses, an improvement was carried out through the three ontology core concept identification strategies [91]. Therefore, the methodology was duly employed to design an ontology for the domain of waste water treatment christened (WaWO) [92].

As partly stated earlier, a robust ontology design is achieved mostly when these methodologies are adapted. Based on literature, some clear cut cases of methodology's modification include the research of Bonanci *et al.* [93] that presents an adapted six-activity ontology design methodology for agriculture knowledge. Also, Aree *et al.* [94] proposed a modified five-activity ontology engineering process for the development of rice crop based ontology. Niu and Issa [95] designed taxonomy for construction domain based on the proposed fused methodology. Gregor *et al.* [96] proposed a logical methodology to design an anticipated intelligence transportation system, ontologically. In the same vein, Dutta *et al.* [97] also presented a ten-activity-based ontology engineering principle for the domain of food. The principle is dubbed as YAMO in their literature. Similarly, Zeb *et al.* [98] equally presented a methodology that consists of ten activities to design ontologies for domains. Another two-activity-based top-down design methodology was also proposed to create ontology in the research of Mezghani *et al.* [99].

Furthermore, there is no doubt that some activities of the existing methodologies are similar. However, in most cases, these activities are combined into three main activities of ontology engineering process [100]. Rayyaan *et al.* [101] examined Methontology and UPON methodologies to develop ontology for the domain of textile supply chain. Based on their results, the researchers pointed out the robustness of UPON for the domain under consideration.

6.4 Way Forward for Robust Ontology Design: The Review

Ontology is said to be robust if it satisfies a set of given design guidelines according to a given standard. More so, the positive outcome of ontology's validity and verification in terms of its content and design largely contributes to the factors that determine the robustness. Primarily, a serious attention has to be paid to the choice of ontology development methodology [102]. However, this is not to state that there are hard rules or guidelines to come up with robust ontology. This is because the research work of [103] emphasizes on collaborative ontology design between ontology engineers and domain experts as one of the reliable means of developing ontology that can stand the test of time. The researchers emphasize on the complete involvement of domain experts at the requirement elicitation level, logical level, and the physical level of ontology design.

John *et al.* [104], in their research, equally faulted the robustness of the existing ontology development methodology. The researchers proposed the hybridization of the procedures or models of software engineering process to the traditional ontological engineering process. Consequently, a software-centric innovative ontology development methodology capable to develop large-scale ontology is targeted. The researchers, in their conclusion, promised to validate the prototype to ascertain the level of accuracy and applicability.

Several methodologies for ontology development have been proposed; however, a robust methodology to address the design of multiple aspect of domain knowledge is largely still in progress. This development propels the research of [105] to propose a four-step methodology that can assist the integration of cross-domain knowledge for multi-aspect ontology design. The major criterion considered by the researchers is on the integration aspect of different domains. A tourism domain is considered to depict the application of the proposed methodology for decision support system based on human machine collective intelligence. Therefore, the researchers aim to experiment the model on multifaceted problems that require knowledge from numerous application domains so as to improve decision making.

The utilization of effective supply chain ontology that aims to ameliorate the interoperability issue often associated with information systems is another aspect of robust ontology design approach to look out for. However, according to the survey work carried out by [106], attention has not been paid to this observation. According to the report, the survey was conducted based on three criteria as a yardstick for measuring the frameworks of six supply chain ontology models. None of the six ontology models considered by the researchers comes without pitfalls. The criteria are follows: scientific paradigm, granularity, and fundamental methodological mechanism. The researchers argued that much effort of ontology development has been concentrated on the organization of human knowledge at the expense of the philosophy of supply chain itself over the year. Therefore, the work suggested that researchers on supply chain ontology should re-channel their efforts on formal ontology. More so, they equally suggested that a more holistic and thorough effort of literature review must be carried out if the dream of robust supply chain ontology methodology has to be achieved.

Furthermore, what gives birth to robust ontology could mean different things to different researchers. The research of [102] strongly argued that if a keen attention is paid to the testing stage of ontology development as a means of measuring quality, a robust ontology would not be an issue. The researchers noted that ontology testing mechanism has the capacity to test the major components of ontology, such as classes, relations, property, and axioms. Consequently, the research was motivated to propose what they called top domain ontology based testing mechanism. Therefore, the efficacy of the proposed mechanism, in terms of semantic matching, would be evaluated.

As noted earlier in the previous section of this work, ontology developer must carefully choose the methodology in order to have effective and efficient ontology that is largely devoid of flaws at minimum. In view of this, the research in [103] proposes a reusable prototype for ontology engineering process, which is premised on the adapted famous NeOn ontology development methodology. Urban Internet of Things was considered as the ontology domain, and the methodology can be applied across other related domains. There are three thematic key points that the method is sitting on, which are level of domain expert participation, logical correctness, and the content performance.

Ontology modeling technique has been identified over time as a reliable means of knowledge management and building of information systems in a contextual form. The commonest technique of modeling ontology is manual, which is very tedious and time consuming; however, the technique is reliable with a good degree of design accuracy. While this strength partly contributes to an effective ontology design, the time spent cannot be easily traded off. Consequently, in order to minimize the human error and design time, Yang *et al.* [107] proposed an ontology learning methodology that can autonomously extract data from extant system engineering standards to form system engineering ontology. The multi-tier methodology consists of a collection of data and pre-processing, natural language preprocessing (NLP) based lexical analysis, and extraction of ontology components. However, the authors submitted that interested researchers have to advance a concerted effort to develop a more robust ontology learning methodology. Currently, the research toward this approach is ongoing as most of the existing prototypes and models suffer from one form of deficiencies or another. For example, they are often incapable to handle implicit terminological and non-terminological relations and data properties.

In a quest to design an ontology that is largely free from errors, the research of [108] described the robustness of ontology as the type of ontology an engineer intends to develop. The researchers argue that most existing domain ontologies are a product of top-level ontology. However, to develop an ontology that is accessible, findable, interoperable, and reusable, a most recent technique called upper ontology alignment is proposed. The technique is an improvement of extreme ontology design methodology, which is based on the ontology design patterns. However, like every other technique, this design pattern equally sought for improvement on the namespace of entity's prefix. Besides, there is a need for robust mechanism to take care of more complex alignment beyond sub-entities (either class or property).

According to the research of [109], there are known limitations among the existing ontology design methodologies; such as vague procedures on how existing ontological and non-ontological collections can be reused. Besides, the issue of usability of finished ontology is yet to attract the needed attention. More importantly, the issue of integration process of various concepts is within a given top-level domain. It implies that ontology's robustness is proportionate to the resources available, and how well can the resources or concepts be integrated within the given domain. Consequently, the researchers proposed a systemic ontology iterative design approach for manufacturing domain herein called *manuService*. The design methodology ranges from requirement analysis to evaluation and feedback. It is a product of popular software development process called rapid application development and extraction of some concepts from existing ontology methodology. The ontology is implemented in OWL using the format of RDF/XML and finally developed using the open source ontology editor, protégé. However, more concepts from the knowledge domain are required to be integrated into the cloud manufacturing based data model.

Similar to the position of [109] on the type of domain, serving as determinant to a methodology to produce robust ontology, Palmirani *et al.* [110] proposed to develop a legal reasoning based ontology named privacy ontology. The researchers on this note hunted for a legal-based design approach called methodology for building legal ontology. This methodology requires ten activities that range from description of the ontology's goal to documentation and collection of feedback. More importantly, the researchers acknowledge the existence of several legal ontologies; however, they lack capacity to integrate with the deontic logic model functioning for legal reasoning. Therefore, the proposed ontology has the capacity for integration. However, the work was reported to be a continuous process as more concepts would be gradually integrated into it.

Yunianta *et al.* [111] argued that the existing ontology design methodologies are not suitable for data integration. Consequently, the research is motivated to develop an enhanced ontology methodology capable for semantic data integration. The enhanced approach called *OntoDI* consists of seven steps, which are categorized in three phases of predevelopment, core development and postdevelopment. The methodology was experimented in the domain of e-learning system. The researchers therefore anticipate that more data would still be integrated; a detailed evaluation would equally be carried out to ascertain the robustness of the methodology.

The fusion of machine learning models with ontology engineering process is gradually receiving attention in the field of knowledge management. Beyond the traditional semantic word representation models, such as word embedding technique (word2Vec), the richness of ontology (especially the OWL-based ontology) has contributed immensely to the robustness of knowledge system. For example, the research of [112] aimed to embed OWL ontology that encrypts the semantic of an ontology by considering its knowledge graph, the lexical knowledge, and the constructors of the knowledge representation model. In other words, each of the ontology's entities such as class, individual, and property would be represented in a vector space – hence termed OWL2Vec. Other ontology-based semantic word representations that literature proposed before the emergence of OWL2Vec are Onto2Vec [113] and OPA2Vec [114]. The methodology for this proposed ontology based on word representation consists of two steps, which are extraction of corpus from ontology and training of word embedding model with the corpus.

Some literature argue that irrespective of the methodology employing to develop ontology, the evaluation technique strongly determines the robustness of ontology. To this end, the research of [115] designed citrus ontology based on the crop production knowledge framework. The researchers employed some standard ontology validation techniques to determine the robustness of the ontology. The evaluation techniques include ontology vocabulary evaluation, structural evaluation based on the eight widely used metrics, antipattern-based evaluation, and ontology competency evaluation. Once a proposed ontology is validated against the technique and the outlook is positive, it implies that such ontology is robust. For example, a high value of schema deepness metric of structural evaluation against the average value of 0.34 indicates that the ontology is deep.

6.5 Proposed Methodology: Determinants for Robust Ontology Design

Presently, the traditional software development methodologies and standards such as waterfall or spiral models are highly deficient to develop an application for some real-life domains such as manufacturing execution systems (MESs). The reason is that most of these domains consist of a complex set of data, and efforts are being made to gear the design standards and functionalities of system-based domains in compliance with the concept of Industry 4.0 requirements [116]. The researchers, in the course of reviewing cutting-edge methodologies and tools to build MES-based Industry 4.0 concept, canvassed for ontology driven based technology. They present the superiority of this technique (that is, the OWL-based ontology-driven approach) over the existing conventional methods of software development. Some of the shortcomings of the conventional approaches highlighted by the authors are as follows: high overhead cost, partially reliable, weak interoperability mechanism, and too much time-consuming exercise.

Conversely, considering the numbers of existing ontologies for domains and continuous growth of heterogeneous data, there is a need to make ontology design process more robust. Therefore, this research aims to employ a concept-based analysis approach to determine the factors that can be responsible for the robustness of ontology design. Table 6.1, based on literature reviewed, summarizes the identified criteria or factors that can lead to the achievement of robust ontology.

Based on the concept-based approach employed in this research, seven criteria are identified to design a robust ontology. More importantly, some literature equally suggested that the goal or objective of ontology design has

Table 6.1 Determinant factors for robust ontology design.

S/N	Determinants for robust ontology
1	Articulate a robust ontology engineering process (for example, modification of existing methodologies)
2	Early collaborative design with domain experts
3	Ontology-based feature learning model (ontology-based word embedding model)
4	Ontology learning
5	Ontology's types: upper ontology alignment (enhancement of extreme ontology design methodology)
6	Standard evaluation techniques
7	Domain types

proportionate effect on the robustness of ontology. This research identifies some of the driven goal of ontology design as data integration [105, 110, 109]; accessible, findable, interoperable, and reusable [108]; granularity, scientific paradigm, and methodological mechanism [106]; testing component [102]. The next section of this article discusses these approaches in detail.

6.6 Discussion and Conclusion

As stated earlier, this section discusses the identified eight criteria as shown in Table 6.1 during the course of this literature review. The criteria are arranged in no particular order of significance.

i Articulate a robust ontology engineering process [102, 110, 111]: It is a common principle that a robust ontology design is a product of a robust ontology engineering process. This is in support of a common computing phenomenon that says garbage in garbage out. Activities or steps of a robust ontology development methodology are to be duly spread across predevelopment, development, and postdevelopment stages. However, since no single existing methodology is self-sufficient, it is expected to modify it. For example, the case of methodologies' hybridization [5].

ii Collaborative design with domain experts [103]: Some literature clearly pointed out that for an ontology engineer to design a robust ontology for any domain, there must be a holistic effort to collaborate with domain experts at the early stage. That is, since ontology developer in most cases has little or no knowledge of the domain, there must be a concerted effort to start the planning with a team of domain experts. Some concepts of real-life domains, such as medicine and agriculture, have a peculiar type of terminologies. Therefore, the experts are well positioned to professionally handle those concepts with their similar lexical concepts in terms of synonyms, hypernyms, hyponyms, meronyms, and holonyms. Consequently, data integration is achieved and expert-based evaluation mechanism would be easily established.

iii Ontology-based learning model: Another angle of literature view robust ontology design via the mechanism of machine learning models. Specifically, the fusion of feature optimization models into the ontology's dataset. For instance, the literature of [112] employs the semantic data representation technique (that is, word embedding technique) nicknamed as OWL2Vec to design a robust ontology-based word embedding system.

iv Ontology learning [107]: Researchers in this aspect of ontology design believe the prospect of this factor in developing a robust ontology. There are efforts to design an automated technique where a given dataset would be autonomously learned by such a technique to model ontology. The motivation behind this goal is that, besides the fact that manual design of ontology is tedious and time consuming, human errors are also unavoidable. Therefore, researchers argue that all these limitations have their consequent effects on a good ontology design.

v Ontology types [108]: Some schools of thought believe that robustness of ontology cannot be ascertained without paying attention to the types of ontologies. This implies that the richness of top-level ontology (upper ontology), for example, cannot in any way be compared with domain ontology (light weight ontology) or task ontology. Therefore, developers are expected to first determine the type of ontology they intend to develop and for what purpose. In some instances, domain ontology is regarded as a rich form of taxonomy system. As a result, the robustness cannot be compared with task or application ontology, which has semantic richness owing to its ability to handle competency questions.

vi Domain's type [109]: The rate of obtaining robustness in ontology design differs across various domains. This is because the rate of obtaining comprehensive data owing to complexity nature of a particular domain differs from another. For example, the data complexity of biomedical domain has to be properly collected and analyzed with a cutting-edge technique in collaboration with experts. Therefore, developers are advised to carry out a thorough feasibility study and collaborative analysis on a proposed domain.

vii Standard evaluation techniques [115]: One of the contending challenges in ontology development has to do with evaluation technique. Consequently, to ascertain the robustness of ontology, some set of evaluation techniques have to be deployed. These include ontology vocabulary evaluation, antipattern-based evaluation, ontology competency evaluation, and structural evaluation. This last technique, the structural-based evaluation, works using the eight widely used metrics. It has an average and median values obtained from a large number of 1413 OWL ontologies as considered by the research of [117].Therefore, the technique is very rich to determine the completeness of ontology. For instance, schema deepness metric one of the eight metrics of the structural based evaluation has the capacity to determine if an ontology design is deep or flat.

Table 6.2 The driven goals for robust ontology design.

S/N	Article details	Proposed methodology	Goal considered	Future work/result
1	Smirnov *et al.* [105]	A four-step methodology	*Integration*	To experiment the model on multifaceted problems domain
2	Sulaeman and Harsono [106]	A review of existing methodology	*Granularity, scientific paradigm, and methodological mechanism*	Suggested a more holistic review to achieve robust supply chain ontology methodology
3	Tebes *et al.* [102]	Reinforcement of Ontology engineering principle	*Testing component*	To evaluate the efficacy of the proposed mechanism in terms of semantic matching
4	Howell *et al.* [103]	Adapted NeOn methodology	*Complete domain expert participation, logical accuracy, and performance*	A reusable engineering process that can be applied in related domain
5	Yang *et al.* [107]	Automation of ontology design (ontology learning)	*Collection of data and preprocessing, NLP-based lexical analysis, and extraction of ontology components*	They are often incapable to handle implicit terminological and non-terminological relations and data properties
6	Dalal [108]	Enhancement of extreme ontology design methodology	*Accessible, findable, interoperable, and reusable*	There should be a robust mechanism to take care of more complex alignment beyond sub-entities (either class or property)
7	Lu *et al.* [109]	Manufacturing ontology approach (combined RAD and existing methodologies' concepts)	*Integration*	It requires more domain concepts to be integrated
8	Palmirani *et al.* [110]	Methodology for building legal ontology (MeLOn)	*Integration*	It requires more domain concepts (deontic logic models) to be integrated
9	Yunianta *et al.* [111]	Enhancement of existing methodology	*Data integration*	More data to be integrated and detailed evaluation technique

Furthermore, this section discusses a fragment of the driven goals behind the development of robust ontology as presented in Table 6.2. Similarly, in order to open up further research drive, it equally presents some areas of research attentions.

Evidently, from the sample of literature presented in Table 6.2 (for instance, S/N 1, 7, 8, and 9), data integration is largely identified as the driven goal to develop robust ontology. It implies that robustness of ontology could be found on its capacity to integrate heterogeneous complex data either within a given ontological domain or across two and more ontological platforms. Therefore, when complex heterogeneous data are modeled for knowledge representation without compromising the sensitivity to contextual meaning of concepts and firmly establishing its relations among concepts, such can be described as robust ontology design. However, an area of research interest open for further work is on the mechanism that can make it possible to integrate more domain concepts without compromising its sensitivity to the factors mentioned earlier. Besides, the existing models for data integration have not been properly experimented. The confirmation to this open problem is the research effort by S/N 3 of Table 6.2 [102] who adapted an ontology engineering process to work on the testing capacity.

Another interesting goal is on the aspect of modifying an existing ontology development methodology to explicitly include the collaboration of domain expert. The work of Howell *et al.* [103] as shown by S/N 4 modified NeOn methodology to pave way for full participation of domain experts. Other driven goals are to define the granularity of the domain concisely, accessibility, and reusability of ontology. The last two goals are very crucial for ontology mapping either for alignment or merging purpose. Robustness of ontology is also obtained when two or more ontologies are mapped.

In conclusion, this review identifies some goals to develop robust ontology and equally points out some areas of research interventions as highlighted by Table 6.1 to design an ontology that can stand the test of time.

References

[1] Viswanath, S., Guntz, S., Dieringer, J., Vaidyaraman, S., Wang, H., & Gounaris, C. (2020) An ontology to describe small molecule pharmaceutical product development and methodology for optimal activity scheduling. *Journal of Pharmaceutical Innovation*, 1-15.

[2] Steinmetz, C., Rettberg, A., Ribeiro, F. G. C., Schroeder, G., Soares, M. S., & Pereira, C. E., (2018) Using Ontology and Standard Middleware for integrating IoT based in the Industry 4.0. *IFAC-PapersOnLine*, *51*(10), 169-174.

[3] Salem, A. B. M. & Parusheva, S. (2018) Developing a web-based ontology for e-business. *International Journal of Electronic Commerce Studies*, *9*(2), 119-132.

[4] Thangaraj, M. & Sujatha, G. (2014) An architectural design for effective information retrieval in semantic web. *Expert Systems with Applications*, *41*(18), 8225–8233. https://doi.org/10.1016/j.eswa.2014.07.017

[5] Aminu, E. F., Oyefolahan, I. O., Abdullahi, M. B. & Salaudeen, M. T. (2020) A Review on Ontology Development Methodologies for Developing Ontological Knowledge Representation Systems for various Domains. *International Journal of Information Engineering and Electronic Business*, *12*(2), 28–39. https://doi.org/10.5815/ijieeb.2020.02.05

[6] Thomas, C. J. (2009) Just what is an ontology, anyway?' *IT Professional,* 11, 22-27.

[7] Wei, Y., Wang, R., Hu, Y., & Wang, X. (2012) From Web Resources to Agricultural Ontology: a Method for Semi-Automatic Construction'*Journalof Integrative Agriculture*, 11(5): 775-783

[8] Meriyem, C., Adil, S., & Hicham, M. (2015) IT Governance ontology building process: example of developing audit ontology. *International Journal of Computer Techniques*, *2*(1), 134-141

[9] Ivanović, M. & Budimac, Z. (2014) An overview of ontologies and data resources in medical domains. *Expert Systems with Applications*, *41*(11), 5158-5166

[10] Barão, A., de Vasconcelos, J. B. Rocha, A., & Pereira, R. (2017) A knowledge management approach to capture organizational learning networks. *International Journal of Information Management*, 37(6), 735-740.

[11] Gruber, T. R., (1993) A Translation Approach to Portable Ontology Specifications. *Knowledge Acquisition,* 5(2): 199-220.

[12] Dou, D., Wang, H. & Liu, H. (2015) Semantic data mining: A survey of ontology-based approaches. In *Semantic Computing (ICSC), 2015 IEEE International Conferenc on* (pp. 244-251). IEEE.

[13] Kang, Y. B., Haghighi, P. D., & Burstein, F. (2014) CFinder: An intelligent key concept finder from text for ontology development. *Expert Systems with Applications*, *41*(9), 4494-4504

[14] Uschold, M. & King, M. (1995) Towards a Methodology for Building Ontologies. In IJCAI95 Workshop on Basic Ontological Issues in Knowledge Sharing. Montreal

[15] Dnyanesh, R., & Rahul, C. A. (2011) Generic ontology development framework for data integration and decision support in a distributed environment. *International Journal of Computer Integrated Manufacturing* Vol. 24, No. 2, 154–170

[16] Wang, R., Nellippallil, A. B. Wang, G., Yan, Y., Allen, J. K. & Mistree, F. (2019) Ontology-based uncertainty management approach in designing of robust decision workflows. *Journal of Engineering Design*, *30*(10-12), 726-757.

[17] Aminu E. F., Oyefolahan I. O., Abdullahi M. B., Salaudeen M. T. (2022) Modeling Competency Questions-Based Ontology for the Domain of Maize Crop: SIMcOnto. In: Mandal J. K., Buyya R., De D. (eds) *Proceedings of International Conference on Advanced Computing Applications*. Advances in Intelligent Systems and Computing, vol 1406. (pp. 751-763). Springer, Singapore. https://doi.org/10.1007/978-981-16-5207-3_61

[18] He, Y., Yu, H., Ong, E., Wang, Y., Liu, Y., Huffman, A., & Smith, B. (2020). CIDO, a community-based ontology for coronavirus disease knowledge and data integration, sharing, and analysis. *Scientific data*, *7*(1), 1-5.

[19] Pratibha, G., Sangeeta, D. & Bhanumurthy, K. (2011). Ontology Development Methods. DESIDOC Journal of Library & Information Technology, Vol. 31, No. 2, pp. 77-83

[20] Agyapong-Kodua, K., Niels, L., Robert, D., & Svctan, R. (2013) 'Review of semantic modeling technologies in support of virtual factory design' *International Journal of Production Research,* Vol. 51, No. 14, 4388–4404

[21] Noy, N. F. and McGuinness, D. L. (2001) 'Ontology Development 101: A guide to Creating Your First Ontology', Stanford University, Stanford, CA, USA, pp. 1-25.

[22] Noy, N. F., & McGuinness, D. L. (2016). Ontology development 101: a guide to creating your first ontology Stanford University.

[23] Walisadeera, A. I., Ginige, A., & Wikramanayake, G. N. (2015) 'User centered ontology for SriLanka farmers' *Ecological Informatics*, 140-150

[24] Bhaskar, K. & Savita S. (2010) 'A Comparative Study Ontology Building Tools for Semantic Web Applications' International journal of Web

& Semantic Technology *(IJWesT)* Vol.1, Num.3, July 2010 DOI : 10.5121/ijwest.2010.1301 1

[25] Ast, M., Glas, M., Roehm, T., & Luftfahrt, V. B. (2014). *Creating an ontology for aircraft design*. Deutsche Gesellschaft für Luft-und Raumfahrt-Lilienthal-Oberth eV

[26] Munir, K., & Anjum, M. S. (2018). The use of ontologies for effective knowledge Modeling and information retrieval. *Applied Computing and Informatics*, *14*(2), 116-126

[27] Bechhofer, S. (2009) "OWL: Web ontology language." *Encyclopedia of database systems*. Springer, Boston, MA, 2009. 2008-2009

[28] Richard F., Patrick H., & Ian H. (2004). OWL-QL—a language for deductive query answering on the Semantic Web. Web Semantics: Science, Services and Agents on the World Wide Web 2 19–29

[29] Cardoso, J., & Pinto, A. M. (2015). The Web Ontology Language (OWL) and its Applications. In *Encyclopedia of Information Science and Technology, Third Edition* (pp. 7662-7673). IGI Global

[30] Sengupta, K., & Hitzler, P. (2014). Web ontology language (OWL). *Encyclopedia of Social Network Analysis and Mining*, 2374-2378.

[31] Arenas, M., Grau, B. C., Kharlamov, E., Marciuška, Š., & Zheleznyakov, D. (2016). Faceted search over RDF-based knowledge graphs. *Journal of Web Semantics*, *37*, 55-74

[32] Calbimonte, J. P., Mora, J., & Corcho, O. (2016, May). Query rewriting in RDF stream processing. In *European Semantic Web Conference* (pp. 486-502). Springer, Cham

[33] Jeen B., Michel K., Stefan D., Dieter F., Frankvan H., & Ian H. (2002). Enabling knowledge representation on the Web by extending RDF Schema. *Computer Networks* 39 609–634

[34] Khan, J. A., & Kumar, S. (2014, October). Deep analysis for development of RDF, RDFS and OWL ontologies with protege. In *Proceedings of 3rd International Conference on Reliability, Infocom Technologies and Optimization* (pp. 1-6). IEEE

[35] Schenner, G., Bischof, S., Polleres, A., & Steyskal, S. (2014, September). Integrating Distributed Configurations With RDFS and SPARQL. In *Configuration Workshop* (Vol.1220, pp. 9-15)

[36] Rodríguez, N. D., Cuéllar, M. P., Lilius, J., & Calvo-Flores, M. D. (2014). A survey on ontologies for human behavior recognition. *ACM Computing Surveys (CSUR)*, *46*(4), 43

[37] Slimani, T. (2015). Ontology development: A comparing study on tools, languages and formalisms. *Indian Journal of Science and Technology*, *8*(24), 1-12

[38] Aminu, E. F. & Adewale, O. S. (2015) 'A Mechanism for Detecting Dead URLs in XTM-Based Ontology Repository' *International Journal of Computer Applications* (0975 – 8887) Volume 111 – No 12

[39] Carroll, J., Herman, I., & Patel-Schneider, P. F. (2015). OWL 2 web ontology language RDF-based semantics. *W3C Recommendation (October 27, 2009)*

[40] Hacherouf, M., Bahloul, S. N., & Cruz, C. (2015). Transforming XML documents to OWL ontologies: A survey. *Journal of Information Science*, *41*(2), 242-259

[41] John, H. G., Mark, A. M., Ray, W. F., Williams, E. G., Monica, C., Henrik, E., Natalya, F. N. & Samson W. T. (2003). The evolution of Protégé: an environment for knowledge-based systems development. *International Journal of Human-Computer Studies*, Vol. 58, Issue1, Pages 89 – 123

[42] Chujai, P., Kerdprasop, N., & Kerdprasop, K. (2014). On transforming the ER model to ontology using protégé OWL tool. *International Journal of Computer Theory and Engineering*, *6*(6), 484

[43] Wohlgenannt, G., Sabou, M., & Hanika, F. (2016). Crowd-based ontology engineering with the uComp Protégé plugin. *Semantic Web*, *7*(4), 379-398

[44] Panita, Y., Dussadee, T., Thanapat, S., Asanee, K., Sachit, R., Margherita, S., & Johannes K. (2008). The AGROVOC Concept Server Workbench: A Collaborative Tool for Managing Multilingual Knowledge. *World Conference on Agricultural Information and IT*

[45] John D., Midori A. H., Melissa Haendel & Suzanna L. (2007). OBO-Edit—an ontology editor for biologists. *Bioinformatics*, Volume 23, Issue 16, 15, Pages 2198–2200

[46] Aditya, K., Bijan, P., Evren, S., Bernardo, C. C. & James, H. (2006) 'Swoop: A Web Ontology Editing Browser' *Web Semantics: Science, Services and Agents on the World Wide Web* Vol. Issue 2, 144–153

[47] Hogan, W. R., Wagner, M. M., Brochhausen, M., Levander, J., Brown, S. T., Millett, N., & Hanna, J. (2016). The Apollo Structured Vocabulary: an OWL2 ontology of Phenomena in infectious disease epidemiology and population biology for use in epidemic simulation. *Journal of biomedical semantics*, *7*(1), 50.

[48] Anikin, A., Litovkin, D., Kultsova, M., Sarkisova, E., & Petrova, T. (2017, September). Ontology visualization: Approaches and software tools for visual representation of large ontologies in learning. In Conference on Creativity in Intelligent Technologies and Data Science (pp. 133-149). Springer, Cham

[49] Alatrish, E. S. (2013) "Comparison some of ontology." *Journal of Management Information Systems* 8.2 (2013): 018-024

[50] García-Peñalvo, F. J., Ordónez de Pablos, P., García, J., & Therón, R. (2014). Using OWL-VisMod through a decision-making process for reusing OWL ontologies. *Behaviour & Information Technology*, *33*(5), 426-442.

[51] Corcho, O., Fernández-López, M., & Gómez-Pérez, A. (2003). Methodologies, tools and languages for building ontologies. Where is their meeting point?. *Data & knowledge engineering*, *46*(1), 41-64.

[52] Nanda, J., Simpson, T. W., Kumara, S. R., & Shooter, S. B. (2006). A methodology for product family ontology development using formal concept analysis and web ontology language. Journal of computing and information science in engineering, 6(2), 103-113.

[53] Gavrilova, T. & Gladkova, M. (2014). Big data structuring: the role of visual models and ontologies. *Procedia Computer Science* 31, 336 – 343

[54] Vigo, M., Bail, S., Jay, C., & Stevens, R. (2014). Overcoming the pitfalls of ontology authoring: Strategies and implications for tool design. *International Journal of Human*-Computer Studies, 72(12), 835-845.

[55] Lenat, D. B., & Guha, R. V. (1990). Building Large Knowledge-based Systems: Representation and Inference in the Cyc Project. Boston, MA: Addison-Wesley.

[56] Dean, J., Trevor, B. C. & Pepijn V. (1998). Methodologies for Ontology Development. 62-75

[57] Fernandez-Lopez, M. Gomez-Perez, A., Pazos-Sierra, A., & Pazos-Sierra, J. (1999) Building a chemical ontology using METHONTOLOGY and the ontology design environment', *IEEE Intelligent Systems & their applications* 4 (1)37–46

[58] Oscar C., Mariano F., and Asuncion G. (2003) 'Methodologies, tools and languages for building ontologies. Where is their meeting point?' *Data & Knowledge Engineering* 46, 41–64

[59] Rizwan, I., & Aida, M. (2013). An experience of developing Quran ontology with contextual information support. *Multicultural Education & Technology Journal* Vol. 7 No. 4, pp. 333-343.

[60] Tan, H., Ismail, M., Tarasov, V., Adlemo, A., & Johansson, M. (2016). Development and evaluation of a software requirements ontology. In *7th International Workshop on Software Knowledge-SKY 2016 in conjunction with the 9th International Joint Conference on Knowledge Discovery, Knowledge Engineering and Knowledge Management-IC3K 2016, November 9-10, 2016, in Porto, Portugal* (pp. 11-18). SciTePress.
[61] Zhou, Z., Goh, Y. M., & Shen, L. (2016). Overview and analysis of ontology studies supporting development of the construction industry. *Journal of Computing in Civil Engineering*, *30*(6), 04016026.Soner, K., Ozgur, A., Orkunt, S., Samet, A., Nihan, K. C. & Ferda, N. A. (2012) 'Ontology-Based Retrieval System using Semantic Indexing' *Information Systems* 37, 294–305
[62] Zhanjun L., Victor R. & Karthik, R. (2007). A Methodology of Engineering Ontology Development for Information Retrieval. *International Conference on Engineering Design, Iced* 07 28 - 31, Paris, France
[63] Obrst, L., Chase, P., & Markeloff, R. (2012, October). Developing an Ontology of the Cyber Security Domain. In *STIDS* (pp. 49-56).
[64] Simperl, E., & Luczak-Rösch, M. (2014). Collaborative ontology engineering: a survey. *The Knowledge Engineering Review*, *29*(1), 101-131.
[65] Stojadinović, S. M., & Majstorović, V. D. (2014). Developing engineering ontology for domain coordinate metrology. *FME Transactions*, *42*(3), 249-255.
[66] Godspower, O. E. and Esingbemi, P. E. (2016). Ontology for Alleviating Poverty among Farmers in Nigeria. *INFOS '16*, May 09-11, 2016, Giza, Egypt
[67] Gruninger, M., & Fox, M. S. (1995). Methodology for the Design and Evaluation of Ontologies. Workshop on Basic Ontological Issues in Knowledge Sharing, Montreal.
[68] Hitzler, P., Gangemi, A., & Janowicz, K. (Eds.). (2016). *Ontology Engineering with Ontology Design Patterns: Foundations and Applications* (Vol. 25). IOS Press.
[69] Kim, H. M., & Laskowski, M. (2018). Toward an ontology-driven blockchain design for supply-chain provenance. *Intelligent Systems in Accounting, Finance and Management*, *25*(1), 18-27.
[70] Hadjar, K. (2016). University Ontology: A Case Study at Ahlia University. In *Semantic Web* (pp. 173-183). Springer, Cham.
[71] Bilgin, G., Dikmen, I., & Birgonul, M. T. (2018). An ontology-based approach for delay analysis in construction. *KSCE Journal of Civil Engineering*, *22*(2), 384-398

[72] El-Diraby, T. E. & Osman, H. (2011). A domain ontology for construction concepts in urban infrastructure products. *Automation in Construction* 20, 1120–1132

[73] Gómez-Pérez, A., Fernández-López, M., & Corcho, O. (2004). Methodologies and methods for building ontologies. *Ontological Engineering: with examples from the áreas of Knowledge Management, e-Commerce and the Semantic Web*, 107-197

[74] Hafedh, N., Mohamed, F., Imed R. F. & Basel S. (2014). A Critical Analysis of lifecycles and Methods for Ontology Construction and Evaluation' *1st International Conference on Advanced Technologies for Signal and Image Processing* - ATSIP'2014

[75] Haghighi, P. D., Burstein, F., Zaslavsky, A., & Arbon, P. (2013). Development and evaluation of ontology for intelligent decision support in medical emergency management for mass gatherings. *Decision Support Systems*, *54*(2), 1192-1204.

[76] Janowicz, K. (2012). Observation-driven geo-ontology engineering. *Transactions in GIS*, *16*(3), 351-374.

[77] Jinsoo, P., Kimoon, S., & Sewon, M. (2008). Developing Graduation Screen Ontology based on the METHONTOLOGY Approach. *Fourth International Conference on Networked Computing and Advanced Information Management.*

[78] Guo, B. H., & Goh, Y. M. (2017). Ontology for design of active fall protection systems. *Automation in Construction*, *82*, 138-153.

[79] Noman, I., Abu, Z. A. and Zubair, A. S. (2010) 'Semantic Web: Choosing the Right Methodologies, Tools and Standards', *Information and Emerging Technologies (ICIET), 2010 International Conference on.* IEEE, 2010.

[80] Rizwan, I., Masrah A. A. M., Aida, M. & Nurfadhlina, M. S. (2013). An Analysis of Ontology Engineering Methodologies: A Literature Review. *Research Journal of Applied Sciences, Engineering and Technology* 6(16): 2993-3000

[81] Gavrilova, T. A., & Leshcheva, I. A. (2015). Ontology design and individual cognitive peculiarities: A pilot study. *Expert systems with Applications*, *42*(8), 3883-3892.

[82] Ibrahim, A. A., Abdul, A. A. G., Wan, N. W. R. & Rodziah, A. (2014). A Comparative Study on Ontology Development Methodologies towards Building Semantic Conflicts Detection Ontology for Heterogeneous Web Services. *Research Journal of Applied Sciences, Engineering and Technology* 7(13): 2674-2679

[83] De Nicola, A., M., Missikoff & Navigli, R. (2009) 'A Software Engineering Approach to Ontology Building' *Inform. Syst.*, 34(2): 258-275.

[84] del Águila, I. M., Palma, J., & Túnez, S. (2014). Milestones in software engineering and knowledge engineering history: A comparative review. *The Scientific World Journal, 2014.*

[85] Jain, V., & Singh, M. (2013). Ontology development and query retrieval using protégé tool. *International Journal of Intelligent Systems and Applications, 9*, 67-75.

[86] Rao, L., Mansingh, G., & Osei-Bryson, K. M. (2012). Building ontology based knowledge maps to assist business process re-engineering. *Decision Support Systems, 52*(3), 577-589.

[87] Serna, E., & Serna, A. (2014). Ontology for knowledge management in software maintenance. *International Journal of Information Management, 34*(5), 704-710.

[88] Tiffani J. B., Furuya E. Y., Gilad J. K., James J. C., & Suzanne B. (2012). Development and evaluation of an ontology for guiding appropriate antibiotic prescribing', *Journal of Biomedical Informatics* 45 120–128

[89] Arp R. (2009) 'Practical steps in building a domain ontology, models and simulations 3: emergence, computation, and reality' Virginia: Charlottesville.

[90] Chen-Huei C., Fatemeh M. Z., & Huimin Z. (2011). Ontology for Developing Web Sites for Natural Disaster Management: Methodology and Implementation. *IEEE Transactions on Systems, Man, and Cybernetics—Part A: Systems and Humans*, Vol. 41, No. 1

[91] Uschold, M & Grüninger, M, (1996). Ontologies: principles methods and applications', *Knowledge Engineering Review*11(2)93–137.

[92] Ceccaroni, L, Cortés, U & Sánchez-Marré, M, (2000) 'WaWO – an ontology embedded into an environmental decision-support system for wastewater treatment plant management' *Workshop on Applications of Ontologies and Problem solving Methods.14th European Conference on ArtificialIntelligence*(ECAI'00)2–1–2–9

[93] Bonanci, R., Nabuco, O. F. & Junior, I. P. (2016). Ontology models of the impacts of agriculture and climate changes on water resources: Scenarios on interoperability and information recovery' *Future Generation Computer Systems*. 54 (2016) 423–434

[94] Aree, T., Asanee, K., Supamard, P. & Uamporn, V. (2009) 'Ontology Development: A Case Study for Thai Rice' *Kasetsart J.* (Nat. Sci.) 43 : 594 – 604

[95] Niu, J., & Issa, R. R. (2015). Developing taxonomy for the domain ontology of construction contractual semantics: A case study on the AIA A201 document. *Advanced Engineering Informatics*, *29*(3), 472-482.

[96] Gregor, D., Toral, S., Ariza, T., Barrero, F., Gregor, R., Rodas, J., & Arzamendia, M. (2016). A methodology for structured ontology construction applied to intelligent transportation systems. *Computer Standards & Interfaces*, *47*, 108-119.

[97] Dutta, B., Chatterjee, U., & Madalli, D. P. (2015). YAMO: yet another methodology for large-scale faceted ontology construction. *Journal of Knowledge Management*, *19*(1), 6-24.

[98] Zeb, J., Froese, T., & Vanier, D. (2015). An ontology-supported asset information integrator system in infrastructure management. *Built Environment Project and Asset Management*, *5*(4), 380-397.

[99] Mezghani, E., Exposito, E., & Drira, K. (2016). A collaborative methodology for tacit knowledge management: Application to scientific research. *Future Generation Computer Systems*, *54*, 450-455.

[100] Keet, C. M., Ławrynowicz, A., d'Amato, C., Kalousis, A., Nguyen, P., Palma, R., ... & Hilario, M. (2015). The data mining OPtimization ontology. *Journal of web semantics*, *32*, 43-53.

[101] Rayyaan, R., Wang, Y., & Kennon, R. (2014). Ontology-based interoperability solutions for textile supply chain. *Advances in Manufacturing*, *2*(2), 97-105.

[102] Tebes, G., Olsina, L., Peppino, D., & Becker, P. (2021). Specifying and Analyzing a Software Testing Ontology at the Top-Domain Ontological Level. *Journal of Computer Science & Technology*, *21*.

[103] Howell, S., Beach, T., & Rezgui, Y. (2021). Robust requirements gathering for ontologies in smart water systems. *Requirements Engineering*, *26*(1), 97-114.

[104] John, S., Shah, N., Stewart, C. D., & Samlov, L. (2017, November). Software Centric Innovative Methodology for Ontology Development. In *KEOD* (pp. 139-146).

[105] Smirnov, A., Levashova, T., Ponomarev, A., & Shilov, N. (2021). Methodology for Multi-Aspect Ontology Development: Ontology for Decision Support Based on Human-Machine Collective Intelligence. *IEEE Access*, *9*, 135167-135185

[106] Sulaeman, M. M., & Harsono, M. (2021). Supply Chain Ontology: Model Overview and Synthesis. *Jurnal Mantik*, *5*(2), 790-799.

[107] Yang, L., Cormican, K., & Yu, M. (2020). Ontology learning for systems engineering body of knowledge. *IEEE Transactions on Industrial Informatics*, *17*(2), 1039-1047.

[108] Dalal, A. (2020). Modular ontology modeling meets upper ontologies: the upper ontology alignment tool.

[109] Lu, Y., Wang, H., & Xu, X. (2019). ManuService ontology: a product data model for service-oriented business interactions in a cloud manufacturing environment. *Journal of Intelligent Manufacturing*, *30*(1), 317-334.

[110] Palmirani, M., Martoni, M., Rossi, A., Bartolini, C., & Robaldo, L. (2019). Pronto: Privacy ontology for legal reasoning. In *International Conference on Electronic Government and the Information Systems Perspective* (pp. 139-152). Springer, Cham.

[111] Yunianta, A., Basori, A. H., Prabuwono, A. S., Bramantoro, A., Syamsuddin, I., Yusof, N., ... & Alsubhi, K. (2019). OntoDI: The methodology for ontology development on data integration. *Int. J. Adv. Comput. Sci. Appl.*, *10*(1), 160-168.

[112] Chen, J., Hu, P., Jimenez-Ruiz, E., Holter, O. M., Antonyrajah, D., & Horrocks, I. (2021). Owl2vec*: Embedding of owl ontologies. *Machine Learning*, *110*(7), 1813-1845.

[113] Smaili, F. Z., Gao, X., & Hoehndorf, R. (2018). Onto2vec: joint vector-based representation of biological entities and their ontology-based annotations. *Bioinformatics, 34*(13), i52–i60.

[114] Smaili, F. Z., Gao, X., & Hoehndorf, R. (2018). OPA2Vec: combining formal and informal content of biomedical ontologies to improve similarity-based prediction. *Bioinformatics, 35*(12).

[115] Wang, Y., & Wang, Y. (2018). Citrus ontology development based on the eight-point charter of agriculture. *Computers and Electronics in Agriculture*, *155*(October), 359–370. https://doi.org/10.1016/j.compag.2018.10.034

[116] Jaskó, S., Skrop, A., Holczinger, T., Chován, T., & Abonyi, J. (2020). Development of manufacturing execution systems in accordance with Industry 4.0 requirements: A review of standard-and ontology-based methodologies and tools. *Computers in industry*, *123*, 103300.

[117] Sicilia, M. A., Rodriguez, D., Garcia-Barriocanal, E., Sanchez-Alonso, S., 2012. Empirical findings on ontology metrics. Expert Syst. Applicat. 39 (8), 6706–6711

7

Semantic Web: An Overview and a .net-based Tool for Knowledge Extraction and Ontology Development

Noman Islam[1], Darakhshan Syed[2], and Zubair A. Shaikh[3]

[1]PAF KIET, Karachi
[2]Iqra University, Karachi
[3]Muhammad Ali Jinnah University, Karachi
E-mail: noman.islam@gmail.com; darakhshan@iqra.edu.pk;
zubair.shaikh@jinnah.edu

Abstract

Semantic web is an evolution of current web that enhances the contents over the web by providing a formal meaning to contents, enabling automated processing of various tasks by means of software agents. Semantic web is a holistic concept comprising a set of languages, tools, and standards. This chapter focuses on challenges pertaining to ontology development in semantic web. Ontology is one of the vital components for realizing the true vision of semantic web. The concept itself comprises issues such as the selection of ontology languages, tools, and methodologies for ontology development. Besides what are discussed above, there are also challenges related to ontology alignment, maintenance, learning, and evolution of ontologies. This research paper offers an updated view on various aspects related to ontology development. First, it provides discussions on various ontology languages, editors, methodologies, and ontology learning strategies. The chapter's second section discusses a .net-based tool for ontology development encompassing features discussed in the first section of the chapter. The main features of the proposed tool are a well-defined web-based user interface, an interface for ontology querying and visualization, and an automated approach for ontology extraction. The chapter also discusses how machine learning can be used for automatic ontology extraction.

Keywords: Ontology editing, ontology extraction, querying, semantic web, visualization.

7.1 Introduction

Semantic web is an evolution of current World Wide Web. It is used to provide the meaning to internet contents such that a machine can process it and enable automation of various tasks. This automation may require understanding the meanings of the contents [1, 2]. The true realization of the semantic web requires addressing a range of issues. Among these issues include the development of an ontology. An *ontology* describes the contents over the web to be processed by machines. Formally, an ontology (C, R, A_o) is a collection of concepts C, relations R ($R \subset C \times C$), and axioms A_o, specified by means of an ontology language. *Ontology languages* provide formal specifications of a domain that are usually stipulated by means of ontology editing tools. An *ontology editor* provides a visual interface to specify domain concepts, define their attributes, and create relationships among these concepts [3]. In addition, different supporting features are also provided to ease the ontology development process. The developed ontology can be finally exported as an ontology document and formally encoded in an ontology language [4].

Besides these issues, there are a number of other aspects related to ontology development. For instance, ontology matching aims at making different ontologies interoperable [5]. Ontology learning aims at automating various tasks of ontology construction [6]. This includes approaches based on linguistic, statistical, and machine learning [7]. Ontology evolution and maintenance is also a difficult task as the domain under consideration might change with the passage of time [8].

The objectives of this book chapter are multifold. It provides a comprehensive overview of semantic web, specifically focusing on aspects related to ontology languages, editing, querying, and ontology extraction. Then, the chapter proposes TODE, i.e., a tool for ontology editing and engineering. The remainder of the chapter is organized as follows. Section 7.2 presents a background study and a review of the literature pertaining to semantic web, ontology languages, and editing tools [9]. A generalized architecture to ontology development for .NET platform is then presented. The proposed architecture provides an extensible approach to development of ontology editor. Different issues inherent in the creation of an ontology editing tool for the NET framework are outlined, and a tool based on the suggested architecture is provided in the subsequent section. The proposed tool TODE provides a web-based highly usable interface and supports querying, visualization, etc. The chapter concludes by outlining the recent work's shortcomings as well as the potential for future research opportunities [10, 11].

7.1.1 Semantic web

Semantic web is an evolution of current web that provides the mechanism for sharing and reuse of data across application, enterprise, and community boundaries. The current World Wide Web has been a universal medium for integration of data and resources. However, the web has been designed for consumption of information by humans and serves only as an information publishing medium. The machines can proficiently parse the web contents for layout and routine processing, but there has been no trustworthy way to savvy the underlying meaning. Realizing the limitations of current web, Tim Berners Lee proposed the concept of semantic web [12], an evolution of current web that enables the machines to understand the meanings of the web contents [13].

Figure 7.1 shows the layered architecture of semantic web. The bottom two layers are the same as conventional web. The *URI* and *Unicode* provide identification of web resources and universal encoding of contents, respectively. Since HTML does not have the capability to completely express semantic information, *XML* has been adopted as serialization syntax by semantic web community. The *RDF* is the building block of semantic web and describes the web resources by means of triples, i.e., subject, object, and property. *RDF schema* or *RDF(S)*, based on RDF, defines relationships among resources and enables taxonomic reasoning. Since the semantics specified by RDF(S) is not useful beyond simple semantic processing, various ontology languages have been proposed. *Ontology* provides complete specification of domain by means of description languages. Simple HTML ontology extension (SHOE), DARPA agent markup language (DAML), web ontology language (OWL), etc., are examples of the widely used ontology languages. The *logic* and *proof* components provide features of first-order logic to semantic web. Together, they enable the provision for proofs of the inferences or conclusions drawn by machines. Finally, the *trust* layer enables agents over the web to interoperate securely using different security technologies, i.e., *signature*, *encryption*, etc.

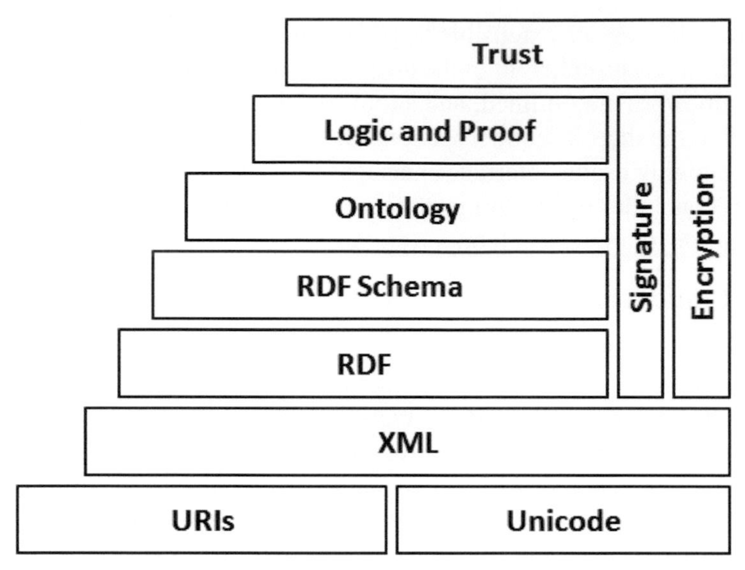

Figure 7.1 Semantic web architecture.

7.1.2 Ontology

Ontology is a formal specification of the knowledge of a domain expressed in the form of concepts, attributes, relations, instances, and axioms. An ontology development requires deliberation on a number of aspects such as ontology languages, editing, development methodology, ontology extraction, etc. [14]. The following paragraphs provide a brief discussion on these aspects.

7.1.3 Ontology languages

There are a number of ontology languages used for ontology development. These languages include simple HTML ontology extension (SHOE), ontology interchange language (OIL), resource description framework (RDF), DARPA agent markup language (DAML), and web ontology language (OWL) [14]. The simplest approach is to use extensible markup language (XML) to describe the domain knowledge [6, 15]. An XML provides a hierarchical representation of concepts in the form of tags that must be uniformly formed. SHOE enables the web pages to be annotated with semantic information that can be processed by machines. The RDF can be used to describe the knowledge in terms of subject, predicate, and object. RDF language does not have the provision for classes and subclasses, which was then addressed in

RDF(S). These languages are extended further with DAML and OWL. OWL is the most widely used language for ontology. It is based on description logic and has three basic versions: OWL Lite, OWL-DL, and OWL-Full [16]. Each of these versions has different power of expressiveness, decidability, and reasoning.

Besides these XML-based languages, there are conventional approaches to ontology representation based on first-order logic and frames. However, these languages often posed intractability problems. Examples of these languages are Ontolingua, KIF, CycL, and Frame Logic (F-Logic) [6, 17]. Recently, JavaScript object notation (JSON) has been used as an ontology language [18]. JSON describes the information in the form of key–value pairs with keys hierarchically nested similar to tags in HTML. The authors have been primarily motivated by JSON schema. The developed ontology can be stored on triples that can be further reasoned. Other languages/formats for representation of ontology include turtle, N3, N-triples, etc. [19]. A summary of ontology languages is provided in Table 7.1.

7.1.3.1 Rule languages

Rule languages can be used to write domain rules that can help in inferencing and reasoning tasks. There are a number of rule languages proposed in

Table 7.1 Summary of ontology languages.

Language	Proposed by	Description	Tools/languages
SHOE	University of Maryland	Annotate web pages with semantic description	PIQ/SHOE Search
RDF/RDFS	W3C	Describe the semantics in terms of subject, predicate, and object	Protégé and TODE
OWL	W3C	Based on description logic	Protégé, TODE, and WebODE
JSON	University of Innsbruck, Austria	A hierarchical representation of concepts based on key/value	
Ontolingua	Stanford University	A collaborative environment to create, edit, and use ontologies	Ontolingua

literature [20] (see Table 7.2). For instance, Rule ML proposed by W3C is written in XML format. Rules are written in atom format. Semantic web rules language (SWRL) enables rules to be embedded in OWL language. It resolves the issues of ontology via enhancing description logic [21]. An extension of SWRL is f-SWRL that is used to model imprecise and vague information. Rule interchange format (RIF) was proposed to provide a standard format for interoperability of rule languages [18]. Rewerse rule markup language (R2ML) is another rule language based on XML for describing rules. A fuzzy temporal extension of SWRL has been presented in [22] called FT-SWRL. The objective is to model imprecise information in SWRL.

7.1.4 Ontology learning

Ontology development is a complex task that requires analysis of collected text, identifying terms, concepts and relations, and encoding them in a formal language such as RDF, RDF(S), and OWL. Ontology development can be performed manually, in a cooperative fashion and/or using any automated ontology construction approaches. Since manually constructing an ontology is a very tedious task, automated or semi-automated ontology construction and knowledge extraction systems have been recommended in literature. However, it should be understood that complete automation of ontology construction is not possible, but the objective is to minimize human effort as much as possible [23].

Table 7.2 Summary of rule languages.

Rule language	Proposed by	Description	Tools/technologies
SWRL	W3C	Rules are written inside OWL using Horn clauses	Protégé , Racer Pro, and Pellet
f-SWRL	University of Aberdeen	To represent imprecise information	Protégé
R2ML	REWERSE Working Group	Exchange of rules between different systems	Xcerpt and visXcept
RIF	W3C	An interchange format between different rule systems	SILK, OntoBroker, fuxi, and Eye
FT-SWRL	Crops For the Future Research Centre (CFFRC), Malaysia	A fuzzy extension of SWRL	

During the past few years with the widespread emergence of unstructured text over the web, automated knowledge extraction/ontology learning has gained a lot of popularity [7]. The life cycle of an ontology learning process comprises extraction, analysis, generation, validation, and evolution [6]. The first step is *extraction*, which is used to acquire the knowledge about the terms such as concepts, attributes, relations, and axioms. Then in the *analysis* phase, integration is performed. Then the ontology is formalized and validated for correctness. The ontology can also evolve with the passage of time. Hence, it also requires careful deliberation.

The authors in [7] have described ontology learning as a layer cake, starting with acquisition of terms and synonyms, then formation of relationships, and finally extraction of axioms. Ontology learning can be classified as linguistic, statistical, and logical approaches. *Linguistic approaches* are based on part of speech (POS) tagging, stemming, lemmatization, parsing, and knowledge extraction. POS is based on annotating a token/word with its corresponding parts of speech. Stemming and lemmatization is used to find the root of a derived word. For instance, the word *doing* can be stemmed to the word *do*. Parsing is the use of any grammar or similar construct to understand the structure of a sentence.

Statistical approaches are based on statistics of underlying corpora without considering the semantics. Among the techniques in this category are C/NC, contrastive analysis, and co-occurrence analysis. A partial multi-diving ontology learning algorithm in this direction has been proposed in [24]. A number of machine learning techniques also fall in this domain such as association rules mining and clustering [25]. For instance, [26] proposed a neural machine translation based approach for ontology learning. Recurrent neural network (RNN) that models the dependency in text data can be used for ontology learning. Finally, *logic-based approaches* are mainly used for extraction of relations and axioms. The automated extraction of axioms is still in infancy stage and a lot of research is required in this domain [23].

Various datasets are also available for ontology learning such as OHSUMED, Genia Corpus, etc. An important aspect in ontology engineering is to minimize the noise term as much as possible. Other important metrics to be considered are precision, recall, *F*-measure, ontological improvement, and ontological loss [23]. A number of frameworks and tools such as GATE, Text2 Onto, and LExO are also proposed in literature [6]. For instance LExO converts textual data into description logic. Text2Onto employs data

Approaches	Linguistic approaches	Stemming	
		Lemmatization	
		POS tagging	
	Statistical approaches	Conventional approaches	C/NC, contrastive analysis, and co-occurrence analysis
		Machine learning	Clustering and association rules mining
		Deep learning	Recurrent neural network, deep belief network, and auto-encoders
	Logic-based approaches	[23]	
Frameworks	GATE, Text2Onto, and LExO		
Datasets	OHSUMED and Genia Corpus		
Metrics	Noise, precision, recall, *F*1-score, ontological improvement, and ontological loss		

Figure 7.2 Summary of ontology learning approaches.

mining and natural language processing techniques for ontology learning [23]. Figure 7.2 shows the summary of various ontology learning approaches.

We end the section with a brief discussion on the approaches for deep learning. Shallow/machine learning, which is the use of data to develop a model for prediction, can be used for ontology learning to improve the results. There is a growing interest toward using deep learning for ontology extraction. There are use of feed forward neural network, recurrent neural network, deep belief network, and auto-encoders being used for ontology extraction [23].

7.1.5 Ontology editor

An ontology editor is a tool that helps designers who develop logics to create ontologies. A good ontology editor provides an *easy-to-use interface* for construction of ontologies. To ease in construction of ontologies, editors adopt different techniques to facilitate developers [8]. Some tools provide a visual drag-and-drop interface to develop ontology, some provide syntax highlighting, and some use natural language techniques to guide users in ontology creation. As ontology development is a highly complicated task, there are some ontology editors that assist the developers in the creation of ontology through wizards and other user interfaces [27].

Once an ontology has been developed, it can be imported and exported in different languages. In order for its acceptance, an editor should support well-known *ontology languages*. An important feature of editors is visualization [28]. *Ontology visualization* provides a handy approach to analyze large

Table 7.3 Requirements for an ontology editor.

Features	Description
Basic editing support	Basic support for editing such as drag and drop, visual error highlighting, imports, and exports should be supported
Easier to use	The editor should have an easy-to-use graphical user interface
Collaboration	The editor should provide support for collaboration among multiple users
Languages	The editor should provide support for a wide variety of languages such as RDF, OWL, and N-triples
Visualization	The editor should provide support for visualization from different perspectives
Inferencing, reasoning, and querying	The editor should provide support for inferencing, reasoning, and querying in various languages such as SPQARQL
Methodology	The editor should support any methodology for ontology editing such as UPON and DILIGENT
Knowledge extraction	The editor should provide support for automated extraction of knowledge from legacy sources

ontologies. It provides a high level view of the overall ontology, facilitate in drilling down a particular aspect of ontology, as well as analyze ontology from various dimensions. An ontology editor should also provide an interface for *querying*. A query written in a specific query language (e.g., SPARQL, RDQL, DQL, OWL-QL, nRQL, etc.) can be posed and relevant information can be extracted from ontology [1, 18]. Finally, the *architecture* of a good ontology editor should be extensible, such that supporting tools can be plugged-in. In addition, it should allow construction of ontologies by multiple users in a *collaborative* fashion. Table 7.3 provides the various requirements for a good ontology editor.

7.1.5.1 Ontology editing tools

A wide variety of ontology editing tools has been discussed in the research. An overview of some of the ontology editing tools can be found in Table 7.4. A survey on ontology editing tools can be seen in [12] and [14]. As can be seen, Java has been the most popular platform for ontology editor development, with the majority of work being performed in open source languages. Interoperability and collaborative effort are supported by some of the technologies. For example, Ontolingua uses a collaborative method to ontology editing. WebODE provides a user interface for collaboratively building ontologies with several users. It also incorporates ontology reasoning and translation facilities. pOWL is an open source multi-user ontology that allows for collaborative ontology building. Another dimension to consider is the web-based ontology editing support. In this direction, a web-based tool has been presented in [29] for ontology editing. An extension of Protégé for web-based ontology development has been proposed in [30]. A further extension in Protégé in the form of an interactive debugging tool has been proposed in [31].

Tools like Protégé and SWOOP provide extensibility by means of plug-ins for visualization, reasoning purposes, and other purposes. However, some of the ontology editors have been developed with special emphasis on *visualization*. Among them are IsaViz and Altova SemanticWorks, which themselves provide a visual interface for ontology development [32]. As far as *ontology languages* are concerned, almost all of the tools support RDF and OWL constructs in their visual interface. In addition, DAML and N-triples are also supported by some of the tools. OBO-Edit, however, is also optimized for the OBO biological ontology file format [33].

Recently, several cloud-based ontology editing tools have emerged. For instance, CLONE is a web-based ontology editor employing service-oriented architecture [34]. It provides most of the feature of conventional ontology editors besides collaboration, role-based access, and maintaining a provenance record.

Besides what was discussed above, other ontology editing tools have also emerged as a result of research efforts in miscellaneous directions. TopBraid Composer is an enterprise-class editor for developing ontologies and semantic web applications [35]. DOE have been created based on linguistic philosophy, with the primary goal of complementing other editors. Similarly, OwlSight is a platform-independent OWL browser, OntoBuilder provides web-based data dictionary creation and editing, Hozo has been developed based on "Role" and "Relationship" model, etc. A library for automated ontology generation has been presented in [36]. A device-independent ontology editing tool has been presented in [37]. An ontology editor for domain experts called PRONTOE has been presented in [38]. The objective is to allow the domain experts to edit ontology without the underlying concerns about technology. An ontology editor for mobile platforms has been proposed in [39]. An ontology editor based on aspect-oriented programming called AVOned has been proposed in [40]. Table 7.4 provides a summary of various ontology editing tools.

Table 7.4 Requirements for an ontology editor.

Editor	Language	Description	Features	Languages
Protégé	Java	Both standalone and web-based version available, extensible by means of various plug-ins	Basic editing, reasoning and visualization support	RDF(S) and OWL
WebODE	Java	Collaborative ontology editing over the web	Collaboration and translation	RDF(S), DAML + OIL, and OWL
Ontolingua	Proprietary	A collaborative environment with the aim to describe the knowledge of different domains	A complete system and collaboration	Ontolingua
Clone	Java	A service-oriented architecture for ontology editing in cloud	Collaborative environment, ontology editing, reasoning, and version control	OWL
PRONTOE	Java	To allow domain experts to construct ontology without worrying about technical details	Ontology editing and reasoning	OWL and PDDL
Zhang [39]	Java	Ontology editing in mobile platforms	Basic ontology editing using a terse user interface, and encryption	OWL
AVOned	Java	Ontology editing based on aspects	Consistency checking, versioning, and visualization	Editable visual language and OWL Lite

7.1.5.2 Ontology editing in .net platforms

We end the section with the note that there are just a few ontology editing tools available for the .NET platform (see Table 7.5). The majority of them have basic RDF parsing capabilities. An empirical evaluation of various ontology editing tools for .net environment has been done in [41]. A quantitative evaluation of two ontology libraries for .net has been done in [42] and [43].

Table 7.5 Summary of .net APIs for ontology editing.

Language	**Tool**	**Features**
RDF	SemWeb	Limited reasoning capabilities, support for RDF, RDF(S), N-triples, XML, and querying in SPQRQL
	Drive	Simple RDF parser
	dotNetRDF	Querying, basic reading and writing RDF
OWL	OWLDotNetAPI	W3C compatible and directed linked graph
	ROWLEX	Basic reasoning support
	OpenAnzo	RDF, OWL, and SPARQL

SemWeb is a library for manipulating RDF statements. It supports ontology manipulation in XML, N3, and RDBMS, among other formats [42]. In addition, limited RDFS and inferencing capabilities are provided. A querying library based on simple graph matching and SPARQL is also available. Drive is a C# RDF parser for the .net framework. It is an open-source RDF API that complies with W3C guidelines [44]. In another work, dotNetRDF application programming interface (API) is provided [45]. The core of the API provides classes for reading/writing RDF and SPARQL querying. There are extensions for integration with other triple stores as well as SPQARQL extensions. Semantics Framework 2.0 is a collection of products for creating semantic web apps with the .net framework [46]. It is equipped with an RDF data storage and SPAQRL querying module.

Apart from RDF tools, there aren't many APIs for manipulating OWL. The OwlDotNetApi is a .NET framework OWL API parser built in C# [47]. It was built using the Drive parser and is completely compatible with the W3C OWL standard [44]. The API builds an underlying directed link graph based on the input ontology. Relaxed OWL Experience (ROWLEX) is a .net package that was created with the goal of alleviating designers who develop logics from the complexities of RDF triples [48]. It provides an abstraction for the creation and browsing of the RDF by means of OWL API. OpenAnzo is a multi-user middleware framework that is available as an open source initiative [49]. It supports RDF and OWL ontologies as well as SPARQL query languages.

7.1.5.3 The need for a .net-based ontology editor

It is evident from the above discussions that the majority of the developments in the semantic web have been made in open source programming technologies, mainly in C/C++, Java, or PHP. Current editors provide good support for different ontology languages as well as equipped with other supporting features. However, there is presently no .NET-based ontology editor accessible. The earlier section's analysis of .net semantic web technologies revealed that there are just a few semantic web frameworks available. So many of these tools are in their infancy, and there is not enough information about them in the existing research. Only a few of the programs include more than basic RDF processing (OWL, inferencing, reasoning, etc.). It can be concluded that there is a serious crisis of semantic web tools for .NET platform. There are no evident plans by Microsoft to support ontology in core of .NET platform. The motivations for development of a .net-based tool are as follows.

- The availability of a tool in a specific platform provides the motivation and momentum for further development of APIs, plug-ins, and supporting tools on that platform.
- An additional motivation behind the tool is the case study of Protégé. The Protégé plug-in library provides a convenient place to develop various types of supporting tools, i.e., API, import, export, reasoner, visualization, and NLP plug-ins, based on its extensible architecture [50]. It is believed that a similar effort in the .NET domain can serve as a catalyst. The extensible architecture of TODE can lead to evolution of various supporting tools and extensions on .NET platform.

Therefore, this study proposes a tool for developing and editing ontologies (TODE) based on our earlier work proposed in [3]. The tool is built on a generalized extendable framework for ontology editors, which will be explained in more detail in the next section.

7.2 A Tool for Ontology Editing in .NET Platform

Figure 7.3 shows the proposed layered architecture for ontology editing. The *application layer* interfaces with end-users and provides access to various functionalities of ontology editor implemented at the server end. Different client technologies can be used at the application layer. Currently, a browser-based HTTP implementation is provided, but one can provide an implementation for Microsoft sliver light, Java applets, Android, HTML 5, etc. The application client interacts with *AJAX engine* [51]. AJAX enables development for highly usable and responsive user interfaces, as we will see in the following paragraphs. The AJAX engine interacts over HTTP with ontology layer, which resides at HTTP server. The *ontology layer* provides implementations of different components (editor, visualizer, and query tool) of TODE. The ontology editor handles basic ontology editing features, while the visualization interface enables viewing ontology from different dimensions. The extensible architecture allows plugging implementations for visualizers by other developers. The query component provides facility to extract relevant information from ontology maintained at the *ontology model* layer. The query can be applied on a base model or a model inferred by reasoners. Currently, we provide default reasoner implementations, i.e., transitive and rule-based reasoners. However, advanced reasoners can also be attached. At the heart of the ontology editor is the graph model that maintains the ontology model in the form of collection of nodes and links. The graphs can be persisted in different formats, i.e., HTML, RDF, OWL, N-triples, relational databases, etc.

The proposed approach provides different salient features such as the following.

- **AJAX-based highly responsive interface:** The proposed approach is based on asynchronous Java script and XML (AJAX). It is a combination of several technologies such as HTML/CSS, document objective model (DOM), JavaScript, etc., to enable development of asynchronous web applications.
- **Single page application:** A single page application provides a fluid interface closer to a desktop application [20, 52]. The overall interface is loaded once, or the appropriate resources can be loaded dynamically in response to THE end-user's actions. The web-interface, however, does not change after the application has been loaded, and the control of the application is not transferred to another page. The advantage of a single page application is smooth and controlled user experience.
- **Extensible architecture:** An extensible architecture is one whose functionality can be extended by other users without recompiling the overall application. The proposed framework recommends an extensible approach for the ontology development, where the default implementations (for reasoning, visualization, storage, etc.) are provided, and extensions can be developed and plugged in.
- **Flexible design:** The proposed architecture recommends a flexible design approach based on model-view-controller (MVC) architecture.

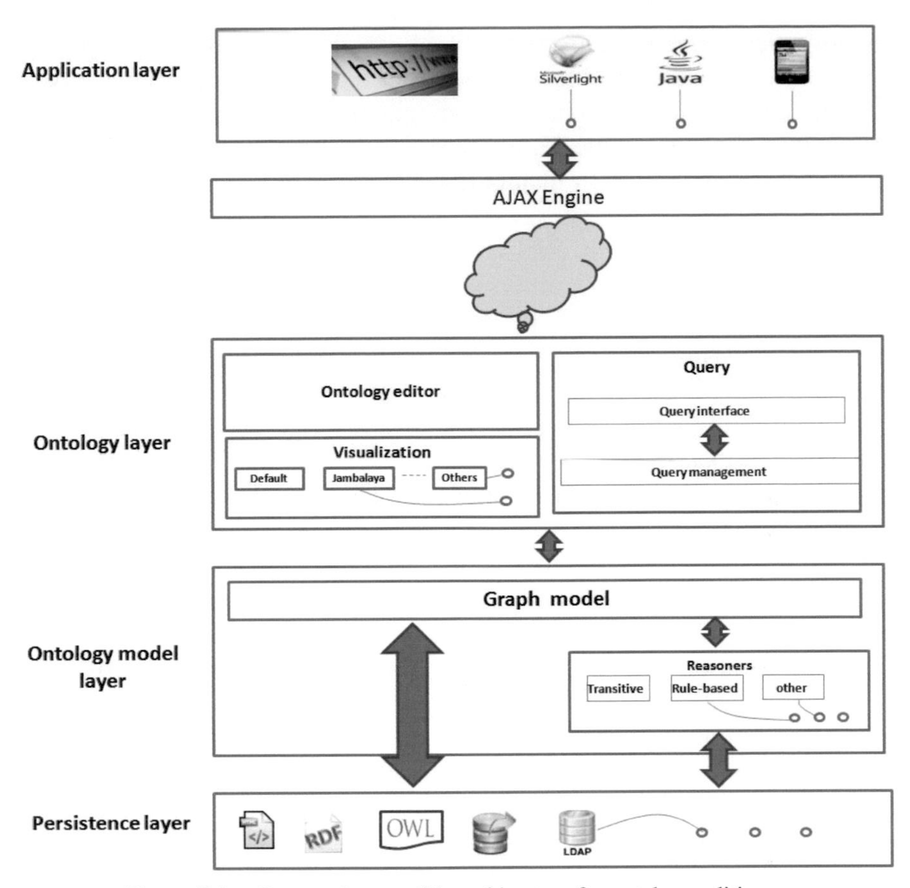

Figure 7.3 Proposed extensible architecture for ontology editing.

The application views and business logic are loosely coupled, allowing developers to modify models or views without affecting each other. This approach enables TODE to be extended for different devices, languages, and features without modifying the overall software.

7.3 Implementation Details

The implementation of an ontology editor requires dealing with a variety of challenges and making numerous implementation decisions. The selection of an ontology library that can allow ontology querying and reasoning, as well as manipulations of ontologies in various languages, is a critical issue in ontology development. Unfortunately, as we described in the previous section, only a few ontology libraries have been built for the .NET framework. Most of these techniques are in their infancy, and there is not enough information about them in the literature. Only a few of the editors are available, which provides more than basic RDF manipulation (OWL, inferencing, reasoning, etc.).

To overcome this problem, JENA, a non-.NET-based ontological framework, was chosen [53]. JENA offers an extensive set of Java classes for ontology editing and reasoning for RDF and OWL ontology languages. Because JENA was created in Java, it cannot be used directly in .NET. As a result, IKVM, a transformation utility, was chosen [54]. The IKVM utility provides the facility of converting a library from Java to the .NET platform. It is to be noted that after converting the toolkit to the .NET platform, one can use this .NET library independently for development of semantic web applications.

An interactive user interface design for ontology generation is an essential element of the ontology editing tool. After extensive research, we have selected *AJAX Control Toolkits* and *Telerik* [55]. These applications come with a suite of modules for creating rich, collaborative, and useable application interfaces.

The .NET framework appears to lack the basic capability as visualizers such as Jamabalaya and OntoViz, among others. We use information connection engine (ICE), a relationship network visualization package, to improve ontology *visualization* easier [56]. ICE is constructed on Microsoft Silverlight and provides a visual representation of relations among data as well as enables interaction with them in real time.

Besides what was discussed above, several other implementation choices have also been made for the development of proposed tool, as summarized in Table 7.6.

7.3.1 Ontology editor

The ontology editor is the core component of the proposed tool. The basic graphical user interface of the ontology editor is shown in Figure 7.4. For ontological editing, TODE provides an easy-to-use online interface that supports a hierarchical structure. It facilitates domain-specific knowledge modeling by utilizing a hierarchy of classes, their properties, associations, and occurrences. The left pane of figure shows that the creation of concepts, properties, and individuals has been provided in a simplified one-tabbed,

Table 7.6 Implementation choices for the proposed tool.

	Issues	**Implementation choice**
1.	Ontology library	JENA
2.	Visualization	Information connection engine (ICE)
3.	Interactive GUI development	ASP.NET, AJAX, and Telerik RAD Control
4.	Querying and reasoning	JENA built-in reasoners
5.	Ontology storage	Files, MS SQL Enterprise
	Other choices	
6.	Development tool	Microsoft Visual Studio
7.	Operating system	Microsoft Windows Server
8.	Browser	Mozilla Firefox, Microsoft Internet Explorer, and Google Chrome

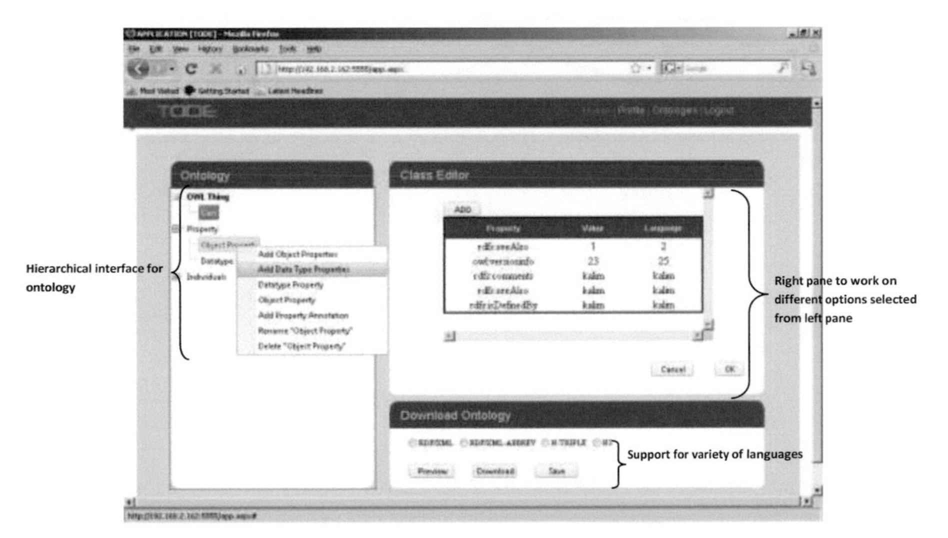

Figure 7.4 TODE interface [3].

AJAX-based, and consistent tree-based web interface. This is in contrast to existing ontology development tools that provides a multi-tabbed or multi-page interface for ontology creation. TODE provides support for all the widely used ontology languages. After the ontology has been constructed, the user can publish it in RDF, OWL, N-triples, and N3 formats by selecting the most appropriate alternatives from the interface's bottom. The developed ontology can also be saved in RDBMS format in an internal database. External ontologies written in various forms can also be imported in the same fashion.

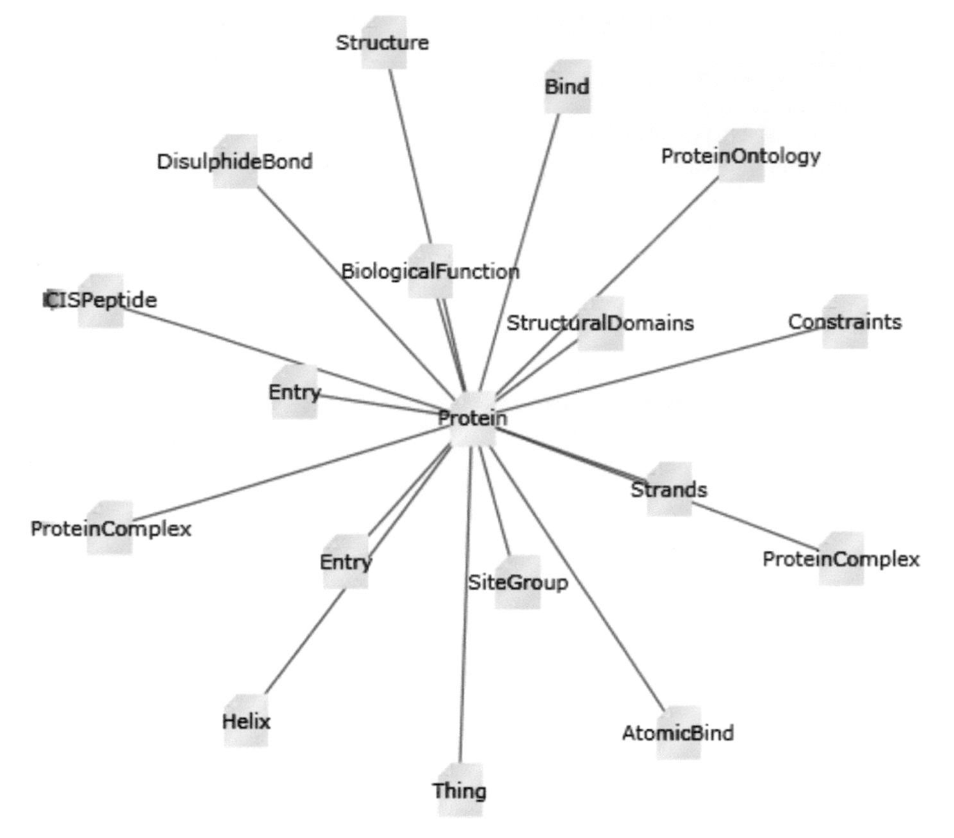

Figure 7.5 TODE visualization interface [3].

7.3.2 Visualizer

To browse any ontology, TODE presents an ICE-based interface. Unlike certain tools that rely on specialized plug-ins for visualization, TODE comes with a built-in modeling tool as a visualizer that can be enhanced further for more advanced features. Figure 7.5 is a screenshot of the proposed visualization tool. Graphical visualizations are used to arrange the various concepts in ontologies. The various concepts of ontologies are organized in the form of graphs. The links among the nodes show the inter-relationships among concepts. The details of concepts can be drilled down by clicking on the corresponding nodes. Clicking on a particular concept shows the object and data type properties of the concepts. This networked view of browsed ontology provides a useful mechanism to get an overall view of ontologies with a single snapshot.

7.3.2.1 Querying interface/reasoning

As we discussed earlier, querying and reasoning are an important component of semantic web stack. The RDF and OWL provide semantics information useful for many information systems. However, the full power of semantic web can only be realized when a reasoning infrastructure is in place to infer further information or conclusions from the existing data. To extract the final information, an appropriate querying interface is required. Most of the ontology editing tools provide an interface for ontology querying, but an external reasoner (Pellet) has to be attached. TODE provides a querying interface where the user can pose a SPARQL query to a model inferred by reasoning engines. The results are displayed in the form of triples.

TODE infers/reasons on the current ontology using the JENA toolkit's built-in rule-based approach. The distributive reasoner, the RDFS rule reasoner, the OWL/OWL Mini/OWL Micro Reasoners, and a generic rule reasoner are all included. TODE's querying interface is shown in Figure 7.6. The user picks the ontology from the left pane before posing the request to the system. The query management software parses the user's question before extracting tuples from inferred information. In the right panel, the query's responses are shown.

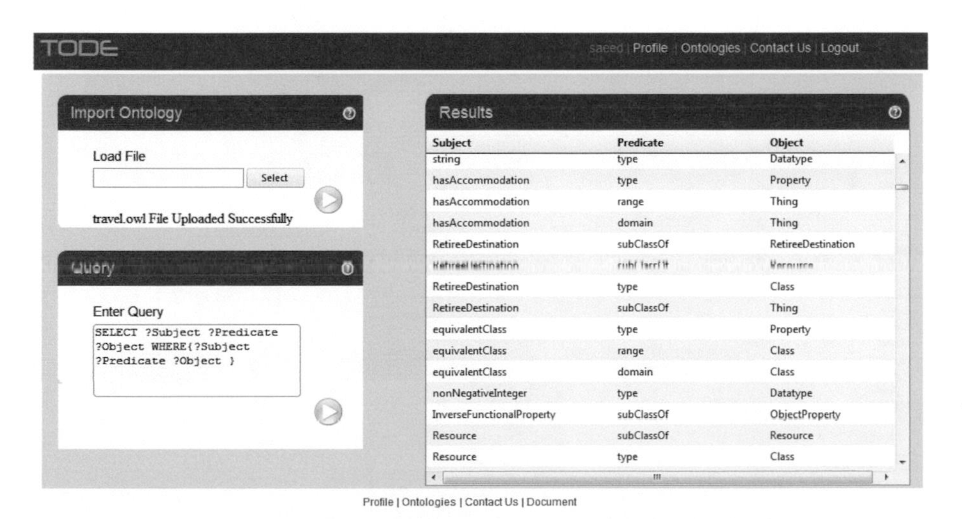

Figure 7.6 TODE querying interface [3].

7.3.2.2 Knowledge extraction interface

One of TODE's most notably interesting characteristics is knowledge extraction. It is developed to capture concepts, properties, and connections from a variety of sources (text documents, databases, web sources, etc.). The knowledge extraction component analyzes the underlying source, identifies and discriminates between the concepts, attributes, and relationships present, allows the user to filter out the choices, and then saves the filtered results in the form of a new ontology. Figure 7.7 shows the TODE knowledge extraction interface. A user can specify any archived file from which concepts are to be extracted. The editor classifies the tokens available in the file as concepts, attributes, or relationships based on its past experience. For this purpose, an ensemble of machine learning classifiers is trained. The model/classifier learns from past data if a token is a concept, attribute, relation, or instance. Users can manually analyze the results of the knowledge extraction component and then save the results as a new ontology.

7.3.2.3 Ontology development methodology

Most of the existing editing tools do not provide guidelines for the creation of ontology. In fact, there are only few tools that are developed specially as a proof-of-concept for a proposed ontology development methodology. One of TODE's benefits is that it offers a simple yet well-defined process for creating ontologies. The proposed technique, together with the TODE user interface, assists an ontology developer through the process of creating an ontology. The adapted methodology is described in depth in the following section.

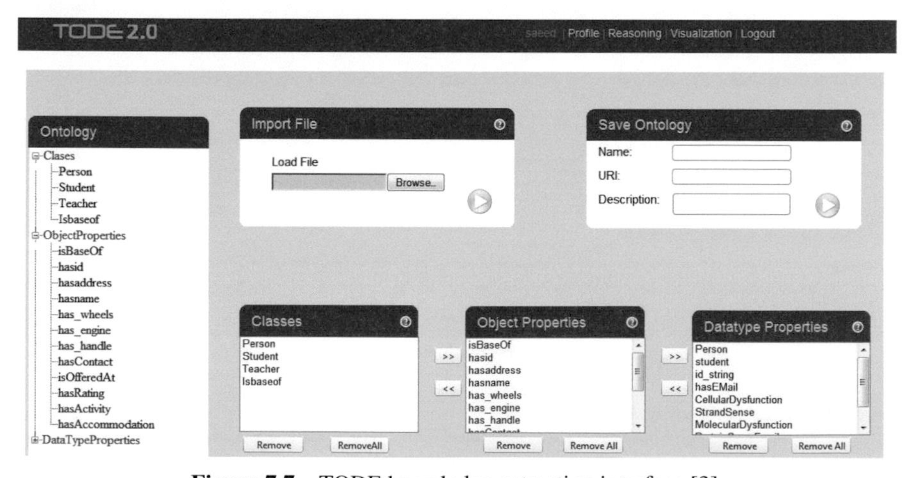

Figure 7.7 TODE knowledge extraction interface [3].

7.4 Ontology Development in TODE

The proposed methodology is inspired by conventional incremental approach to software development. TODE recommends development of ontology in increments. Figure 7.8 shows the six basic steps of ontology creation, and how various steps of ontology development can be performed by the proposed tool.

Knowledge extraction: Similar to [57], TODE recommends knowledge extraction as the first step toward ontology creation. The basic objective is to extract concepts and properties from different sources, i.e., archived files, websites, etc. The data of these archived sources can be input as a text file to knowledge extraction interface as shown in Figure 7.8 (b).

Metadata specification/modification: The next step is the specification of metadata about the ontology being created. This involves providing the name

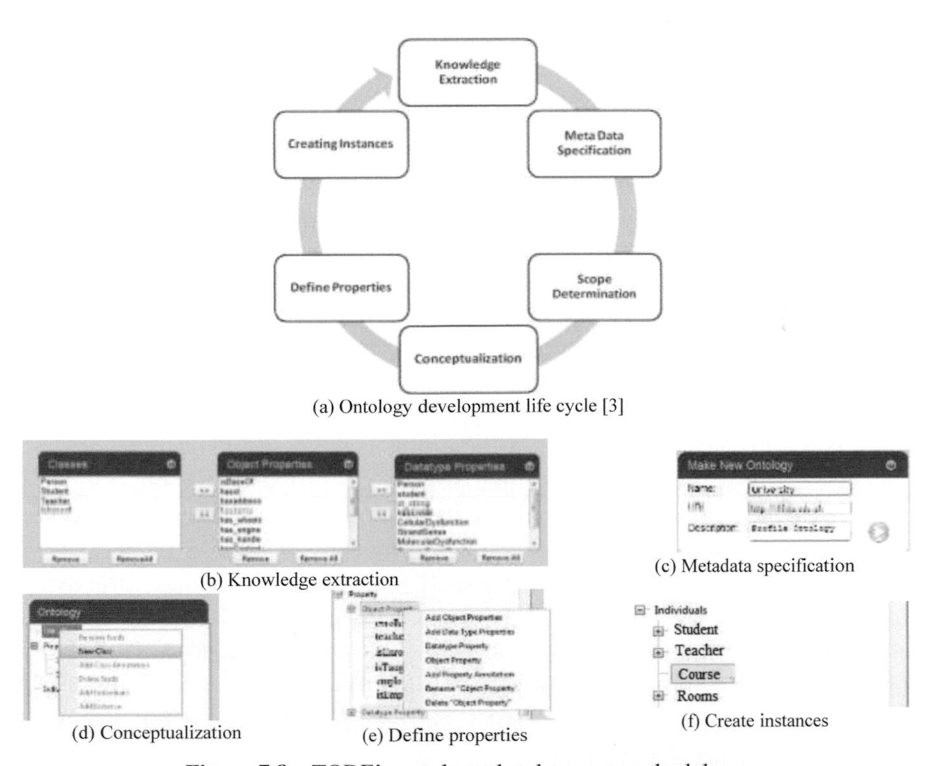

(a) Ontology development life cycle [3]

(b) Knowledge extraction

(c) Metadata specification

(d) Conceptualization

(e) Define properties

(f) Create instances

Figure 7.8 TODE's ontology development methodology.

of the ontology being created, base URI, further details, etc. Figure 7.8 (c) shows how this step can be achieved using TODE.

Scope determination: Having extracted concepts from archived sources and metadata specified, the user can now start editing the ontology using the web-based interface. The scope of ontology is specified (or refined), which can be used in the next steps to extract concepts and properties. The scope of the ontology can be identified by answering several basic questions about the domain of ontology, purpose of ontology, the target user of the ontology, etc. Once the scope of the problem is identified, it can be written as a statement of scope in few lines.

Conceptualization: The next stage is to determine which ontology notions exist. The determination of nouns from the scope declaration written in the previous stage is the first step in conceptualizing. These nouns serve as ontology concepts. Consequently, idea generalization and specialization are carried out. The TODE ontology editor provides a hierarchical interface for construction of classes, attributes, and relationships, as shown in Figure 7.8 (d). From the left pane of editing interface, one can select the corresponding node and then add the new concept.

Define properties: Properties of the concepts are identified. There are two types of properties, i.e., data type properties and object properties. The former can be obtained by looking for adjectives in the statement of scope. The object properties are the relationships among the concepts found in scope. Figure 7.8 (e) shows how data type and object properties can be defined using the proposed tool.

Instances creation: The occurrences of the ideas are recognized in this stage. The editor is used to produce instances and specify the values of data type characteristics. Figure 7.8 (f) depicts some of the TODE-created occurrences.

We close this section with a comparison of the proposed methodology with some of the ontology development methodologies available in recent studies. Presently, there are a plethora of approaches to choose from. For comparison, we restrict our self to some well-known methodologies. It is to be noted that due to the diversity in the proposals presented in the existing research, it is not possible to align them on a simple comparison table. Table 7.7 simply highlights the features provided by the proposed methodology in comparison to other available approaches. Therefore, interested readers can see further details in [14].

Table 7.7 A comparison of TODE's methodology with currently available methodologies.

Methodology	Major ontology development activities	Life cycle	Integral processes				Collaboration	Supporting tools
			Knowledge acquisition	Verification/validation/evaluation	Documentation	Training		
Ontology 101 [62]	Scoping, reuse ontology, enumerate items, conceptualization, define properties and facets, and instantiation	×	×	×	×	×	×	Protégé-2000, Ontolingua, and Chimaera
UPON [58]	Inception, elaboration, construction, and transition	✓	✓	✓	×	×	×	Athos and Ontolearn
TOVE [63]	Scenarios, informal competency questions, FOL terms, formal competency questions, FOL axioms, and completeness theorem	✓	✓	✓	✓	✓	✓	
DILIGENT [64]	Build, local adaptation, analysis, revision, and local update	✓	✓	✓	✓	✓	✓	KAON and SWAP
TODE	Knowledge extraction, metadata specification/modification, scope determination, conceptualization, define properties, and instances creation	✓	✓	✓	✓	✓	✓	TODE

Major ontology development methodologies: The methodologies available in literature proposed different steps for development of ontology. UPON [58], for instance, proposed steps similar to software development. Similarly, DILIGENT [59] provides steps to cater for distributed ontology development. TODE proposes basic steps for ontology development similar to Noy and McGuinness' method [60].

Well-defined life cycle: Some of the methodologies are inspired by the conventional life cycle model of software development. For example, UPON [61] is based on rational unified process model. Hence, the activities it proposes are similar to the life cycle of software. These activities include inception, elaboration, construction, and release. In each of these activities, different sub-activities are performed. TODE's ontology development methodology is an inspiration of incremental model. The ontologies are developed in small increments. Each increment comprises different steps outlined in Figure 7.8 and refines the current ontology.

Integral processes: Besides the basic ontology development activities, some methodologies also provide different integral processes, e.g., knowledge acquisition, documentation, training, etc. We can see that TODE's methodology does not talk too much about integral processes besides only the knowledge acquisition step.

Distributed development/collaboration: Collaboration is also one of the important aspects of ontology development. Different users can work during ontology development and can later on merge their ontologies. To address different types of issues that can arise during distributed ontology development, some collaboration mechanism is required. Besides DILIGENT, none of the methodologies listed in Table 7.7 addresses the issues for distributed construction of ontologies.

Supported tools: It is also interesting to see which tools provide the support for the proposed development methodology. It has been observed that there is no standard methodology for ontology development that can be supported by all tools. In fact, there are supporting tools developed for supporting a development methodology or vice-versa.

It can be concluded that TODE's ontology methodology is not very comprehensive and matured. However, it provides a basic foundation for ontology development for novice users.

7.5 Conclusion

This chapter provides a brief introduction to semantic web. It presents a discussion on ontology languages, rules, editors, ontology engineering, and ontology editing. Finally, a .net-based tool has been presented. Semantic web is a holistic concept comprising languages, tools, and standards. We have briefly discussed these techniques. The proposed tool TODE provides a wide range of new features. It is planned to develop a mobile based front end for the tool. Also, deep learning is expected to be incorporated to provide extensive support for ontology extraction and learning.

References

[1] Islam, N., & Laeeq, K. (2019). Salaat Ontology: A domain ontology for modeling information related to prayers in Islam. *Indian Journal of Science and Technology, 12*, 31.

[2] Noman Islam, D. S., Mariz Zafar & Asif Raza. (2022). Visualizing Chatbot Knowledge Graph using RDF. In N. C. D. Archana Patel, Bharat Bhushan (Ed.), *Semantic Web Technologies, Research and Applications*.

[3] Islam, N., & Shaikh, Z. A. (2019). *Towards ontology editing, querying and visualization in. net environment.* Paper presented at the 2019 8th International Conference on Information and Communication Technologies (ICICT).

[4] O' Conner, J. (2007). http://www.oracle.com/technetwork/articles/javase/extensible-137159.html.

[5] Trojahn, C., Vieira, R., Schmidt, D., Pease, A., & Guizzardi, G. (2021). Foundational ontologies meet ontology matching: A survey. *Semantic Web*(Preprint), 1-20.

[6] Khadir, A. C., Aliane, H., & Guessoum, A. (2021). Ontology learning: Grand tour and challenges. *Computer Science Review, 39*, 100339.

[7] Asim, M. N., Wasim, M., Khan, M. U. G., Mahmood, W., & Abbasi, H. M. (2018). A survey of ontology learning techniques and applications. *Database, 2018*.

[8] Braun, G. A., Estevez, E. C., & Fillottrani, P. (2019). A reference architecture for ontology engineering web environments.

[9] Gómez-Pérez, A. (2004). Ontology evaluation *Handbook on ontologies* (pp. 251-273): Springer.

[10] Nielsen, J. (1994). Heuristic evaluation. *Usability inspection methods*.

[11] Brooke, J., Jordan, P., Thomas, B., Weerdmeester, B., & McClelland, I. (1996). Usability evaluation in industry.

[12] Siricharoen, W. V. (2018). *Ontology editors approach for ontology engineering.* Paper presented at the 2018 4th International Conference on Control, Automation and Robotics (ICCAR).

[13] Pietriga, E. (2002). *Isaviz, a visual environment for browsing and authoring rdf models.* Paper presented at the Eleventh International World Wide Web Conference Developers Day, 2002.

[14] Islam, N., Abbasi, A. Z., & Shaikh, Z. A. (2010). *Semantic Web: Choosing the right methodologies, tools and standards.* Paper presented at the 2010 International Conference on Information and Emerging Technologies.

[15] Pulido, J., Ruiz, M., Herrera, R., Cabello, E., Legrand, S., & Elliman, D. (2006). Ontology languages for the semantic web: A never completely updated review. *Knowledge-Based Systems, 19*(7), 489-497.

[16] Hogan, A. (2020). Web ontology language *The web of data* (pp. 185-322): Springer.

[17] Farquhar, A., Fikes, R., & Rice, J. (1997). The ontolingua server: A tool for collaborative ontology construction. *International journal of human-computer studies, 46*(6), 707-727.

[18] Angele23, K., & Angele, J. (2021). JSON towards a simple Ontology and Rule.

[19] Patel, A., & Jain, S. (2021). Present and future of semantic web technologies: a research statement. *International Journal of Computers and Applications, 43*(5), 413-422.

[20] Mehla, S., & Jain, S. (2019). Rule languages for the semantic web *Emerging Technologies in Data Mining and Information Security* (pp. 825-834): Springer.

[21] Bechhofer, S., Horrocks, I., Goble, C., & Stevens, R. (2001). *OilEd: a reason-able ontology editor for the semantic web.* Paper presented at the Annual Conference on Artificial Intelligence.

[22] Lawan, A., & Rakib, A. (2019). FT-SWRL: A Fuzzy-Temporal Extension of Semantic Web Rule Language. *arXiv preprint arXiv:1911.12399.*

[23] Al-Aswadi, F. N., Chan, H. Y., & Gan, K. H. (2020). Automatic ontology construction from text: a review from shallow to deep learning trend. *Artificial Intelligence Review, 53*(6), 3901-3928.

[24] Gao, W., Guirao, J. L., Basavanagoud, B., & Wu, J. (2018). Partial multi-dividing ontology learning algorithm. *Information Sciences, 467*, 35-58.

[25] Konys, A. (2018). Knowledge systematization for ontology learning methods. *Procedia Computer Science, 126*, 2194-2207.
[26] Petrucci, G., Rospocher, M., & Ghidini, C. (2018). Expressive ontology learning as neural machine translation. *Journal of Web Semantics, 52*, 66-82.
[27] Fortuna, B., Grobelnik, M., & Mladenic, D. (2007). *Ontogen: Semi-automatic ontology editor.* Paper presented at the Symposium on Human Interface and the Management of Information.
[28] Bernstein, A., & Kaufmann, E. (2006). *GINO–a guided input natural language ontology editor.* Paper presented at the International semantic web conference.
[29] Mendonça, F. M., Emygdio, J. L., de Castro, L. P., & Felipe, E. R. Onto4ALLEditor: a Graphic Web Ontology Editor for Information Science.
[30] Horridge, M., Gonçalves, R. S., Nyulas, C. I., Tudorache, T., & Musen, M. A. (2019). *Webprotégé: A cloud-based ontology editor.* Paper presented at the Companion Proceedings of The 2019 World Wide Web Conference.
[31] Schekotihin, K., Rodler, P., & Schmid, W. (2018). *Ontodebug: Interactive ontology debugging plug-in for Protégé.* Paper presented at the International Symposium on Foundations of Information and Knowledge Systems.
[32] Kalyanpur, A., Parsia, B., Sirin, E., Grau, B. C., & Hendler, J. (2006). Swoop: A web ontology editing browser. *Journal of Web Semantics, 4*(2), 144-153.
[33] Day-Richter, J., Harris, M. A., Haendel, M., Group, G. O. O.-E. W., & Lewis, S. (2007). OBO-Edit—an ontology editor for biologists. *Bioinformatics, 23*(16), 2198-2200.
[34] Preventis, A., & Petrakis, E. G. (2021). CLONE: Collaborative Ontology Editor as a Service in the Cloud. *Procedia Computer Science, 184*, 275-282.
[35] COMPOSER, T. (2007). Features and getting Started Guide Version 1.0, created by TopQuadrant, US.
[36] Jackson, R. C., Balhoff, J. P., Douglass, E., Harris, N. L., Mungall, C. J., & Overton, J. A. (2019). ROBOT: a tool for automating ontology workflows. *BMC bioinformatics, 20*(1), 1-10.
[37] Wiens, V., Lohmann, S., & Auer, S. (2018). *WebVOWL Editor: Device-Independent Visual Ontology Modeling.* Paper presented at the International Semantic Web Conference (P&D/Industry/BlueSky).

[38] Bell, S., Bonasso, P., Boddy, M., Kortenkamp, D., & Schreckenghost, D. (2013). *PRONTOE: An Ontology Editor for Domain Experts.* Paper presented at the International Joint Conference on Knowledge Discovery, Knowledge Engineering, and Knowledge Management.

[39] Zhang, M. (2015). *A mobile web-based ontology editor for smartphone.* Paper presented at the Proceedings of the 2nd International Workshop on Materials Engineering and Computer Sciences.

[40] Hallay, F., Hartmann, S., Kewitz, N., & Mertens, R. (2017). *An aspect-oriented visual ontology editor with edit-time consistency checking.* Paper presented at the 2017 IEEE 11th International Conference on Semantic Computing (ICSC).

[41] Mahoro, L. J., & Fonou-Dombeu, J. V. (2019). *An Empirical Evaluation of dot NET-Based Tools for OWL/RDF Ontologies Processing.* Paper presented at the 2019 International Conference on Advances in Big Data, Computing and Data Communication Systems (icABCD).

[42] Mahoro, L. J., & Fonou-Dombeu, J. V. (2020). *A Comparative Study of dotNetRDF And Semweb. NET Semantic Web Libraries.* Paper presented at the 2020 International Conference on Artificial Intelligence, Big Data, Computing and Data Communication Systems (icABCD).

[43] Haase, P., Broekstra, J., Eberhart, A., & Volz, R. (2004). *A comparison of RDF query languages.* Paper presented at the International Semantic Web Conference.

[44] .net., D.-A. R. P. f., Availablefrom:http://sourceforge.net/projects/drive -rdf/support.,lastaccessedonMarch,2022.

[45] *dotNetRDF - Semantic Web/RDF Library for C#/.Net. www.dotnetrdf.org.* . last accessed on Feb 2022.

[46] 2.0., S. P.,http://www.intellidimension.com/products/semantics-platfo rm/,. last accessed on Feb 2022.

[47] API., O. D.,http://code.google.com/p/owldotnetapi/.. Last accessed on March 2022.

[48] *ROWLEX API, http://rowlex.nc3a.nato.int/.*

[49] Standards., O.-S. W., www.w3.org/2001/sw/wiki/OpenAnzo. , Last accessed on March 2022.

[50] Library., P. P., http://protegewiki.stanford.edu/wiki/Protege_Plugin_Lib rary. last accessed on Feb, 2022.

[51] Garrett, J. J., *Ajax: A new approach to web applications.* 2005.

[52] paradigm., I.-b.c.t.b.n., https://www.opcnlinksw.com/wcblog/kidc hen@openlinksw.com/127/index.vspx?page=&id=86&cmf=1,. last accessed on March 2022.
[53] JENA, A., https://jena.apache.org/ last accessed on March 2022.
[54] https://github.com/jessielesbian/ikvm last accessed on March 2022.
[55] Spalding, D. R., *Telerik WPF Controls Tutorial*. 2014: Packt Publishing.
[56] (ICE)., I. C. E., http://icedotnet.codeplex.com/.
[57] Fernández-López, M., Gómez-Pérez, A., & Juristo, N. (1997). Methontology: from ontological art towards ontological engineering.
[58] Nicola, A. D., Missikoff, M., & Navigli, R. (2005). *A proposal for a unified process for ontology building: UPON.* Paper presented at the International Conference on Database and Expert Systems Applications.
[59] P Pinto, H. S., Staab, S., & Tempich, C. (2004). *DILIGENT: Towards a fine-grained methodology for DIstributed, Loosely-controlled and evolvInG Engineering of oNTologies.* Paper presented at the ECAI.
[60] Noy, N. F., & McGuinness, D. L. (2001). Ontology development 101: A guide to creating your first ontology. *Knowledge Systems Laboratory*, 01-05.
[61] De Nicola, A., Missikoff, M., & Navigli, R. (2005). A proposal for a Unified Process for ONtology building: UPON. *Lecture Notes in Computer Science, 3588*, 655.
[62] Noy, N. F. a. D. L. M. (2001). Ontology development 101: A guide to creating your first ontology. Knowledge Systems Laboratory, . p. 01-05.
[63] Grüninger, M., & Fox, M. S. (1995). Methodology for the design and evaluation of ontologies.
[64] Pinto, H. S., Staab, S., & Tempich, C. (2004). *DILIGENT: Towards a fine-grained methodology for DIstributed, Loosely-controlled and evolvInG Engineering of oNTologies.* Paper presented at the ECAI.

8

Aedes Ont: Ontology for Aedes Mosquito Vectors to Predict Semantic Relations of Biocontrol Agents

G. Jeyakodi, Trisha Agarwal, and P. Shanthi Bala

Department of Computer Science, Pondicherry University, India
E-mail: rjeyakodi02@gmail.com; trisha.agarwal27@gmail.com; shanthibala.cs@gmail.com

Abstract

Mosquitoes belonging to the genus *Aedes*, such as *Aedes aegypti* and *Aedes albopictus* species, are the prime vectors of several arboviral diseases such as dengue, zika, yellow fever, and chikungunya. Every year, numerous cases of dengue infections occur throughout the world. Vector control is an efficient method to control these arboviral diseases. However, the onset of insecticide resistance has failed in controlling the vectors. The continuous synthetic insecticide may cause toxicity to human health through the food chain. The alternate strategy of vector control can be achieved with the use of biological control by microbes or medicinal plants which are harmless to human health. Most of the biological agents control the immature species in its oviposition, larvae, or pupae stages. The semantic relations among the mosquitoes and biocontrol agents help to control the species harmlessly. Still, the semantic relations among the *Aedes* mosquito vectors and their biocontrol agents are not explored well. Learning semantic relations is challenging due to the lack of semantic representation for *Aedes* mosquito vectors. Ontology helps to provide the semantic information that may be used to extract meaningful relations. This chapter discusses the role of *Aedes* ontology and the issues in extracting domain specific semantic relations.

Keywords: *Aedes*, biocontrol agent, mosquito, microbes, ontology, plant, semantic relation.

8.1 Introduction

Vector-borne diseases (VBD), such as dengue, malaria, zika, yellow fever, lymphatic filariasis, and chikungunya are the infections caused in humans or animal hosts wherein the pathogens use vectors as a host for their survival or transmission. These diseases spread 17% of the global infections. *Aedes* species mosquitoes, *Aedes aegypti* and *Aedes albopictus*, are the most common vectors for VBDs like chikungunya, dengue, West Nile infection, and zika [1]. Dengue has been listed as one of the ten high-priority diseases by the World Health Organization (WHO) posing global public health threats [2]. It is a mosquito-borne disease caused by four genetically distinct strains of the dengue virus (DENV) serotype belonging to the Flaviviridae family. Mosquito bite by female *Aedes* mosquito vectors transmits dengue to humans [3]. The global burden of the disease study reported that there was a major increase in the number of dengue cases as well as the mortality rates globally [4]. Dengue is prevalent in the urban and semi-urban areas of tropical and subtropical countries in the regions of South Asia and Southeast Asia. However, now the disease is spreading beyond the tropics and has been found to be fatal in Florida with infected people who are having a travel history of Cuba and also from some locally infected cases [5]. More than half of the world's population is now susceptible to dengue virus infection [6]. The severity of dengue virus infection ranges from asymptomatic mild fever to fatal diseases like dengue hemorrhagic fever and dengue shock syndrome. Although initial dengue infection can create an immune system against the same serotype DENV infection, the subsequent DENV infections over a prolonged period are observed to be more fatal [7].

It is important to understand the environmental and biological factors for controlling the development and movement of *Ae. aegypti* mosquitoes [8]. It has been observed that the climatic factors influence the *Aedes* mosquito population densities in metropolitan regions of Brazil [9]. Vector control is the primary method of controlling VBDs, especially at the larval stage, which is crucial for reducing the spread of VBDs [10]. Dispersion of pathogens can be controlled by observing the skip oviposition behavior of female *Ae.* aegypti mosquitoes. Controlling the environment of the breeding sites can also be an important control mechanism for the transmission of VBDs. However, it has been observed that wiping off the breeding sites can have adverse effects on the disease spread [11].

Pyrethroid insecticide has played a pivotal role in vector control. However, the development of resistance among the mosquito vectors against this insecticide has been reported in Bangladesh, with studies conducted to determine the resistance mechanism [12]. Similarly, the development of resistance for permethrin insecticide among the *Aedes* vectors in Thailand was reported, which prove the inefficiency of DDT for vector control [13]. Systematic comparison of the effectiveness of chemical and biological vector control mechanisms show that the chemical control does not help in controlling vector incidences over a prolonged period. On the other hand, biological control has a long-term vector as well as disease incidence control [14].

The inefficiency of synthetic insecticides on account of insecticide resistance and concern for the environment as well as human health has led to the need for the development of alternate sustainable biological vector control strategies. The biological agents target the development of mosquitos at different stages of their lifecycle [15]. These agents work on the behavioral aspects of adult mosquitoes and inhibit growth and reproduction at larval stages [16]. The toxicity of plant-based natural extracts also provides alternative to chemical insecticides [17]. A study conducted in French Guiana also exposed that the plant-based extracts increased the larval mortality by 50% that includes pyrethroid-resistant population [18]. Thus, finding correlations between the biocontrol agents and dengue virus vectors is an important factor in vector-borne disease management.

Ontology is an explicit, formal specification (i.e., machine-readable) of a shared conceptualization of a domain of interest. It represents a set of concepts within a domain and the relationships that exist between them. Ontology-based information extraction has recently emerged as a subfield of information extraction from natural language through a mechanism that is guided by ontologies and also explores the semantic information from it. The ontology data model can be applied to a set of individual facts to create a knowledge graph as a collection of entities, in which the types and relationships between them are expressed by nodes and edges, respectively. The ontology helps to construct the knowledge graph. This knowledge graph is required for understanding the relationship between the *Aedes* vector and its biocontrol agents to develop an alternate strategy against vector-borne diseases [19]. This knowledge graph can be reused through ontology enrichment to get desired information for the development strategies of vector-borne disease management.

8.2 Aedes Mosquito Vector

Aedes mosquitoes, both *Ae. aegypti* and *Ae. albopictus*, which are carriers of VBDs, are of great medical and veterinary importance. The arthropod-borne viruses (arboviruses), such as zika virus (ZIKV) and DENV, are transmitted by *Aedes* mosquito bites and pose a great threat to human and animal lives globally. The vector gets infected with DENV through the dengue-infected human after an incubation period of 8–10 days. With the infection being inherited by the offspring, i.e., transovarial transmission, vector control has become more difficult. Dengue has now become endemic in India with the rapid increase in cases as well as recurring outbreaks reported in more than 10 states [20]. The global expansion of arboviral diseases has been primarily associated with trade and travel, with *Aedes* occurrences now also being recorded in Asia, America, Africa, as well as Europe. Since the maximum occurrence is recorded in Asia [8], it becomes important to study and understand the behavior of *Aedes* mosquitoes and the ways to control their spread.

The most important variables affecting the spread of *Ae. aegypti* are the precipitation and population density. The factors such as temperature and rainfall affect the spread of *Ae. albopictus* in an urban area. The elongated proboscis and piercing mouthparts render female *Aedes* mosquitoes to transmit pathogens and diseases among humans as well as animals. In India, *Ae. aegypti* is considered as a primary vector and *Ae. albopictus* as a secondary vector for dengue transmission [21]. However, *Ae. albopictus* plays a primary role in chikungunya transmission in Kerala. *Ae. aegypti* is observed to be endophilic in nature, while *Ae. albopictus* prefers resting outdoors showing exophilic nature. They both are strong daytime biters with peak biting in the morning and late afternoon before sunset. Female *Ae. aegypti* bites, multiple hosts, to complete one blood meal and has a flight range of 450–600 m. On the other hand, female *Ae. albopictus* is a weak flyer and prefers any region between the ankle and knees.

8.2.1 Aedes life cycle

The lifespan of *Aedes* mosquitoes ranges from two weeks to less than a month depending on the environmental conditions, which has a higher impact on the longevity of *Ae. albopictus* than *Ae. aegypti*. Due to the high tolerance level of *Ae. albopictus* in the cold climatic conditions, it was spread to more than 32 states in America and a majority section of the European region [22]. In India, the breeding sources of both *Aedes* species are found to be increased in post-monsoon and peak during October and November. The *Ae. aegypti* mosquitoes prefer container inhibitors for their breeding such as mud pots and plastic containers, and *Ae. albopictus* mosquitoes prefer forest breeding such as leaf axils and bamboo stumps. Since *Ae. aegypti* has been adapted well to man-made containers, it has become a major threat in densely populated urban regions. *Ae. albopictus*, on the other hand, breeding in densely vegetated areas, poses threat to rural life. *Ae. aegypti* mosquitoes are primarily anthropophilic, while *Ae. albopictus*, initially zoophilic, has started exhibiting anthropophilic nature with evolved host preferences. *Ae. albopictus* is an opportunistic feeder with blood hosts including humans as well as domestic and wild animals. Apart from DENV and ZIKV, *Ae. aegypti* has shown strong vector competency for chikungunya virus, yellow fever virus, as well as West Nile virus, while *Ae. albopictus* has been a competent vector experimentally for 22 other arboviruses. *Aedes* mosquitoes undergo four different stages during their life cycle as shown in Figure 8.1 and Table 8.1.

8.2.2 Insecticide resistance behavior

The development of insecticide resistance among *Aedes* mosquitoes has been a major setback in the control of VBDs. The study conducted in the Northern districts of Bengal stated that metabolic detoxification is the primary reason for resistance against four major insecticides for the dengue vectors [23]. *Ae. aegypti* from Ceará, Brazil has shown resistance against temephos and organophosphates, the most common insecticides in Brazil. It has also been observed that resistance reversal at the larval stage is very slow and may take more than seven years. Detoxifying enzymes seem to be a more probable reason for the temephos resistance rather than the target site mechanism [24]. The knockdown resistances (kdr) among *Ae. albopictus* mosquitoes against DDT and pyrethroid insecticides are the most affecting efficacy of chemical insecticides [25]. Climate changes influence the evolution of mosquitoes to develop insecticide resistance [26]. Insecticide Resistance Management (IRM) under the Insecticide Resistance Action Committee (IRAC) has helped in creating resistance awareness by building IRM tools and schemes like the mode of action (MoA) classification scheme [27]. This insecticide resistance

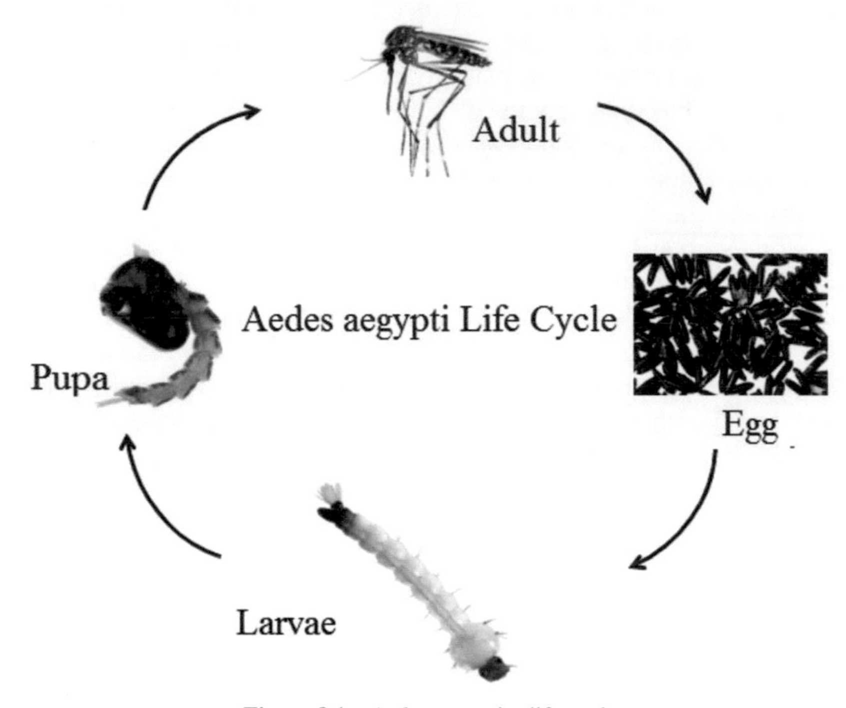

Figure 8.1 *Aedes* mosquito life cycle.

behavior leads the researchers to identify the alternate strategies for vector control.

8.2.3 Aedes mosquito variants

Insect transgenesis is a novel tactic to detect vector competency in *Aedes* mosquitoes. Behavioral studies of *Aedes* mosquitoes are used to understand the molecular genetics progress of a vector control strategy. Systemic immunity has been successfully activated in stable transgenic female *Ae. aegypti* mosquito fat body by a blood meal. Thus, vitellogenin (Vg) – Defensis A (DefA) insertion in *Aedes* mosquitoes is a pioneering work in the field of genetic engineering to inhibit pathogen transmission [28]. Restricting the population of *Ae. albopictus* through the insertion of transposable piggyBac has been a novel attempt in enhancing the Sterile Insect Technique (SIT) to reduce the fertile mating [29]. The piggyBac transposon vectors promoted with antiviral hammerhead ribozymes (hRzs), #9 and #14 effector genes

Table 8.1 Various stages of *Aedes* mosquito life cycle.

Sl. no.	Mosquito stage	Description
1	Egg	• Adult female mosquitoes lay black ovoidal eggs on the inner walls of containers, at varying distances, just above the waterline, on damp mud leaf litter of pools damp surfaces of tree holes, clay pots, and rock pools. They stick to the walls of containers and can survive without water for eight months or occasionally over a year. These eggs hatch as soon as they come in contact with water while some may require prolonged immersion in water. Even in favorable climatic conditions, it is often the case when these eggs entering into the resting period and require various ecological disturbances like variation in temperature, daytime, and the oxygen content in water to break the diapause stage.
2	Larvae	• Larvae hatch from eggs when they are flooded with water. The physical characteristics of *Aedes* larvae isolate it from the rest of culicine genera except from South American Haemagogus larvae. Its development is mainly affected by temperature variations.
3	Pupae	• Pupae come out of the skin of larva at the end of the fourth larval instar. Swims in a kicking motion but, in the last few minutes, loses its ability to swim and floats on the water body surface.
4	Adults	• In female adults, the abdomen is a pointed tip that adds to a distinctive physical feature. White scales on the top surface of the thorax form conspicuous patterns such as violin or lyre of the adult mosquito. *Ae. aegypti* and *Ae. albopictus* have a white stripe down the middle of the top of the thorax. Males are comparatively smaller than females.

being most effective, have been constructed to suppress CHIKV in *Aedes* mosquitoes [30]. However, vector genes with tissue and specific properties are not easily available. Hence, direct genetic manipulation of the vector itself is a complex approach with a very limited number of laboratories that may perform gene transformation successfully [31]. The fitness of transgenic mosquitoes in natural habitats is also a major setback. Hence, developing transgenic symbionts of the vector to eliminate the development of pathogens in the host has helped in moving forward in the genetic control approach. The introduction of transgenic *wolbachia* symbiont is a major achievement in inhibiting pathogen development in the target organs of *Aedes* mosquitoes. The density of *wolbachia* in male and female mosquitoes has been studied and found that female *Aedes* has a higher density as compared to males because of vertical transmission. Region and temperature are also important factors in determining *wolbachia* density [32].

8.3 Vector Control Techniques

Vector control can be broadly classified into chemical, genetic, environmental, immunological, and biological as shown in Figure 8.2. Chemical agents are the most common and cost-effective vector control measure. However, the adverse effects of insecticides on the environment, biomagnification of chemicals in the ecosystem, as well as the development of insecticide resistance in mosquitoes help to the development of tools in the field of vector control agents. The vector control through genetic engineering can develop deliberate tools and produce satisfactory results. However, this vector control strategy is very costly with a low success rate on specific conditions. Hence, it becomes important to explore various natural extracts from plants and microbes that can help in developing bioinsecticides for dengue control. Bioinsecticides successfully serve as an alternate strategy to insecticide resistance with almost negligible environmental hazards.

8.3.1 Environmental control

Environmental control aims at reducing mosquito breeding sources and making the environment detrimental to larval development. This method seeks both short-term and long-term changes. The irrigation system for draining breeding sites to prevent larval development is an example of short-term changes. It also includes the removal of water plants and other preferred artificial *Aedes* habitats such as water storage containers, flower pots, mud pots, coconut shells, tree holes for rubber milk collection, and plastic containers. Intermittent irrigation, the long-term changes are used for controlling vector-borne diseases effectively. It also suppresses the rice irrigation's favorable conditions for breeding by supplying limited water, draining, and flooding shallow water periodically. This method of draining and flooding the fields helps in preventing mosquito reproduction. In addition to this, the

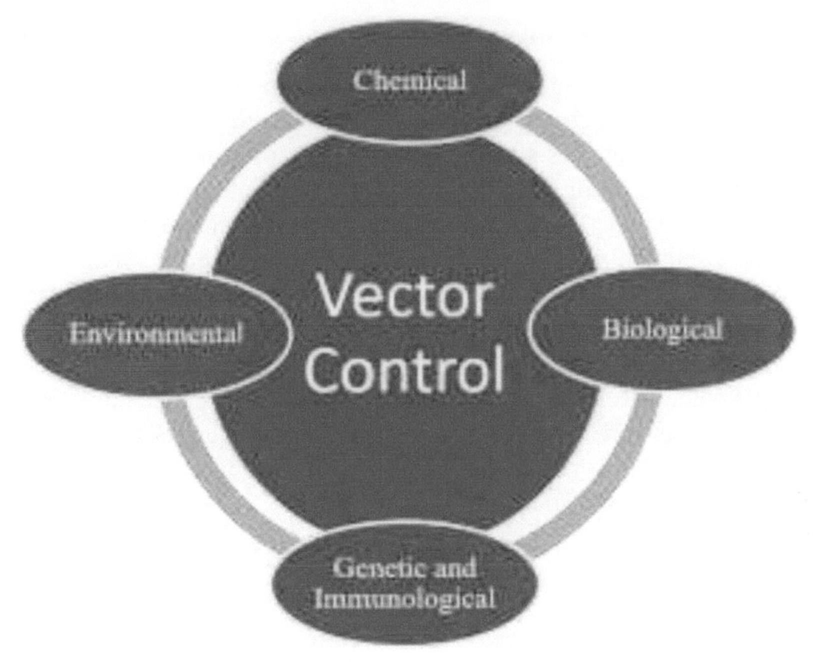

Figure 8.2 Vector control techniques.

construction of ditches to prevent water stagnating in salt marshes, improving sanitation systems by installing sewerages are some of the other effective methods for controlling mosquito vector species.

8.3.2 Chemical control

Chemical control is extensively used for larvicides and adulticides for immediate control using insecticides and repellents. Indoor residual spraying, space spraying, and insecticide treated mosquito nets are used for controlling adult *Aedes* mosquitoes. The effectiveness of insecticide-treated materials benefited in a 70% reduction in dengue-positive cases [14]. The chemical insecticide pyrethroid, an adulticide, is refrained over water bodies due to high toxicity against aquatic and cold-blooded animals. In general, larvicides are used in powdered form. Ovitraps are also used for larvae control by keeping them in the vector breeding sites. It is an effective method for a smaller scale and expensive for a larger scale. Repellents and mosquito nets are used for personal protection against mosquito species.

8.3.3 Genetic and immunological control

Genetically modified mosquitoes have been developed to overcome the limitations of existing vector control strategies. Genetics aims to provide environment-friendly, species-specific methods for mosquito control. This is achieved by either reducing the mosquito population or introducing harmless novel traits for the species. The release of genetically modified mosquitoes is a self-limiting vector control technique. It requires monitoring and maintaining the release of modified vectors; otherwise, the modified vector may vanish. Thus, a strategic approach to spread the modified mosquito in the wild population is required to ensure the reduction of overall vector competency. Oxitec's Release of Insects carrying a Dominant Lethal (RIDL) technique for *Ae. aegypti*, based on Sterile Insect Technique (SIT), is one of the successful strategies to reduce the *Aedes* population by the periodic mass release of sterilizing mosquitoes [33].

8.3.4 Biological control

The need of the hour is to develop eco-friendly alternative strategies to control dengue vectors. Biological control agents such as bacteria, nematodes, predators like fishes, copepods, tadpoles, microbes, and plant extracts have a substantial influence on reducing mosquito density. The release of larvivorous fishes in water bodies has proved to be one of the most economical vector control strategies. Among the many larvivorous fishes available locally, mosquitofish (*Gambusia affinis*) is one of the most widely used biocontrol agents [34]. It mainly preys on *Ae. aegypti* at the third instar larvae. Plant extracts, phytochemicals, and their nanoformulations can be used as eco-friendly ovipositional attractants, insecticides, and repellents. The nectar of *Impatiens walleriana*, the common garden impatiens, was found to be preferred by more than 80% of both male and female *Aedes* mosquitoes on a 24-hour feeding activity behavior surveillance [35]. The development of another vector control tool wherein anti-mosquito proteins can be delivered through the nectar of this plant. The *Azolla pinnata* plant extract has shown effective larvicidal activities for *Aedes* mosquito control. The bioactive molecules of *A. pinnata* can be used to develop insecticide affecting the fourth instar larvae of *Aedes* mosquito [36].

The inhibitory effects of *Ipomoea cairica* crude extracts on the physical and biological development of both *Ae. aegypti* and *Ae. albopictus* have made *I. cairica* a suitable candidate for the development of plant-based insecticide

Table 8.2 Biocontrol by plant extracts.

Sl. no.	Stage	Name	Type
1	Eggs	*Hedychium larsenii* and *Piper purusanum*	Essential oil [40]
		Lpomoea cairica	Leaf [25]
2	Larvae	*Hedychium larsenii*, Trichodesmium, *Ayapana triplinervis*, *Eucalyptus camaldulensis*, *Conyza* (asteracease), *Baccharis reticularia*, *Tetradium glabrifolium*, *Lantana montevidensis*, *Vitex negundo* L.	Essential oil [40–48]
		Feronia limonia, *Cupresaceae* species, *Ocimum gratissimum*, *Croton linearis* Jacq, *Pinus kesiya*, and *Heracleum sprengelianum*	Leaf essential oil [49–54]
		Azadirachta indica, Neem	Neem oil [55, 56]
		Toddalia asiatica Lam, *Angelica dahurica*, *Angelica pubescentis*, and *Saussurea lappa*	Root oil [57–59]
		Cnidoscolus phyllacanthus and *Ricinus communis*	Vegetable oil [60]
		Pterocarpus santalinoides and *Tropical Manihot* species	Seed oil [61]
		Couroupita guianensis Aubl and *Solanum xanthocarpum*	Fruit [62, 63]
		Ipomoea cairica (L.) or railway creeper	Flower [64]
		Ipomoea cairica (L.) or railway creeper	Stem [64]
		Derris trifoliata, *Tephrosia vogelii*, *Zornia diphylla*, *Ricinus communis*, *Lpomoea cairica*, and *Murraya koenigii*	Leaves [65–71]
		Couroupita guianensis Aubl and *Ipomoea cairica* (L.) or railway creeper	[62–64]
		Jatropha curcas and *Ricinus communis*, *Ricinus communis*, *Millettia pinnata*, *Terminalia chebula* Retz., and *Clitoria fairchildiana*	Seed [72–75]
3	Pupae	*Cnidoscolus phyllacanthus* and *Ricinus communis*	Vegetables oils [60]
4	Adult	*Origanum scabrum*	Essential oil [76]

Table 8.3 Biocontrol by microbes.

Sl. no.	Stage	Name	Type
1	Egg	*Beauveria bassiana*	Fungus [77]
		Aedes densonucleosis virus (AeDNV)	Virus [78]
2	Larvae	*Bacillus sphaericus*, *B. thuringiensis* H-14, *Lysinibacillus sphaericus*, *Bacillus thuringiensis* var. *israelensis* (Bti), *Steinernema*, and *Heterorhabditis nematodes*	Bacteria [79–82]
		cylindrosporum Kal, *Leptolegnia chapmanii*, and *Isaria tenuipes*	Fungus [83–85]
		C6/36DNV	Virus [86]
3	Pupae	*Bacillus marisflavi*	Bacteria [87]
4	Adult	*Metarhizium anisopliae*	Fungus [88]
		Wolbachia endosymbiont	Bacteria [89]

[37]. Plant leaf extracts of *Glycosmis pentaphylla*, collected from Kolli Hills, India, especially Acetone extracts, have shown to be effective at both larval and adult stages of *Aedes* mosquito. These bioactive molecule ingredients can help in developing eco-friendly insecticides for dengue control [38]. The findings of the high mortality rate of *Aedes* against exposure to *Areca catechu* nut extracts have led to the development of potential larvicide against early fourth stage *Aedes* larvae [39]. Similarly, various plant extracts have been discovered that affect the development of the *Aedes* mosquito at different life stages as shown in Table 8.2. Mosquitocidal microbes that include bacteria, fungi, and viruses also provide non-toxic vector control solutions. These microbes and their effects on mosquito development stages have been specified in Table 8.3. The efficacy of these biocontrol agents is largely based on the results of small-scale field operations. The expense of raising the biocontrol organisms, difficulty in production, and limited applicability are some of the limitations of biocontrol measures.

8.4 Role of Ontologies in Vector Control

Information on mosquito vectors in terms of pathogens (virus, bacteria, and nematode), vector-borne diseases (dengue, malaria, chikungunya, and lymphatic filariasis), insecticides (Dichloro Diphenyl Trichloroethance (DDT), deltamethrin, malathion, pyrethroid, etc.), and biocontrol agents such as medicinal plants and microorganisms are generated continuously from scientific research and public health sectors. Standardization of this terminology is required for understanding the domain-related technical terms and their dependency. This terminology helps the researchers for the control of mosquito vectors, surveillance, and management, the alternate strategy for controlling vector-borne diseases (VBDs). There is an urge on ontologies for VBDs to describe respective diseases in terms of clinical, therapeutics, and pathogenic transmissions. The vector's physiological processes, epidemiology, and prevention control measures are the other major essential attributes of VBDs.

Ontology is the standard method for representing unambiguous domain-related terms and their relationships in an easily understandable form. They are the powerful decision support system that helps to take decisions in various scenarios and provide semantic knowledge. It also helps in integrating and managing data to enhance complex search queries and facilitate the reuse of existing ontologies. The ontologies follow the rules of Open Biological and Biomedical Ontologies (OBO) Foundry or follow the basic formal ontology (BFO) format. The common ontology relationships are "is_a" relationship and "has" relation. The "is_a" relation is used to define the hierarchical structure for organizing the contents. For example, dengue is_a disease, mosquito is_a insect, leaf is_a part, etc. Ontologies can also include additional relationships to provide knowledge using the specified relation. Examples of such relations are *Aedes* has larvae stage, biocontrol agent has mosquito control, arbovirus has transmitted pathogens, etc. This reveals the necessity of the development of ontologies on mosquito vectors.

8.4.1 Existing ontologies for vector control

The vector surveillance and management ontology (VSMO) is the ontology for arthropod species focused on pathogenic microorganism transmitters of infectious disease to vectors such as mosquitoes, fleas, bugs, and flies. VSMO provides details of the vector, its continent, the pathogen it transmits, and the disease it spreads [90]. The Infectious Disease Ontology (IDO) is the ontology for the world's global health infectious disease, malaria. IDO consists of limited information on the malaria control process such as malaria treatment, malaria vectors, and their control. It has been updated as IDOMAL with malaria disease additional features such as clinical manifestations, disease biology including immunology, therapeutic approaches, and epidemiology. IDOMAL consists of 3158 classes and 16 properties for malaria control and is freely available on BioPortal website, a repository of biomedical ontologies [91]. The functional ontology of male and female *Ae. aegypti* mosquitoes provides information based on the biological process, molecular function, and cellular components that may help to block the viral transmissions. This ontology also identifies the abundance of mosquitoes related to the proteins in male and female mosquitoes [92].

Mosquito Isecticide Resistance Ontology (MIRO) consists of information on insecticide resistance-related databases focused on fieldwork and monitoring. The MIRO ontology succeeded by developing the core insecticide resistance database (IRbase) and was adopted immediately by the Regional Office for Africa of the World Health Organization (WHO-AFRO). The IRbase has become the repository for storing field studies conducted or supported by WHO-AFRO. MIRO consists of 3790 terms related to mosquito population, field catch studies, chemical resistance, and species identification method [93, 99]. Biomedical ontologies have been developed to represent the pandemics of infectious diseases such as COVID-19, malaria, dengue, and so on. IDODEN is an ontology for dengue fever, the world infectious disease that is transmitted through the dengue virus to humans and animals by the bite of *Aedes* genus mosquitoes. It is freely available in BioPortal and consists of 5019 classes and 18 properties to describe the biology of dengue disease, the entomological and epidemiological data on dengue fever, fever control, diagnosis, and therapeutic procedure, treatment, and the chemicals and repellents used for mosquito control. In addition to this, spatio-temporal information, such as the development stage, is also provided [94].

The gene ontology (GO) is the widely used biological ontology that is used to represent the genes and gene products of various organisms concerned with cellular localization, molecular function, and biological processes in genome databases. GO terms are an important representation in many experiments such as microarray experiments, protein sequence similarity analysis, and various bioinformatics experimental analyses. GO is the base for developing many other biological ontologies to describe biological notions, scientific disciplines, and inter-related biomedical terms. The limitation of GO is that the pathogen–host interaction details are not provided that may be required for the exhaustive knowledge of vector biology. Hence, the VectorBase database funded by the National Institute of Allergy and Infectious Diseases (NIAID) started to construct the ontologies for the vectors of mosquitoes and ticks to include physiology using the existing anatomy ontologies [95].

The ontology for Infectious Disease Diagnosis and Antibiotic Prescription (IDDAP) is an antibiotic decision supporting system developed for suggesting valid prescriptions for patients who need an immediate consultation and could not contact doctors directly. IDDAP consists of 1,267,004 classes, 7,608,725 axioms, and 1,266,993 subclasses, constructed from the existing ontologies of infectious disease diagnosis, antibiotic therapy, syndromes, microbes, and drugs. Diseases such as candidiasis, balantidiasis, tonsillitis, and zika fever have been modeled based on the factors such as germ family, carrier, reservoir, the microorganism's entry and exit, and disease using IDDAP ontology. For example, zika fever is modeled by the factors such as its pathogen (zika virus), vector (*Aedes* mosquito), reservoir (unknown), mode of transmission (mosquito bite), and the infectious agent exit portal (urogenital tract) [96]. Rajapakse *et al.* [97] developed a dengue ontology to provide unstructured information that is scattered in literature and web resources as a generic infrastructure with a reasoning facility by a visual query tool. This ontology helps to share the domain knowledge on dengue viral disease that infects millions of tropical and subtropical region people every year. The ontology is constructed by using 12 blogs and 14 news feeds to provide knowledge on dengue serotype trends, clinical information, diagnosis, remedy, phenotype description, disease manifestation, proteins, symptoms, mosquito vector breeding site, virus description, and so on [97].

Ontology for Infectious Disease Risk (IDR) prediction system is used to examine the risk of infectious diseases in humans based on environmental factors such as geographical region, location features, and climatic conditions like temperature, humidity, etc. IDR represents epidemiological knowledge on infectious diseases and help entomologists to enrich their job. The system is evaluated for infectious diseases such as dengue fever and tuberculosis in Indonesia and cholera disease in India [98]. Table 8.4 shows the list of existing ontologies for vector control and maintenance. Ontology combined with the Information Technology (IT) tools such as Geographic Information System (GIS), databases, and Android applications have a vital role in the control of mosquito vectors that transmit vector-borne diseases [99]. Mobile phones have been used as sensors for monitoring the mosquito populations in the field by recording the time and location of mosquito wingbeat sounds with the help of mobile microphones [100]. The survey concluded that even

Table 8.4 List of existing ontologies for vector control and maintenance.

Sl. no	Vector ontology	Description
1	IDO	Infectious disease ontology
2	IDOMAL	Infectious disease ontology for malaria disease
3	Functional ontology	Ontology for *Ae. aegypti* male and female mosquitoes based on molecular and biological studies
4	MIRO	Ontology for mosquito insecticide resistance
5	TGMA	Ontology for mosquito gross anatomy
6	CARO	Common anatomy reference ontology
7	TADS	Tick anatomy ontology
8	IDODEN	Ontology for dengue fever
9	GO	Gene ontology consists of annotated genes and gene products of organisms
10	IDDAP	Ontology for infectious disease diagnosis and antibiotic prescription
11	Dengue Ontology	Ontology centric combination and exploration of dengue literature
12	IDR	Ontology for infectious disease risk

though various ontologies are available for mosquito vector control, no specific ontology is preferred for *Aedes* mosquito vectors concerning biological control agents.

8.4.2 Need of ontology for aedes mosquito

The difficulties in controlling the *Aedes* mosquito vectors motivate the researchers to work on biological control agents to identify the harmless biological component that can act as a bio-insecticide for *Aedes* mosquito control. This mandates the complete knowledge of existing biocontrol agents such as microbes and medicinal plants, and the relations among them. The knowledge is scattered in the research articles and there is a necessity for composing all the information in a structured way to identify the semantic relations. Ontology is the best-suited technique for representing the information on *Aedes* mosquito vectors in terms of biological control agents. The remaining part of the chapter focuses on the development of an ontology for *Aedes* mosquito vectors concerned with its medicinal plants and microorganisms to express the various mosquito development stages, plant name and its part used for the control, the organ it disturbs, and microbes. The Resource Description Framework Schema (RDFS) has been used for *Aedes* mosquito ontology development.

8.5 Aedes Mosquito Vector Ontology

The development of an ontology includes various phases such as defining the classes, arranging the classes in a taxonomic (subclass–superclass) hierarchy, defining properties, and assigning the property values for instance. The classes are the collection of objects and play a major role in ontology development. It describes the concepts in the domain. For example, a class of mosquitoes represents all the different genera of mosquitoes. Specific genera are instances of this class. A class can have subclasses to represent the concepts that are more specific than the superclasses. For example, the class of mosquitoes can be divided into *Aedes*, *Anopheles*, *Culex*, and *Mansonia*. Alternatively, the class of *Aedes* genera can be subdivided into a class of male and female. Finally, the knowledge base can be generated by merging the individual instances of the classes by assigning the appropriate property values.

The knowledge base in the form of ontology is represented through machine-readable languages such as Resource Description Framework Schema (RDFS), Web Ontology Language (OWL), etc. Ontology construction is difficult and time-consuming as there is no standard method for developing an ontology. Tools such as Text2Onto, OntoGain, ASIUM system, the Mo'K Workbench, OntoLearn, or OntoLT are available to support ontology construction from a given set of textual data. Text2Onto is the most preferred tool for two important features such as Probabilistic Ontology Models (POMs) and Data Driven Change Discovery (DDCC). POM represents the results by attaching a probability to them and the data-driven change discovery is responsible for detecting changes in the corpus [101].

This chapter aims in developing an ontology to represent the knowledge that is hidden in articles by extracting terms, identifying named entities or concepts, and their relationships from existing texts using minimal domain knowledge. This will help in the development of a domain-specific ontology for knowledge representation and its subsequent utilization.

8.5.1 Aedes ontology development

The development of *Aedes* ontology consists of various phases such as data collection, data preprocessing, semantic relation extraction, *Aedes* ontology construction, and visualization as shown in Figure 8.3.

Data collection:

The ontology development starts with collecting the relevant articles from the knowledge bases such as Science Direct, PubMed, and IEEE databases.

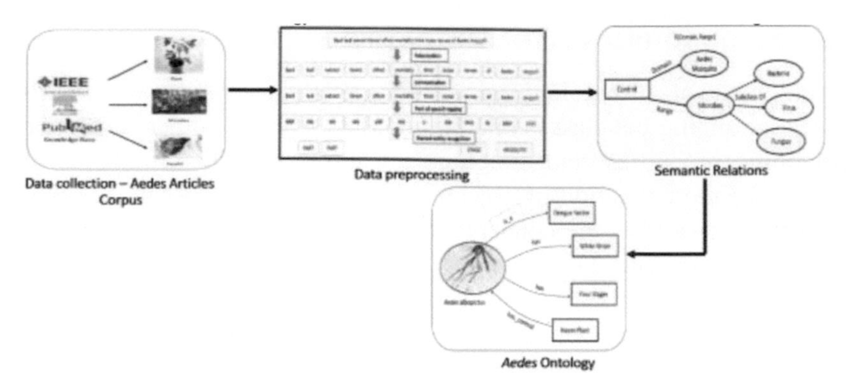

Figure 8.3 *Aedes* ontology development phases.

The domain-related search keywords are used to download the related articles to frame the corpus, a comprehensive collection of data. The corpus is used for implementing tokenization and other data preprocessing steps.

Data preprocessing:

Data preprocessing begins with tokenization, the task of dividing an input text sequence into smaller parts called tokens that include words, terms (keywords and phrases), sentences, and symbols. It also removes unwanted characters such as punctuations. In some contexts, tokens are used in a group to form a semantic unit. The Part Of Speech (POS) tagging, stemming, and lemmatization are the other preprocessing steps required for facilitating relation extractions. POS tagging is used to map the linguistic words to their lexical term such as noun, verb, adverb, determiner, etc. Stemming extracts the base words by removing both the suffixes and prefixes attached to the word. For example, the stem word for identified is "identifi," identifies is "identifi," and identifying is "identify." But the lemmatization extracts the root words by morphological analysis. The lemmatized words are called a lemma. For example, the lemma of the words identifying, identifies, and identified is identify.

Semantic relation extraction:

Relations are defined as the associations among the words that are used to express the semantic associations among them. Before relation extraction, the primitives such as classes or concepts, instances, subclass_of, and instance_of need to be categorized. The Named Entity Recognition (NER) uses various algorithms to extract and categorize the primitives. For example, in the sentence "chikungunya is the widespread vector-borne disease," chikungunya

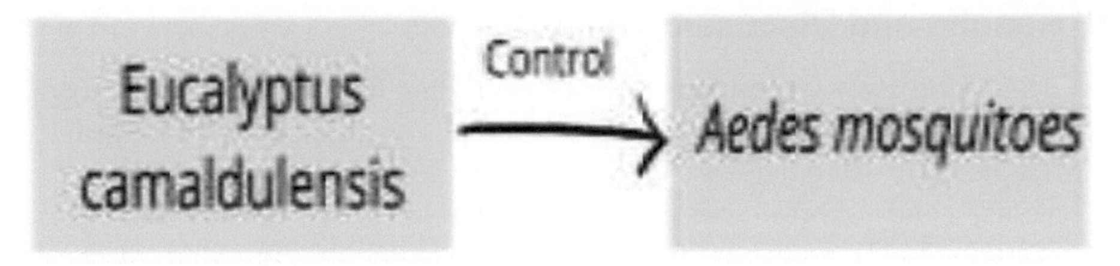

Figure 8.4 Semantic relation representation.

and vector-borne diseases are classes/concepts and chikungunya is recognized as a subclass of vector-borne disease. Relation extraction algorithms are used for extracting the relations and expressing them in the form of R (domain, range). The extracted relations occur in triplet form, subject–predicate–object, as shown in Figure 8.4, which represents the relation (control) between the medicinal plant and *Aedes* mosquitoes.

The accuracy of the predicted relationship is evaluated by a statistical measure such as accuracy and by the domain experts. Accuracy is calculated by dividing the number of predicted domain-specific semantic relations by the total number of predicted relations. The domain experts identify the domain-specific relations by manual analysis.

Ontology construction and visualization:

The extracted concepts and relations are expressed in Resource Description Framework Schema (RDFS) format, the language for ontology development. RDFS is an extension to the Resource Description Framework (RDF) that identifies the things using web identifiers (URIs) and describes the resources with properties and property values. The combination of a resource, a property, and a property value forms a triplet known as the subject, predicate, and object of a statement. RDF schema provides the framework to describe application-specific classes and properties. Classes in RDF schema are much like classes in object-oriented programming languages. This allows the resources to be defined as instances of classes and subclasses of classes. The generated RDFS file is expressed as a graphical representation for visualization and easy understanding of the relations in a specific domain. Protégé, Ontoplot, Jambalaya, Knoocks, and OntoViewer are the common tools for ontology visualization [102]. Cytoscape, a free software for visualizing network interactions, can also be used for ontology visualization [103].

8.6 Results and Discussion

The articles on *Aedes* mosquito vectors regarding the biocontrol agents such as plants, microbes, and aquatic organisms were downloaded from PubMed database using the search keywords "*Aedes* and biocontrol," "*Aedes* and plants," and "*Aedes* and microbes." The abstracts of the last 10 years (2012–2022) articles were downloaded using the PubMed filters "Text Availability" and "Publication Date." Around 250 abstracts were downloaded in a text file and the corpus was generated. The Text2Onto, a free online tool, was used to extract the primitives such as concepts, instances, subclass_of, instance_of, and relations from the corpus. Text2Onto used a Java-based tool, General Architecture for Text Engineering (GATE), for tokenization, sentence splitting, and part of speech tagging. Table 8.5 shows the various algorithms used for executing the primitives and the total number of primitives extracted. Figure 8.5 shows the probabilistic ontology model obtained from Text2Onto.

Among the predicted relations, 56 relations were identified as domain-specific semantic relations by the manual analysis and listed in Table 8.6. The proposed system achieved 46.7% of domain-specific semantic relations. Text2Onto used WordNet lexical database for mapping entities. The relations such as exist_in, reduce_in, occur_as, affect, prevent, restrict, etc., were identified. Table 8.6 shows the extracted relations. The sub-classes of concepts such as dengue, vector, mosquito, arthropod, plant, biocontrol agent, etc., were identified. Figure 8.6 shows the ontology generated from Text2Onto as RDFS [S1]. The Cytoscape tool was used for visualizing the concepts, subclasses, and relations, which is shown in Figure 8.7.

Table 8.5 Number of primitives identified from Text2Onto.

Sl. no	Primitives	algorithm used	Number of primitives
1	Concept extraction	Term frequency-inverse document frequency	5854
2	Instance extraction	Term frequency-inverse document frequency	6371
3	Subclass of	Pattern concept classification	108
4	Instance of	Pattern instance classification	44
5	Relation extraction	Sub-categorization framework (SubCat)	120

8.7 Conclusion

This chapter explored the need of an ontology for *Aedes* mosquito vectors to recognize the semantic relations between *Aedes* and biocontrol agents. The *Aedes* mosquito vector biology, their resistance behavior with synthetic insecticides, *Aedes* mosquito variants, vector control methods, and their issues were discussed. The biocontrol agents such as medicinal plants, microbes, and aquatics used as adulticides, larvicides, and ovicides were discussed to provide the importance of harmless biocontrol agents. The role of ontologies in vector control strategy and the existing ontologies for vector control and management were discussed. This chapter endorsed the formal method for the development of *Aedes* mosquito vectors ontology to explore various biocontrol agents with mosquito development stages. The ontology was constructed using Text2Onto. The ontology was visualized using Cytoscape to empathize the relations of *Aedes* mosquito with biocontrol agents. The *Aedes* ontology is used for generating association rules to represent the complex semantic associations among the *Aedes* mosquitoes and biocontrol agents. The user dictionaries and logical axioms can be used for further enhancement of domain-specific semantic relation predictions. The combination of user dictionaries and deep learning neural network algorithms such as recurrent neural network, long short-term memory, and bi-directional long short-term memory may help to provide more semantic associations for any applications.

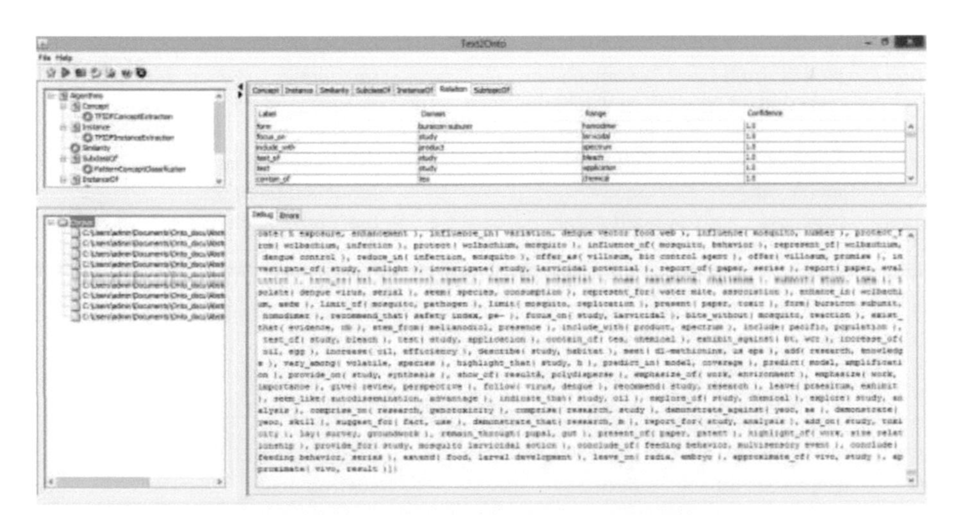

Figure 8.5 Generated probabilistic ontology model.

Table 8.6 Domain-specific relations extracted from Text2Onto.

Sl. no.	Relation	Entity	Entity
1	suggest	evidence	wmelpop
2	cause	virus	development
3	induce	wolbachium	fitness
4	prefer	aegyptus	oviposition site
5	need	novel alternative	investigation
6	restrict	wolbachia pipienti	rna
7	produce	wolbachium	rna
8	prevent	wolbachium	*aedes*
9	lengthen	wmel	EIP
10	confirm	study	wolbachium
11	absorb	anisoplia	hemolymph sugar
12	reduce	infection	ability
13	possess	wolbachium	characteristic
14	affect	infection	population dynamics
15	exhibit	capsid	feature
16	rely_on	strategy	factor
17	suggest_that	bti-bioinsecticide	wolbachium
18	exist_in	virus	transmission cycle
19	occur_as	infection	result
20	affect_of	infection	host
21	reduce_of	aegyptus mosquito	survival
22	influence_in	variation	dengue vector food web
23	influence_of	mosquito	behavior
24	predict_that	theory	wolbachium
25	enhance	wolbachium	infection
26	protect_from	wolbachium	infection
27	represent_of	wolbachium	dengue control
28	reduce_in	infection	mosquito
29	offer_as	villosum	bio control agent
30	investigate	study	larvicidal potential
31	have_as	kal	biocontrol agent
32	limit_of	mosquito	pathogen
33	limit	mosquito	replication
34	pose	resistance	challenge
35	form	bursicon subunit	homodimer
36	bite_without	mosquito	reaction
37	increase	oil	efficiency
38	exhibit_against	bt	wcr
39	show_of	result	polydisperse
40	emphasize_of	work	environment

Table 8.6 (Continued.)

41	increase_of	oil	egg
42	follow	virus	dengue
43	remain_through	pupal	gut
44	increase	oil	efficiency
45	describe	study	habitat
46	focus_on	study	larvicidal
47	bite_without	mosquito	reaction
48	stem_from	melianodiol	presence
49	include_with	product	spectrum
50	include	pacific	population
51	indicate_that	study	oil
52	provide_for	study	mosquito larvicidal action
53	extend	food	larval development
54	leave_on	radia	embryo
55	conclude_of	feeding behavior	multisensory event
56	conclude	feeding behavior	series

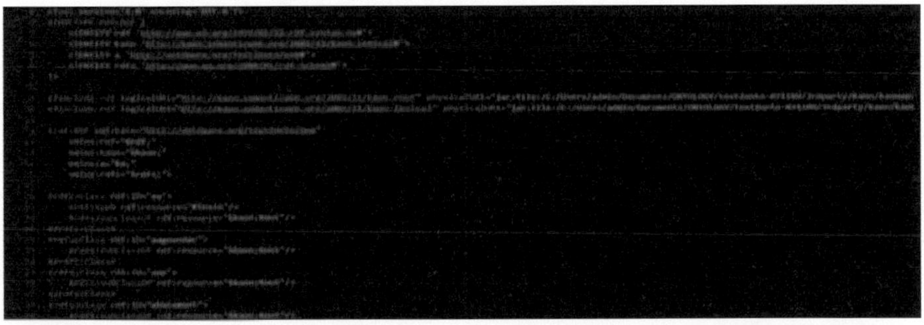

Figure 8.6 Ontology in RDFS.

Supporting File

[S1] – The RDFS file generated from Text2Onto tool.

Acknowledgement

The authors have not received any financial support to carry out this work.

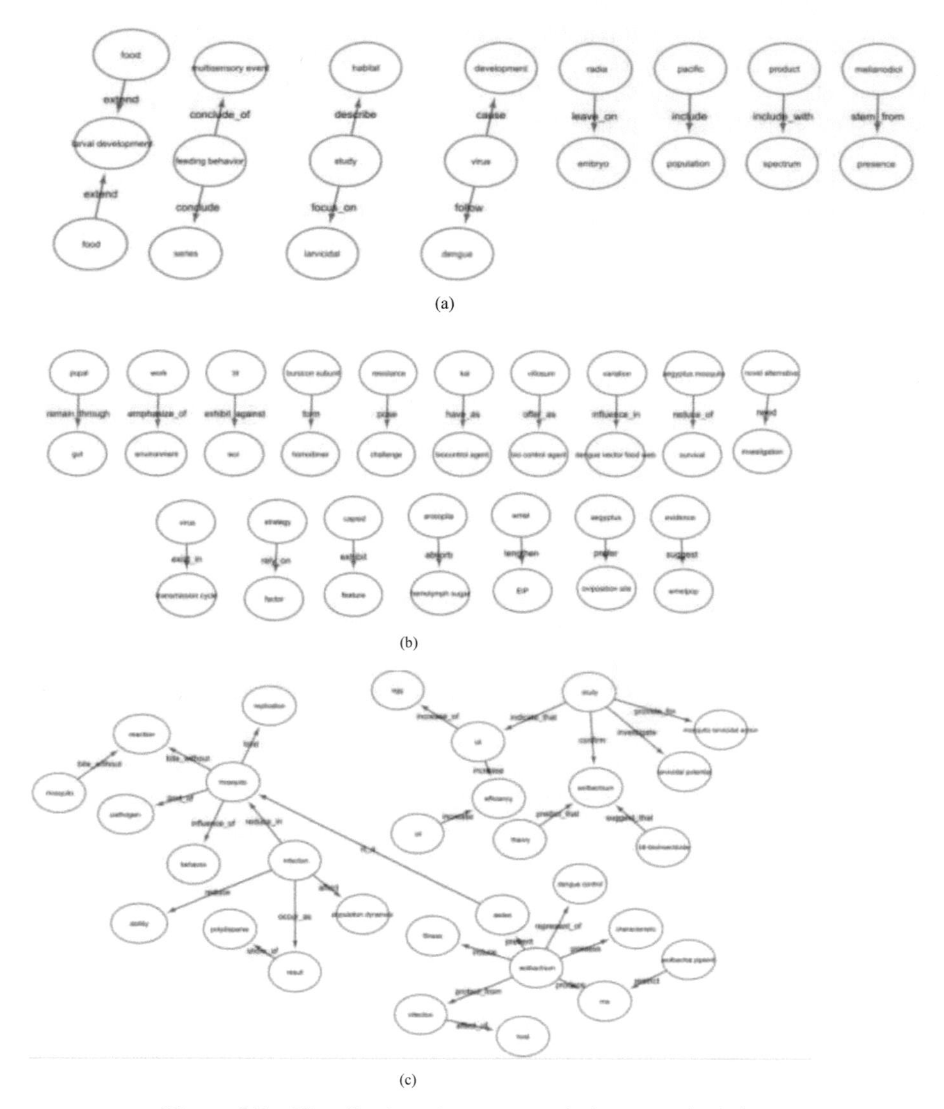

Figure 8.7 Visualization of concepts, subclasses, and relations.

References

[1] Huntington, M. K., Allison, J., & Nair, D. (2016). Emerging Vector-Borne Diseases. American family physician, 94(7), 551–557.

[2] Kuthyar, S., Anthony, C. L., Fashina, T., Yeh, S., & Shantha, J. G. (2021). World Health Organization High Priority Pathogens: Ophthalmic Disease Findings and Vision Health Perspectives. Pathogens (Basel, Switzerland), 10(4), 442. https://doi.org/10.3390/pathogens10040442

[3] Islam, M. T., Quispe, C., Herrera-Bravo, J., Sarkar, C., Sharma, R., Garg, N., Fredes, L. I., Martorell, M., Alshehri, M. M., Sharifi-Rad, J., Daştan, S. D., Calina, D., Alsafi, R., Alghamdi, S., Batiha, G. E., & Cruz-Martins, N. (2021). Production, Transmission, Pathogenesis, and Control of Dengue Virus: A Literature-Based Undivided Perspective. BioMed research international, 2021, 4224816. https://doi.org/10.1155/2021/4224816

[4] Zeng, Z., Zhan, J., Chen, L., Chen, H., & Cheng, S. (2021). Global, regional, and national dengue burden from 1990 to 2017: A systematic analysis based on the global burden of disease study 2017. EClinicalMedicine, 32, 100712.

[5] Sharp, T. M., Morris, S., Morrison, A., de Lima Corvino, D., Santiago, G. A., Shieh, W. J., ... & Stanek, D. (2021). Fatal Dengue Acquired in Florida. New England Journal of Medicine, 384(23), 2257-2259.

[6] Bennett, S. N., Holmes, E. C., Chirivella, M., Rodriguez, D. M., Beltran, M., Vorndam, V., Gubler, D. J., & McMillan, W. O. (2003). Selection-driven evolution of emergent dengue virus. Molecular biology and evolution, 20(10), 1650–1658. https://doi.org/10.1093/molbev/msg182

[7] Guzmán, M. G., Kourí, G., Valdés, L., Bravo, J., Vázquez, S., & Halstead, S. B. (2002). Enhanced severity of secondary dengue-2 infections: death rates in 1981 and 1997 Cuban outbreaks. Revista Panamericana de Salud Pública, 11, 223-227.

[8] Kraemer, M. U., Sinka, M. E., Duda, K. A., Mylne, A. Q., Shearer, F. M., Barker, C. M., Moore, C. G., Carvalho, R. G., Coelho, G. E., Van Bortel, W., Hendrickx, G., Schaffner, F., Elyazar, I. R., Teng, H. J., Brady, O. J., Messina, J. P., Pigott, D. M., Scott, T. W., Smith, D. L., Wint, G. R., ... Hay, S. I. (2015). The global distribution of the arbovirus vectors Aedes aegypti and Ae. albopictus. eLife, 4, e08347.https://doi.org/10.7554/eLife.08347

[9] Xavier, L. L., Honório, N. A., Pessanha, J., & Peiter, P. C. (2021). Analysis of climate factors and dengue incidence in the metropolitan region of Rio de Janeiro, Brazil. PloS one, 16(5), e0251403. https://doi.org/10.1371/journal.pone.0251403

[10] Baglan, H., Lazzari, C., & Guerrieri, F. (2017). Learning in mosquito larvae (Aedes aegypti): Habituation to a visual danger signal. Journal of insect physiology, 98, 160–166. https://doi.org/10.1016/j.jinsphys.2017.01.001

[11] Abreu, F. V., Morais, M. M., Ribeiro, S. P., & Eiras, Á. E. (2015). Influence of breeding site availability on the oviposition behavior of Aedes aegypti. Memorias do Instituto Oswaldo Cruz, 110(5), 669–676. https://doi.org/10.1590/0074-02760140490

[12] Al-Amin, H. M., Johora, F. T., Irish, S. R., Hossainey, M., Vizcaino, L., Paul, K. K., Khan, W. A., Haque, R., Alam, M. S., & Lenhart, A. (2020). Insecticide resistance status of Aedes aegypti in Bangladesh. Parasites & vectors, 13(1), 622. https://doi.org/10.1186/s13071-020-04503-6

[13] Paeporn, P., Supaphathom, K., Srisawat, R., Komalamisra, N., Deesin, V., Ya-umphan, P., & Leeming Sawat, S. (2004). Biochemical detection of pyrethroid resistance mechanism in Aedes aegypti in Ratchaburi province, Thailand. Tropical biomedicine, 21(2), 145–151.

[14] Bouzid, M., Brainard, J., Hooper, L., & Hunter, P. R. (2016). Public Health Interventions for Aedes Control in the Time of Zikavirus-A Meta-Review on Effectiveness of Vector Control Strategies. PLoS neglected tropical diseases, 10(12), e0005176.https://doi.org/10.1371/journal.pntd.0005176.

[15] T. R. Gruber, "A translation approach to portable ontology specifications" in *Knowledge acquisition*, 1993, 5(2), pp. 199-220.

[16] Borovsky, D., & Meola, S. M. (2004). Biochemical and cytoimmunological evidence for the control of Aedes aegypti larval trypsin with Aea-TMOF. Archives of insect biochemistry and physiology, 55(3), 124–139. https://doi.org/10.1002/arch.10132

[17] Silvério, M., Espindola, L. S., Lopes, N. P., & Vieira, P. C. (2020). Plant Natural Products for the Control of Aedes aegypti: The Main Vector of Important Arboviruses. Molecules (Basel, Switzerland), 25(15), 3484. https://doi.org/10.3390/molecules25153484

[18] Falkowski, M., Jahn-Oyac, A., Odonne, G., Flora, C., Estevez, Y., Touré, S., Boulogne, I., Robinson, J. C., Béreau, D., Petit, P., Azam, D., Coke, M., Issaly, J., Gaborit, P., Stien, D., Eparvier, V., Dusfour, I., & Houël, E. (2020). Towards the optimization of botanical insecticides research: Aedes aegypti larvicidal natural products in French Guiana. Acta tropica, 201, 105179. https://doi.org/10.1016/j.actatropica.2019.105179

[19] Topalis, P., Dialynas, E., Mitraka, E., Deligianni, E., Siden-Kiamos, I., & Louis, C. (2011). A set of ontologies to drive tools for the control of vector-borne diseases. Journal of biomedical informatics, 44(1), 42–47. https://doi.org/10.1016/j.jbi.2010.03.012

[20] World Health Organization. (2014). Fact sheets on vector borne diseases in India. Available at https://www.who.int/docs/default-source/searo/india/health-topic-pdf/vbd-fact-sheets.pdf?sfvrsn=c1908b04_2

[21] G. Rezza, "Dengue and chikungunya: long-distance spread and outbreaks in naïve areas", in *Pathogens and global health*, 2014 Dec 1;108(8), pp. 349-55.

[22] World Health Organization.(2022). Dengue and severe dengue. Available at https://www.who.int/news-room/fact-sheets/detail/dengue-and-severe-dengue

[23] M. Bharati, and D. Saha, D, "Assessment of insecticide resistance in primary dengue vector, Aedes aegypti (Linn.) from Northern Districts of West Bengal, India" in *Acta tropica*, 2018, 187, pp. 78–86.

[24] Lima, E. P., Paiva, M. H. S., de Araújo, A. P., da Silva, É. V. G., da Silva, U. M., de Oliveira, L. N., ... & de Melo Santos, M. A. V. (2011). Insecticide resistance in Aedes aegypti populations from Ceará, Brazil. Parasites & vectors, 4(1), 1-12.

[25] Auteri, M., La Russa, F., Blanda, V., & Torina, A. (2018). Insecticide Resistance Associated with kdr Mutations in Aedes albopictus: An Update on Worldwide Evidences. BioMed research international, 2018, 3098575. https://doi.org/10.1155/2018/3098575

[26] Pu, J., Wang, Z., & Chung, H. (2020). Climate change and the genetics of insecticide resistance. Pest management science, 76(3), 846–852. https://doi.org/10.1002/ps.5700

[27] Sparks, T. C., Storer, N., Porter, A., Slater, R., & Nauen, R. (2021). Insecticide resistance management and industry: the origins and evolution of the Insecticide Resistance Action Committee (IRAC) and the mode of action classification scheme. Pest management science, 77(6), 2609–2619. https://doi.org/10.1002/ps.6254

[28] Kokoza, V., Ahmed, A., Cho, W. L., Jasinskiene, N., James, A. A., & Raikhel, A. (2000). Engineering blood meal-activated systemic immunity in the yellow fever mosquito, Aedes aegypti. Proceedings of the National Academy of Sciences of the United States of America, 97(16), 9144–9149. https://doi.org/10.1073/pnas.160258197

[29] Labbé, G. M., Nimmo, D. D., & Alphey, L. (2010). piggybac- and PhiC31-mediated genetic transformation of the Asian tiger mosquito,

Aedes albopictus (Skuse). PLoS neglected tropical diseases, 4(8), e788. https://doi.org/10.1371/journal.pntd.0000788

[30] Mishra, P., Furey, C., Balaraman, V., & Fraser, M. J. (2016). Antiviral Hammerhead Ribozymes Are Effective for Developing Transgenic Suppression of Chikungunya Virus in Aedes aegypti Mosquitoes. Viruses, 8(6), 163. https://doi.org/10.3390/v8060163

[31] Coutinho-Abreu, I. V., Zhu, K. Y., & Ramalho-Ortigao, M. (2010). Transgenesis and paratransgenesis to control insect-borne diseases: current status and future challenges. Parasitology international, 59(1), 1–8. https://doi.org/10.1016/j.parint.2009.10.002

[32] Mohanty, I., Rath, A., Swain, S. P., Pradhan, N., & Hazra, R. K. (2019). Wolbachia Population in Vectors and Non-vectors: A Sustainable Approach Towards Dengue Control. Current microbiology, 76(2), 133–143. https://doi.org/10.1007/s00284-018-1596-8

[33] Carvalho, D. O., Nimmo, D., Naish, N., McKemey, A. R., Gray, P., Wilke, A. B. B., Marrelli, M. T., Virginio, J. F., Alphey, L., Capurro, M. L. Mass Production of Genetically Modified *Aedes aegypti* for Field Releases in Brazil. *J. Vis. Exp.* (83), e3579, doi:10.3791/3579 (2014).

[34] Sarwar, M.H. (2015). Reducing Dengue Fever Through Biological Control of Disease Carrier Aedes Mosquitoes (Diptera : Culicidae).

[35] Pruett, G., Hawes, J., Varnado, W., Deerman, H., Goddard, J., Burkett-Cadena, N., & Kearney, C. (2020). The readily transformable Impatiens walleriana efficiently attracts nectar feeding with Aedes and Culex mosquitoes in simulated outdoor garden settings in Mississippi and Florida. Acta Tropica, 210, 105624.

[36] Ravi, R., Zulkrnin, N., Rozhan, N. N., Nik Yusoff, N. R., Mat Rasat, M. S., Ahmad, M. I., Ishak, I. H., & Amin, M. (2018). Chemical composition and larvicidal activities of Azolla pinnata extracts against Aedes (Diptera:Culicidae). PloS one, 13(11), e0206982. https://doi.org/10.1371/journal.pone.0206982

[37] Zuharah, W. F., Ahbirami, R., Dieng, H., Thiagaletchumi, M., & Fadzly, N. (2016). EVALUATION OF SUBLETHAL EFFECTS OF Ipomoea cairica LINN. EXTRACT ON LIFE HISTORY TRAITS OF DENGUE VECTORS. Revista do Instituto de Medicina Tropical de Sao Paulo, 58, 44. https://doi.org/10.1590/S1678-9946201658044

[38] Ramkumar, G., Karthi, S., Muthusamy, R., Suganya, P., Natarajan, D., Kweka, E. J., & Shivakumar, M. S. (2016). Mosquitocidal Effect of Glycosmis pentaphylla Leaf Extracts against Three Mosquito Species

(Diptera: Culicidae). PloS one, 11(7), e0158088. https://doi.org/10.1371/journal.pone.0158088

[39] Bharathithasan, M., Ravindran, D. R., Rajendran, D., Chun, S. K., Abbas, S. A., Sugathan, S., Yahaya, Z. S., Said, A. R., Oh, W. D., Kotra, V., Mathews, A., Mohd Amin, M. F., Ishak, I. H., & Ravi, R. (2021). Analysis of chemical compositions and larvicidal activity of nut extracts from Areca catechu Linn against Aedes (Diptera: Culicidae). PloS one, 16(11), e0260281. https://doi.org/10.1371/journal.pone.0260281

[40] AlShebly, M. M., AlQahtani, F. S., Govindarajan, M., Gopinath, K., Vijayan, P., & Benelli, G. (2017). Toxicity of ar-curcumene and epi-β-bisabolol from Hedychium larsenii (Zingiberaceae) essential oil on malaria, chikungunya and St. Louis encephalitis mosquito vectors. Ecotoxicology and environmental safety, 137, 149-157

[41] Suroowan, S., Mahomoodally, F., & Ragoo, L. (2016). Management and treatment of Dengue and Chikungunya-natural products to the rescue. Combinatorial Chemistry & High Throughput Screening, 19(7), 554-564.

[42] Rodrigues, A. B. L., Lopes, R. M., Rabelo, É. M., Tomazi, R., Santos, L. L., Brandão, L. B., ... & Galardo, A. K. R. (2020). Development of nano-emulsions based on Ayapana triplinervis for the control of Aedes aegypti larvae. bioRxiv.

[43] Sierra, I., Latorre-Estivalis, J. M., Traverso, L., Gonzalez, P. V., Aptekmann, A., Nadra, A. D., ... & Ons, S. (2021). Transcriptomic analysis and molecular docking reveal genes involved in the response of Aedes aegypti larvae to an essential oil extracted from Eucalyptus. PLoS Neglected Tropical Diseases, 15(7), e0009587.

[44] Hoi, T. M., Huong, L. T., Chinh, H. V., Hau, D. V., Satyal, P., Tai, T. A., ... & Setzer, W. N. (2020). Essential oil compositions of three invasive Conyza species collected in Vietnam and their larvicidal activities against Aedes aegypti, Aedes albopictus, and Culex quinquefasciatus. Molecules, 25(19), 4576..

[45] Botas, G. D. S., Cruz, R. A., De Almeida, F. B., Duarte, J. L., Araújo, R. S., Souto, R. N. P., ... & Fernandes, C. P. (2017). Baccharis reticularia DC. and limonene nanoemulsions: promising larvicidal agents for Aedes aegypti (Diptera: Culicidae) control. Molecules, 22(11), 1990.

[46] Liu, X. C., Liu, Q., Chen, X. B., Zhou, L., & Liu, Z. L. (2015). Larvicidal activity of the essential oil from Tetradium glabrifolium

fruits and its constituents against Aedes albopictus. Pest management science, 71(11), 1582-1586.

[47] Blythe, E. K., Tabanca, N., Demirci, B., Tsikolia, M., Bloomquist, J. R., & Bernier, U. R. (2016). Lantana montevidensis essential oil: chemical composition and mosquito repellent activity against Aedes aegypti. Natural Product Communications, 11(11), 1934578X1601101122.

[48] Balasubramani, S., Rajendhiran, T., Moola, A. K., & Diana, R. K. B. (2017). Development of nanoemulsion from Vitex negundo L. essential oil and their efficacy of antioxidant, antimicrobial and larvicidal activities (Aedes aegypti L.). Environmental Science and Pollution Research, 24(17), 15125-15133.

[49] Senthilkumar, A., Jayaraman, M., & Venkatesalu, V. (2013). Chemical constituents and larvicidal potential of Feronia limonia leaf essential oil against Anopheles stephensi, Aedes aegypti and Culex quinquefasciatus. Parasitology research, 112(3), 1337-1342.

[50] Giatropoulos, A., Pitarokili, D., Papaioannou, F., Papachristos, D. P., Koliopoulos, G., Emmanouel, N., ... & Michaelakis, A. (2013). Essential oil composition, adult repellency and larvicidal activity of eight Cupressaceae species from Greece against Aedes albopictus (Diptera: Culicidae). Parasitology research, 112(3), 1113-1123

[51] Sumitha, K. V., & Thoppil, J. E. (2016). Larvicidal efficacy and chemical constituents of O. gratissimum L.(Lamiaceae) essential oil against Aedes albopictus Skuse (Diptera: Culicidae). Parasitology research, 115(2), 673-680.

[52] Amado, J. R. R., Prada, A. L., Diaz, J. G., Souto, R. N. P., Arranz, J. C. E., & de Souza, T. P. (2020). Development, larvicide activity, and toxicity in nontarget species of the Croton linearis Jacq essential oil nanoemulsion. Environmental Science and Pollution Research, 27(9), 9410-9423.

[53] Govindarajan, M., Rajeswary, M., & Benelli, G. (2016). Chemical composition, toxicity and non-target effects of Pinus kesiya essential oil: an eco-friendly and novel larvicide against malaria, dengue and lymphatic filariasis mosquito vectors. Ecotoxicology and Environmental Safety, 129, 85-90.

[54] Govindarajan, M., & Benelli, G. (2016). Eco-friendly larvicides from Indian plants: effectiveness of lavandulyl acetate and bicyclogermacrene on malaria, dengue and Japanese encephalitis mosquito vectors. Ecotoxicology and environmental safety, 133, 395-402.

[55] Chandramohan, B., Murugan, K., Panneerselvam, C., Madhiyazhagan, P., Chandirasekar, R., Dinesh, D., ... & Benelli, G. (2016). Characterization and mosquitocidal potential of neem cake-synthesized silver nanoparticles: genotoxicity and impact on predation efficiency of mosquito natural enemies. Parasitology research, 115(3), 1015-1025.

[56] Gomes, S. A., Paula, A. R., Ribeiro, A., Moraes, C. O., Santos, J. W., Silva, C. P., & Samuels, R. I. (2015). Neem oil increases the efficiency of the entomopathogenic fungus Metarhizium anisopliae for the control of Aedes aegypti (Diptera: Culicidae) larvae. Parasites & vectors, 8(1), 1-8.

[57] Liu, X. C., Dong, H. W., Zhou, L., Du, S. S., & Liu, Z. L. (2013). Essential oil composition and larvicidal activity of Toddalia asiatica roots against the mosquito Aedes albopictus (Diptera: Culicidae). Parasitology research, 112(3), 1197-1203.

[58] Tabanca, N., Gao, Z., Demirci, B., Techen, N., Wedge, D. E., Ali, A., ... & Baser, K. H. C. (2014). Molecular and phytochemical investigation of Angelica dahurica and Angelica pubescentis essential oils and their biological activity against Aedes aegypti, Stephanitis pyrioides, and Colletotrichum species. Journal of agricultural and food chemistry, 62(35), 8848-8857.

[59] Liu, Z. L., He, Q., Chu, S. S., Wang, C. F., Du, S. S., & Deng, Z. W. (2012). Essential oil composition and larvicidal activity of Saussurea lappa roots against the mosquito Aedes albopictus (Diptera: Culicidae). Parasitology research, 110(6), 2125-2130.

[60] Candido, L. P., Cavalcanti, M. T., & Beserra, E. B. (2013). Bioactivity of plant extracts on the larval and pupal stages of Aedes aegypti (Diptera, Culicidea). Revista da Sociedade Brasileira de Medicina Tropical, 46, 420-425.

[61] Adeleke, M. A., Popoola, S. A., Agbaje, W. B., Adewale, B., Adeoye, M. D., & Jimoh, W. A. (2009). Larvicidal efficacy of seed oils of Pterocarpus santalinoides and Tropical Manihot species against Aedes aegypti and effects on aquatic fauna. Tanzania Journal of Health Research, 11(4).

[62] Vimala, R. T. V., Sathishkumar, G., & Sivaramakrishnan, S. (2015). Optimization of reaction conditions to fabricate nano-silver using Couroupita guianensis Aubl.(leaf & fruit) and its enhanced larvicidal effect. Spectrochimica Acta Part A: Molecular and Biomolecular Spectroscopy, 135, 110-115.

[63] Mahesh Kumar, P., Murugan, K., Kovendan, K., Panneerselvam, C., Prasanna Kumar, K., Amerasan, D., ... & Nataraj, T. (2012). Mosquitocidal activity of Solanum xanthocarpum fruit extract and copepod Mesocyclops thermocyclopoides for the control of dengue vector Aedes aegypti. Parasitology Research, 111(2), 609-618.

[64] AhbiRami, R., Zuharah, W. F., Thiagaletchumi, M., Subramaniam, S., & Sundarasekar, J. (2014). Larvicidal efficacy of different plant parts of railway creeper, Ipomoea cairica extract against dengue vector mosquitoes, Aedes albopictus (Diptera: Culicidae) and Aedes aegypti (Diptera: Culicidae). Journal of Insect Science, 14(1).

[65] Kumar, V. A., Ammani, K., Jobina, R., Subhaswaraj, P., & Siddhardha, B. (2017). Photo-induced and phytomediated synthesis of silver nanoparticles using Derris trifoliata leaf extract and its larvicidal activity against Aedes aegypti. Journal of Photochemistry and Photobiology B: Biology, 171, 1-8.

[66] Li, W., Huang, C., Wang, K., Fu, J., Cheng, D., & Zhang, Z. (2015). Laboratory evaluation of aqueous leaf extract of Tephrosia vogelii against larvae of Aedes albopictus (Diptera: Culicidae) and non-target aquatic organisms. Acta tropica, 146, 36-41.

[67] Govindarajan, M., Rajeswary, M., Muthukumaran, U., Hoti, S. L., Khater, H. F., & Benelli, G. (2016). Single-step biosynthesis and characterization of silver nanoparticles using Zornia diphylla leaves: a potent eco-friendly tool against malaria and arbovirus vectors. Journal of Photochemistry and Photobiology B: Biology, 161, 482-489.

[68] Ahbirami, R., Zuharah, W. F., Yahaya, Z. S., Dieng, H., Thiagaletchumi, M., Fadzly, N., ... & Bakar, S. A. (2014). Oviposition deterring and oviciding potentials of Ipomoea cairica L. leaf extract against dengue vectors. Trop Biomed, 31(3), 456-465.

[69] Sogan, N., Kapoor, N., Singh, H., Kala, S., Nayak, A., & Nagpal, B. N. (2018). Larvicidal activity of Ricinus communis extract against mosquitoes. Journal of vector borne diseases, 55(4), 282.

[70] Rodzay, R., & Zuharah, W. F. (2021). RESEARCH ARTICLE The determination of effective concentration of acethonilic Ipomoea cairica leaves extract against laboratory and field strains of Aedes albopictus and Aedes aegypti mosquito larvae. Tropical Biomedicine, 38(3), 446-452.

[71] Suganya, A., Murugan, K., Kovendan, K., Mahesh Kumar, P., & Hwang, J. S. (2013). Green synthesis of silver nanoparticles using

Murraya koenigii leaf extract against Anopheles stephensi and Aedes aegypti. Parasitology Research, 112(4), 1385-1397.

[72] Nuchsuk, C., Wetprasit, N., Roytrakul, S., & Ratanapo, S. (2012). Larvicidal activity of a toxin from the seeds of Jatropha curcas Linn. against Aedes aegypti Linn. and Culex quinquefasciatus Say. Tropical Biomedicine, 29(2), 286-296.

[73] Waris, M., Nasir, S., Abbas, S., Azeem, M., Ahmad, B., Khan, N. A., ... & Mahboob, S. (2020). Evaluation of larvicidal efficacy of Ricinus communis (Castor) and synthesized green silver nanoparticles against Aedes aegypti L. Saudi journal of biological sciences, 27(9), 2403-2409.

[74] Perumalsamy, H., Jang, M. J., Kim, J. R., Kadarkarai, M., & Ahn, Y. J. (2015). Larvicidal activity and possible mode of action of four flavonoids and two fatty acids identified in Millettia pinnata seed toward three mosquito species. Parasites & vectors, 8(1), 1-14.

[75] Thanigaivel, A., Vasantha-Srinivasan, P., Senthil-Nathan, S., Edwin, E. S., Ponsankar, A., Chellappandian, M., ... & Kalaivani, K. (2017). Impact of Terminalia chebula Retz. against Aedes aegypti L. and non-target aquatic predatory insects. Ecotoxicology and environmental safety, 137, 210-217.

[76] Govindarajan, M., Kadaikunnan, S., Alharbi, N. S., & Benelli, G. (2016). Acute toxicity and repellent activity of the Origanum scabrum Boiss. & Heldr.(Lamiaceae) essential oil against four mosquito vectors of public health importance and its biosafety on non-target aquatic organisms. Environmental Science and Pollution Research, 23(22), 23228-23238.

[77] Darbro, J. M., Johnson, P. H., Thomas, M. B., Ritchie, S. A., Kay, B. H., & Ryan, P. A. (2012). Effects of Beauveria bassiana on survival, blood-feeding success, and fecundity of Aedes aegypti in laboratory and semi-field conditions. The American journal of tropical medicine and hygiene, 86(4), 656.

[78] De Valdez, M. R. W., Suchman, E. L., Carlson, J. O., & Black IV, W. C. (2010). A large scale laboratory cage trial of Aedes densonucleosis virus (AeDNV). Journal of medical entomology, 47(3), 392-399.

[79] J. A. Obeta, "Effect of inactivation by sunlight on the larvicidal activities of mosquitocidal Bacillus thuringiensis H-14 isolates from Nigerian soils" in *The Journal of Communicable Diseases*, 1996, 28(2), pp. 94-100.

[80] Rojas-Pinzón, P. A., & Dussán, J. (2017). Efficacy of the vegetative cells of Lysinibacillus sphaericus for biological control of insecticide-resistant Aedes aegypti. Parasites & vectors, 10(1), 1-7.

[81] Harwood, J. F., Farooq, M., Turnwall, B. T., & Richardson, A. G. (2015). Evaluating liquid and granular Bacillus thuringiensis var. israelensis broadcast applications for controlling vectors of dengue and chikungunya viruses in artificial containers and tree holes. Journal of medical entomology, 52(4), 663-671.

[82] Subkrasae, C., Ardpairin, J., Dumidae, A., Janthu, P., Meesil, W., Muangpat, P., ... & Vitta, A. (2022). Molecular identification and phylogeny of Steinernema and Heterorhabditis nematodes and their efficacy in controlling the larvae of Aedes aegypti, a major vector of the dengue virus. Acta Tropica, 106318

[83] Serit, M. A., & Yap, H. H. (1984). Comparative bioassays of Tolypocladium cylindrosporum Gams (Californian strain) against four species of mosquitoes in Malaysia. The Southeast Asian journal of tropical medicine and public health, 15(3), 331-336.

[84] Muniz, E. R., Catão, A. M., Rueda-Páramo, M. E., Rodrigues, J., Lastra, C. C. L., García, J. J., ... & Luz, C. (2018). Impact of short-term temperature challenges on the larvicidal activities of the entomopathogenic watermold Leptolegnia chapmanii against Aedes aegypti, and development on infected dead larvae. Fungal biology, 122(6), 430-435.

[85] Karthi, S., Vasantha-Srinivasan, P., Ganesan, R., Ramasamy, V., Senthil-Nathan, S., Khater, H. F., ... & Krutmuang, P. (2020). Target activity of isaria tenuipes (Hypocreales: Clavicipitaceae) fungal strains against dengue vector aedes aegypti (linn.) and its non-target activity against aquatic predators. Journal of Fungi, 6(4), 196.

[86] Wei, W., Shao, D., Huang, X., Li, J., Chen, H., Zhang, Q., & Zhang, J. (2006). The pathogenicity of mosquito densovirus (C6/36DNV) and its interaction with dengue virus type II in Aedes albopictus. The American journal of tropical medicine and hygiene, 75(6), 1118-1126.

[87] Thelma, J., & Balasubramanian, C. (2021). Ovicidal, larvicidal and pupicidal efficacy of silver nanoparticles synthesized by Bacillus marisflavi against the chosen mosquito species. PloS one, 16(12), e0260253.

[88] Carolino, A. T., Paula, A. R., Silva, C. P., Butt, T. M., & Samuels, R. I. (2014). Monitoring persistence of the entomopathogenic fungus Metarhizium anisopliae under simulated field conditions with the aim

of controlling adult Aedes aegypti (Diptera: Culicidae). Parasites & vectors, 7(1), 1-7.

[89] Audsley, M. D., Seleznev, A., Joubert, D. A., Woolfit, M., O'Neill, S. L., & McGraw, E. A. (2018). Wolbachia infection alters the relative abundance of resident bacteria in adult Aedes aegypti mosquitoes, but not larvae. Molecular Ecology, 27(1), 297-309.

[90] Lozano-Fuentes, S., Bandyopadhyay, A., Cowell, L. G., Goldfain, A., & Eisen, L. (2013). Ontology for vector surveillance and management. *Journal of medical entomology*, *50*(1), 1-14.

[91] Topalis, P., Mitraka, E., Bujila, I., Deligianni, E., Dialynas, E., Siden-Kiamos, I., ... & Louis, C. (2010). IDOMAL: an ontology for malaria. *Malaria Journal*, *9*(1), 1-11.

[92] Shettima, A., Joseph, S., Ishak, I. H., Abdul Raiz, S. H., Abu Hasan, H., & Othman, N. (2021). Evaluation of Total Female and Male Aedes aegypti Proteomes Reveals Significant Predictive Protein–Protein Interactions, Functional Ontologies, and Differentially Abundant Proteins. *Insects*, *12*(8), 752.

[93] Topalis, P., Dialynas, E., Mitraka, E., Deligianni, E., Siden-Kiamos, I., & Louis, C. (2011). A set of ontologies to drive tools for the control of vector-borne diseases. *Journal of biomedical informatics*, *44*(1), 42-47.

[94] Mitraka, E., Topalis, P., Dritsou, V., Dialynas, E., & Louis, C. (2015). Describing the breakbone fever: IDODEN, an ontology for dengue fever. *PLoS neglected tropical diseases*, *9*(2), e0003479.

[95] Topalis, P., Lawson, D., Collins, F. H., & Louis, C. (2008). How can ontologies help vector biology?. *Trends in parasitology*, *24*(6), 249-252.

[96] Shen, Y., Yuan, K., Chen, D., Colloc, J., Yang, M., Li, Y., & Lei, K. (2018). An ontology-driven clinical decision support system (IDDAP) for infectious disease diagnosis and antibiotic prescription. *Artificial intelligence in medicine*, *86*, 20-32.

[97] Rajapakse, M., Kanagasabai, R., Ang, W. T., Veeramani, A., Schreiber, M. J., & Baker, C. J. (2008). Ontology-centric integration and navigation of the dengue literature. *Journal of biomedical informatics*, *41*(5), 806-815.

[98] Vinarti, R. A., & Hederman, L. M. (2019). A personalized infectious disease risk prediction system. *Expert Systems with Applications*, *131*, 266-274.

[99] Dialynas, E., Topalis, P., Vontas, J., & Louis, C. (2009). MIRO and IRbase: IT tools for the epidemiological monitoring of insecticide resistance in mosquito disease vectors. *PLoS Neglected Tropical Diseases*, *3*(6), e465.

[100] Mukundarajan, H., Hol, F. J. H., Castillo, E. A., Newby, C., & Prakash, M. (2017). Using mobile phones as acoustic sensors for high-throughput mosquito surveillance. *elife*, *6*, e27854.

[101] Cimiano, P., & Völker, J. (2005). A framework for ontology learning and data-driven change discovery. In *Proceedings of the 10th International Conference on Applications of Natural Language to Information Systems (NLDB)* (pp. 227-238).

[102] Yang, Y., Wybrow, M., Li, Y. F., Czauderna, T., & He, Y. (2019). OntoPlot: a novel visualisation for non-hierarchical associations in large ontologies. IEEE Transactions on Visualization and Computer Graphics, 26(1), 1140-1150.

[103] Smoot, M. E., Ono, K., Ruscheinski, J., Wang, P. L., & Ideker, T. (2011). Cytoscape 2.8: new features for data integration and network visualization. Bioinformatics, 27(3), 431-432.

Biography

G. Jeyakodi is a Research Scholar with the Department of Computer Science, Pondicherry University, Puducherry, India. She is an innovative and knowledgeable professional having 10 years of teaching experience in various recognized institutes like Pondicherry University, Puducherry and Govt. Arts & Science College, Tumkur, Karnataka. She has five years of scientific research experience with ICMR – Vector Control Research Centre, Puducherry. Her research interest includes bioinformatics, text mining, database development, and ontology. She has published various books and book chapters with well-reputed publishers. She also participated and presented papers in national and international conferences. She has also published her research articles in SCOPUS and SCI indexed Journals.

Trisha Agarwal is currently working toward the Master of Computer Applications degree with the Department of Computer Science, Pondicherry University, Puducherry, India. She is experienced in developing android applications and currently working on deep learning algorithms for image classification. She also has well exposure in python programming language and web development language.

P. Shanthi Bala received the Ph.D. degree from Pondicherry University, Puducherry, India. She is presently working as an Assistant Professor with the Department of Computer Science, Pondicherry University. She has around 16 years of teaching and 12 years of research experience. Her research interests include knowledge engineering, artificial intelligence, ontology and networks, and IoT. She has published more than 50 papers in international journals and conferences.

9

Paradigms for Integration of Biomedical Knowledge with Patients' Records: Brief Trajectory and Roles of Ontology

Xia Jing[1,*], James J. Cimino[2,*], Xuequn Pan[3], and Shijuan Li[4]

[1]Department of Public Health Sciences, College of Behavioural, Social and Health Sciences, Clemson University, USA
[2]Informatics Institute, School of Medicine, The University of Alabama at Birmingham, USA
[3]Changsha Medical University, China
[4]Department of Information Management, Peking University, China
*Xia Jing and James J. Cimino contributed to the chapter equally; therefore, they share the first authorship.
E-mail: xjing@clemson.edu; jamescimino@uabmc.edu; xuequn_pan@hotmail.com; shijuan.li@pku.edu.cn

Abstract

Integrating biomedical knowledge with patients' health record systems is a common approach to helping clinicians make more informed decisions. We aim to explore and compare different integration mechanisms and identify integration paradigms to enlighten future improvement. We searched PubMed (n = 471) and set up the initial inclusion/exclusion criteria, and the initial paradigms via a pilot classification (n = 61). We then conducted brainstorming sessions to achieve consensus. We then compared, discussed, refined the results, and formed the integration paradigms. We then reviewed, selected, and classified the remainder of the publications (n = 410). There exist at least five knowledge integration paradigms: knowledge inscription, knowledge catalog, knowledge agent, expert systems, and knowledge modeled as an ontology among 73 publications. Integration mechanisms, scalability, maintenance, and reusability of the knowledge sources were reviewed and compared among paradigms. The knowledge modeled as an ontology paradigm has advantages in scalability, reusability, and maintainability. However, computational efficiency can be a challenge. Our review provides an initial summary of approaches to knowledge integration with patients' records. The work demonstrates the quick transitions among different paradigms when integrating knowledge with patient records and highlights the roles of ontology applications in healthcare computing.

Keywords: Clinical decision making, Electronic health record, knowledge bases, knowledge integration paradigms, ontology, personalized information, semantic web technologies.

9.1 Introduction

For decades, researchers and clinicians have sought ways to integrate biomedical knowledge with patients' records to help clinicians make more informed decisions at the point of care [1]. These more informed decisions can play a critical role in keeping patients safe and delivering high-quality and consistent care. Such efforts include medication dosage calculators, abnormal laboratory test results flags, and automated clinical decision support systems (CDSSs), which often contain explicit rules for proactively delivering alerts and reminders [2]. These forms of knowledge delivery exist within both paper-based and electronic health records (EHRs) [1].

Previous publications have described decision support for clinicians and patients, using services external to the EHR (i.e., not a component within an EHR) [3–5]. We refer to an EHR as an "electronic knowledge record" (EKR) when the EHR ***incorporates*** knowledge and displays it alongside the patient's data as an integrated component within an EHR; and the knowledge meets the following criteria:(1) it can be customized according to patient and usage context; and (2) it can be used for education, training, decision-making, or interpretation of test results. The main advantage of an EKR versus an EHR is that an EKR can provide more relevant individual information in addition to the patient's existing data, which can be found routinely in the EHR, and the additional information may help clinical users to make more informed decisions. The EKR potentially allows users to make better-evidenced decisions during clinical care. The *knowledge integration paradigm* refers to a conceptual model or a design pattern that can guide the development and implementation of knowledge integration with EHR systems. Knowledge integration refers to external knowledge that is incorporated into, or incorporated with, patients' records to be displayed next to patients' data.

Providing the right amount and type of information in the right situation to clinicians poses significant challenges. Ellsworth and colleagues' study shows that clinicians seek the support of the literature for approximately one-quarter of patient interactions [6]. Both Smith [7] and Mickan and Askew [8] noted that to avoid information overload, patient information and general biomedical knowledge need to be selected based on their relevance to the individual patient.

In some instances, solutions for knowledge integration that initially appeared successful for a particular problem have often failed to ***scale*** up to cover a practical subset for a specific domain. In addition, systems have proven difficult to ***maintain*** to stay current. Finally, the work required to integrate knowledge with patients' records has been directed toward specific purposes, without consideration for knowledge ***reuse***. This chapter, however, focuses on the knowledge integrated with a patient database or the user interface to be displayed with the patient's data.

In this chapter, we use ***biomedical knowledge*** to refer to the information that is either external to the EHR or organized and integrated within it (i.e., EKR), both of which can answer "how come" or "what" questions. We do not intend to define biomedical knowledge universally. Instead, we want to pragmatically distinguish knowledge from the typical patient's data in EHR. Greenes' viewpoint paper [9] reviews all aspects of clinical decision support (CDS): from integrating CDS into clinical workflow to evaluating CDS. Greenes summarizes existing theoretical CDS models and proposes a comprehensive framework that serves as a high-level navigation map for research to improve CDS utility. Our chapter focuses more on the following aspects of that framework: knowledge management, interoperability, and sharing. We examine the integration mechanisms between knowledge facts and a patient's record. Guise's paper has a more comprehensive description of external evidence that is outside of EHR and knowledge within EHR and their roles in a learning health system [10]. Alavi [11] and Floridi [12] provide more general definitions of knowledge without specifying application fields. The sources of knowledge may include human expertise, empirical observations, or biomedical literature that can be used to support clinicians and patients in their decision making. We use the term ***rules*** to refer to (1) clinical rules, i.e., the rules are used to express biomedical domain knowledge in a natural or machine-processable language; and (2) machine-executable rules, i.e., the rules are used to decide how programs function and these rules can be executed via rule engines. We use ***logic*** to refer to the relationships among groups of events within one or multiple tables, which can be achieved by combining a set of rules. In this chapter, ***rules*** can provide finer granularity levels of content than ***logic***. We prepared this chapter to illustrate and compare different integration mechanisms and identify integration paradigms that can be utilized as a baseline for future design and improvement.

9.2 Methods

These are the questions we hope to answer in this chapter: (1) What are the existing paradigms by which knowledge is integrated with patients' records? (2)What are the differences in scalability, reusability, maintenance, and explicitness of logic of knowledge sources among the integration paradigms? The first question was utilized to search, screen, and review literature; the second question was used to guide the selected publications' analysis, comparisons, and synthesis.

To collect the reported paradigms to integrate knowledge with patients' records, we follow the PRISMA-ScR [13] guidelines to conduct and report this chapter as a scoping review. We report 20 items in PRISMA-ScR. The PRISMA-ScR checklist provides the details of the included items. The study methods were not registered in advance because the method development was part of the study. No IRB approval was needed due to the nature of the study. We separate the research methods into three stages: literature search, literature review, results analysis, and synthesis.

Stage 1 — literature search:

a) All authors discussed, developed, and revised the literature search strategies. After several trial searches, we searched for knowledge, knowledge bases, databases, and CDSS integration with EHR and electronic medical record (EMR) systems. After iterative tests and revisions, we decided to use the following search strategy in PubMed.
("Electronic Health Records"[Mesh] OR "Medical Records Systems, Computerized"[Mesh]) AND (("Knowledge Bases"[Mesh]) OR (("Decision Support Systems, Clinical"[Mesh]) AND ("Systems Integration"[Mesh])) OR (knowledg"[Title] AND ((integr*[Title]) OR (incorpor*[Title]))))

b) There were no restrictions on publication types, publication dates, or languages. All the records were aggregated; then, all the duplicates based on titles were removed before the review started.

Stage 2 – literature review:

a) All authors served as reviewers to conduct a pilot review (n = 61) of the relevance of each publication to form the initial paradigms; we then conducted a consensus process about disagreements and refined inclusion and exclusion criteria during the process. These were the first 61 records of the literature search results, which were the most recent publications at the time.
b) We also conducted brainstorming sessions by ***incorporating the authors' own experience*** and discussion and consensus processes to form the initial paradigms in addition to the pilot review. We itemized ideas about integration individually on paper or whiteboard first; we then aggregated the items and put similar items together, discussed, and conceptualized the paradigms.
c) At least two reviewers reviewed the remainder of the records (n = 410) based on the refined criteria.
d) We compared manual review results and discussed iteratively to achieve consensus.

Inclusion criteria:

a) The publication has to report integration between knowledge and patient's record (EHR or EMR) via a demonstration or conceptual design.
b) Mechanism of integration should be included in the publication.
c) At least one of the following items should be mentioned in the publication: integration mechanism at the technical level, function descriptions of the integration, or the knowledge customization.
d) The publication can be original research or a literature review.
e) Only English publications were included.

Exclusion criteria:

a) The introduction of pure natural-language-processing tools without a demonstration or a design of integrating knowledge with EHR/EMR via the tools.
b) A publication indicates the potential for future usages without an integration mechanism, a demonstration, or a design of the integration. The publication may cover the following topics:

i A knowledge base or ontology.
ii Integration at the governance level, organizational level, or policy level, not on a technical level.
iii Information retrieval within EHR without explicitly including the mechanism or design of the knowledge integration.
iv Utilizing EHR data and knowledge bases for secondary purposes without knowledge integration.
v The publication focuses on health information exchange or data transfer without knowledge integration.

c) Application tools of EHR without direct knowledge involvement of end-users. For example, a study explores improving CDSS usage without specifying how knowledge is integrated at the backend.
d) Publications that are comments, editorials, or a literature review about functionalities of a computerized physician order entry (CPOE) or CDSS without knowledge integration.

Stage 3 – result analysis and synthesis:

We summarized the review results in separate Excel documents and classified them into different paradigms; we then compared and discussed the differences iteratively; we also discussed, analyzed, conceptualized, and revised diagrams iteratively for each paradigm before we synthesized and presented the final results. We used thought experiments to compare four dimensions of each paradigm subjectively (scalability, maintainability, reusability, and the explicitness of logic) to compare the different paradigms. For example, for scalability, we used an example of increasing knowledge facts from 50 to 5000; how much effort would be needed? Maintainability refers to the attempt to update (e.g., modify or delete) a specific knowledge fact. Reusability is how easily knowledge facts can be used in a different system without substantial effort. We mainly used easy or challenging to demonstrate relative assessment of the dimensions among different paradigms.

The workflow of the methodology is presented in Figure 9.1 [13], with modification from PRISMA [14] to reflect the processes in this study. The actual review process was not linear; there were many back and forth and many revisions in the process. The paradigms become clear gradually after many discussions. The process of classification, review, forming of inclusion and exclusion criteria, discussions, and revisions reinforce the forming and refinement of the paradigms.

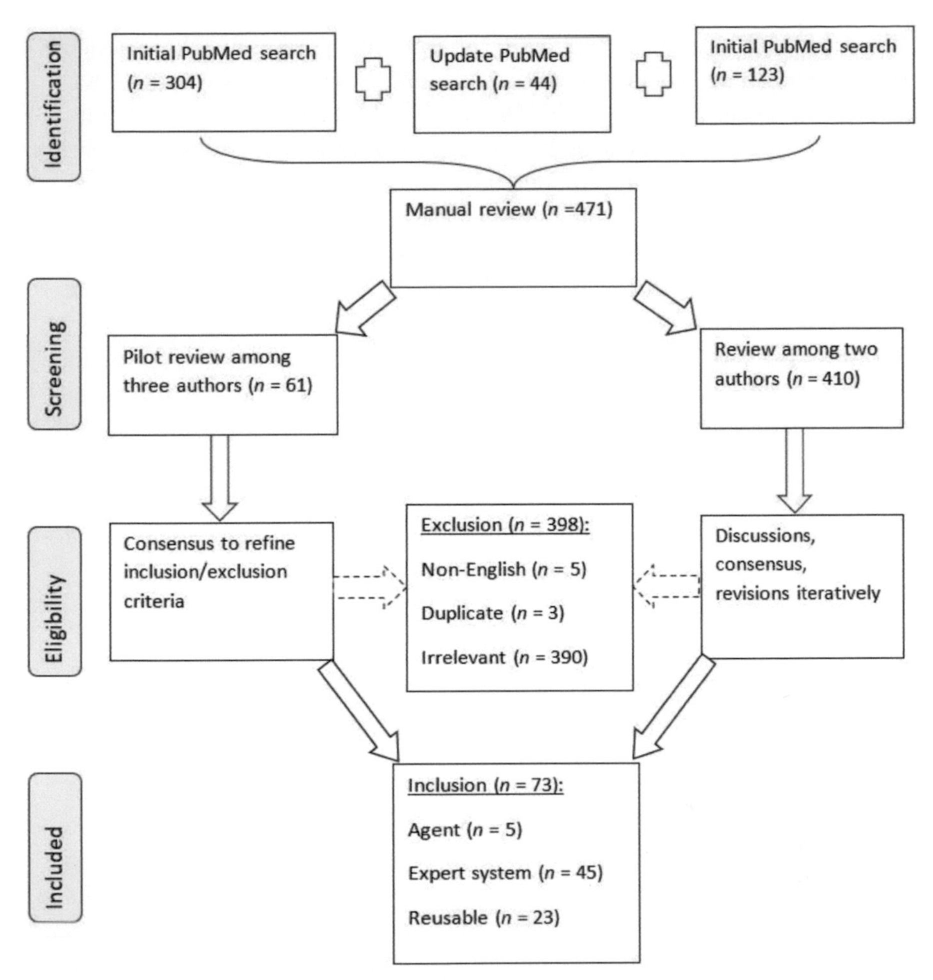

Figure 9.1 Workflow of literature search, review, and classification (the solid arrows annotate the direction of the work; the dashed arrows show contribution).

9.3 Results

The first PubMed search was conducted on September 12, 2016, from which we found 304 unique records; the update search was conducted on June 27, 2017, from which we found 44 unique records. The most recent update was additional 123 unique records retrieved on October 9, 2020.

There exist at least five knowledge integration paradigms: knowledge inscription, knowledge catalog, knowledge agent, expert systems, and knowledge modeled as an ontology. We classified the selected 73 relevant publications into three paradigms of knowledge integration: knowledge agent, expert system, and knowledge modeled as an ontology. Three publications from the authors' reference collection can be classified into the knowledge catalog. Knowledge inscription is most commonly recognizable; although no relevant publication from our search can be classified into knowledge inscription, it has been ubiquitous in patient records (electronic and paper) for decades. We formed the paradigms according to how knowledge facts are integrated into patients' records (i.e., EKRs) regarding the ***forms*** of knowledge, ***mechanisms*** of integration, and the ***relationships*** between the individual knowledge facts and components of patients' records, such as the patient database, system logic, and user interface. In the following sections, we summarize the typical paradigms and their integration mechanisms first and then provide the publication list for each paradigm.

9.3.1 Knowledge inscription

Long before EHRs existed, biomedical knowledge was included in paper-based health records in the form of statements in clinicians' notes, such as published evidence of best treatment options. A more consistent form of knowledge can be found in laboratory reports (e.g., normal ranges). This type of information represents knowledge added to the raw clinical data to help users to comprehend the raw data and to apply these data appropriately. In this case, the external knowledge is inserted into the record next to the patient's data and displayed to the user via a piece of paper or on a computer screen in a *static manner* (Figure 9.2). We refer to this as "knowledge inscription."

One advantage of this approach is that the knowledge is readily available to the user of the record with straightforward display logic. Thus, increase in knowledge does not affect the process needed to integrate it into the record and does not require changes to the logic or rules required to display it. We, therefore, consider this approach to be easily scalable with increases in knowledge, including the addition of new knowledge domains.

A disadvantage of this paradigm is less maintainable than are other approaches. For example, there is the possibility of reinterpretations of a test result. The knowledge already stored in the patients' records may become inaccurate, outdated, and difficult to locate and correct.

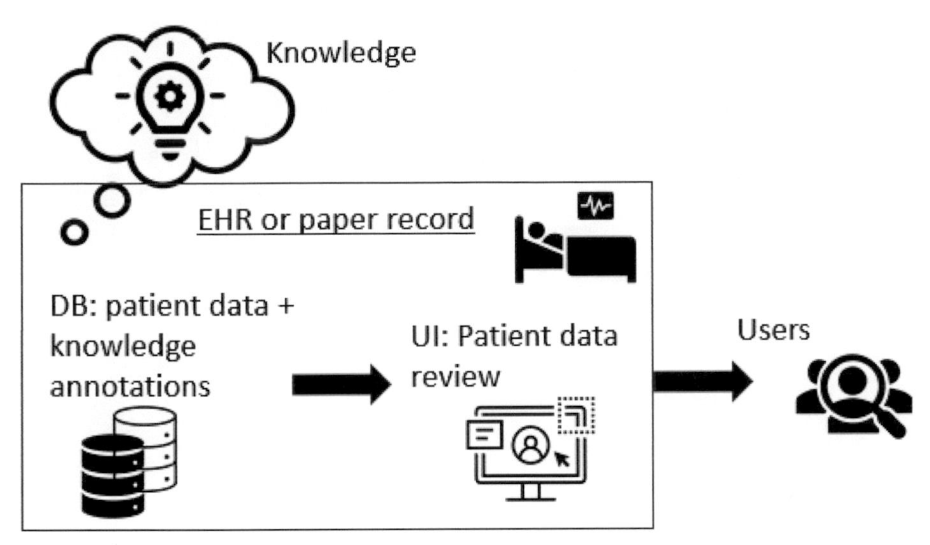

Figure 9.2 Knowledge inscription approach: The knowledge is explicitly included *next to* patients' data. EHR, electronic health record; DB, database; UI, user interface.

9.3.2 Knowledge catalog

Another approach is to select from a predefined list of known facts, such as drug formulary or available laboratory tests, which can be compiled into a reference source ("catalog") designed to address specific information needs dynamically. For example, a user who plans to prescribe a medication may request to see the proper dosing information for that medication. The knowledge can be represented as a table or an indexed set of documents (Figure 9.3). The logic in this approach consists of simply retrieving the appropriate facts; the knowledge itself is not used in any inferencing process. Several studies [2, 15, 16] provide examples of this approach.

The knowledge catalog approach shares a similar knowledge integration mechanism with the knowledge inscription approach. However, knowledge inscription inserts the external knowledge into the patients' records, making updates and maintenance difficult. This is in contrast to the knowledge catalog, in which the knowledge is either embedded within or accessible via EHR applications, separate from the patients' data. Maintenance of the knowledge catalog is therefore easier than for the knowledge inscription approach. Like the knowledge inscription approach: (1) the knowledge catalog approach has the advantage of straightforward logic, i.e., to retrieve the facts and to display; (2) this approach is readily scalable. Comparing to knowledge inscription, the

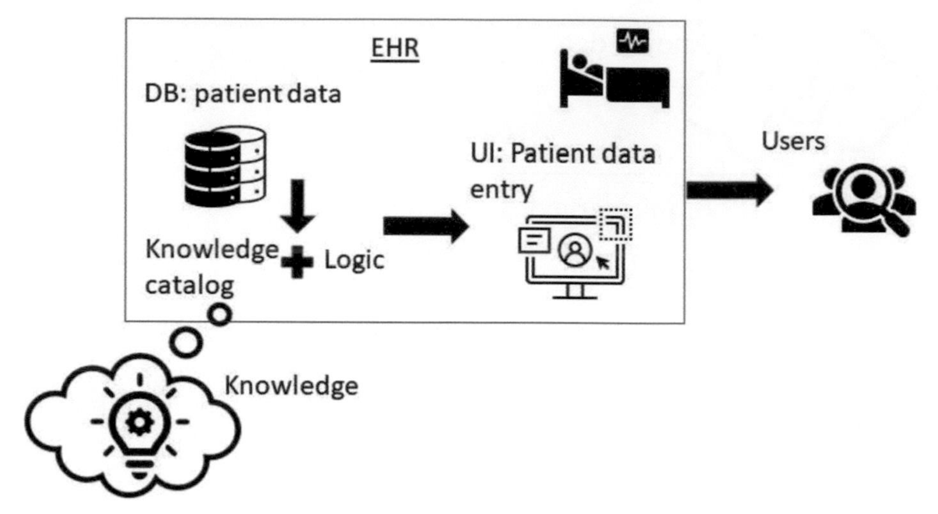

Figure 9.3 Knowledge catalog approach: The knowledge is included in the EHR system, *external* to the patient data, and accessible as needed dynamically. EHR, electronic health record; DB, database; UI, user interface.

knowledge catalog approach has the potential advantage of having relatively reusable knowledge.

9.3.3 Knowledge agent

One method for delivering knowledge to an EHR user is to pass patient information to external knowledge resources to retrieve the relevant information. A particular resource can be "hardwired" to the EHR, or the selection of the resources is determined dynamically based on the characteristics of the patient and the EHR user. The context-specific links to outside resources can be generated automatically. These go by a variety of names such as *infobuttons* [17] (an HL7 standard) and KnowledgeLink [18, 19]. The technology has been adopted by different healthcare delivery systems in the USA as a CDS to meet clinicians' information needs at the point of care [20].

The logic for this approach is concerned with identifying the most appropriate resources that match the user's context. This approach uses a knowledge base, one that is concerned with facts about information needs, relevant resources, and the mechanisms for retrieving information from those resources, rather than with biomedical domain knowledge (Figure 9.4 (A)). This approach's logic and knowledge base do not have to be integrated into

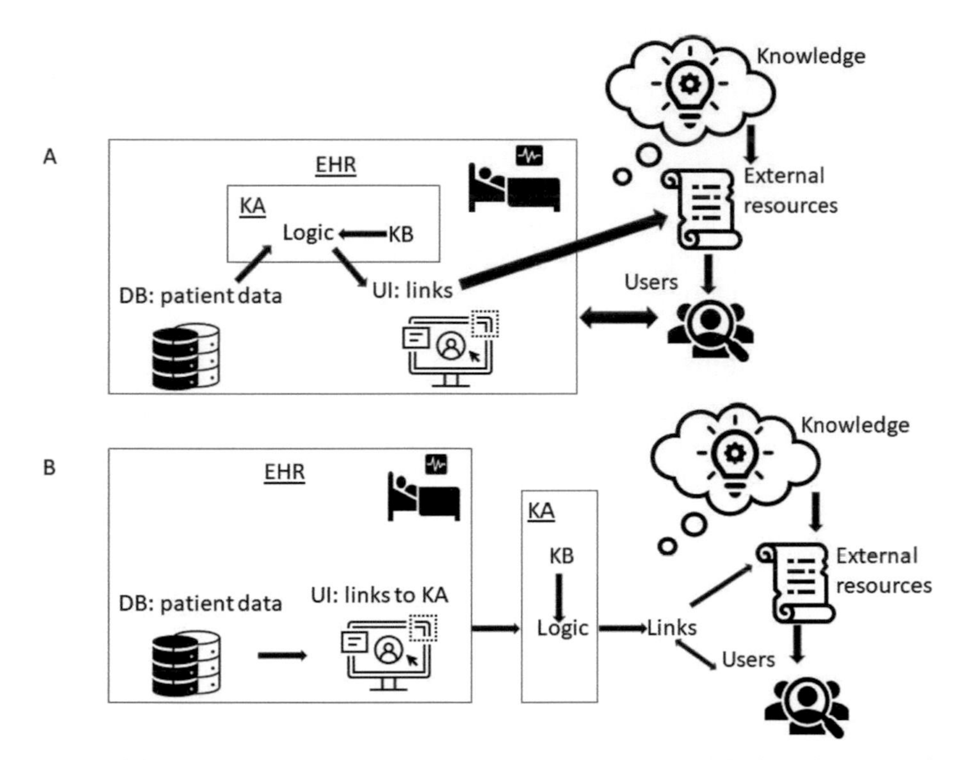

Figure 9.4 Knowledge agent (KA). In (A), the KA uses internal logic and an internal knowledge base to determine the links to external knowledge resources (the characteristics of information needs and of the external biomedical knowledge resources to address those needs). In (B), the logic and knowledge needed for selecting context-appropriate resource links are external to the EHR in an outside KA; the internal "logic" determines a link to access an external resource, such as an infobutton manager. KA, knowledge agent; KB, knowledge base; EHR, electronic health record; DB, database; UI, user interface.

the EHR; they can be invoked as an external, independent service. Del Fiol and colleagues provide Open Infobutton [20, 21] as an independent service. Open Infobutton can process HL7 compatible and non-HL7 compatible XML knowledge requests [22] and respond in an XML format. Multiple healthcare delivery systems' EHRs may generate the requests, and the responses can be displayed in the EHRs (Figure 9.4 (B)).

As with the knowledge catalogue, the biomedical knowledge used by the "knowledge agent" is external to the EHR; therefore, its scalability and maintainability are attributes of the external knowledge resources. Further, the domains that knowledge covers can readily be addressed by simply adding links to additional resources. The logic used by the EHR has the advantage of simplicity, especially when an outside service, such as an infobutton manager, is employed.

A disadvantage of the approach is that the rules are not readily available for reuse because the biomedical knowledge in this approach resides in external resources. It would be challenging for an internal reminder system (described in the next section) to use the rules found in some external rule sets without substantial effort. Five publications are in this paradigm (Table 9.1).

9.3.4 Expert systems

Expert systems, sometimes in the form of automatic alerts, reminders, or more complicated inferred suggestions, are, in effect, a dynamic version of knowledge inscription. The alert can be categorized as the straightforward CDSS compared to the more sophisticated CDSS, such as case-based reasoning or differential diagnosis purposes [1, 23, 24]. This paradigm is broader than a traditional expert system (e.g., MYCIN [25]) as long as the system is integrated with an EHR (versus a stand-alone system). Users of an EHR can get an alert or reminder as soon as the individual patient's data or user's action triggers a rule. CLIPMERGE PGx [26], for example, provides order sets for medications or dosages based on a patient's genomic data. The CDSS and the CDSS architecture and deployment architecture are described in Kam's [27] and Cho's [28] papers. Most of the CDSS described in the two taxonomy papers authored by Wright [23] and Berlin [29] can be classified into this category. Forty-five publications are in this paradigm (Table 9.1).

Knowledge here refers to the biomedical domain (i.e., content) knowledge, such as harmful interactions among different medications. In this approach, the domain knowledge is organized and authored as rules in a ***rule base***. A domain-independent ***rules engine*** is needed to execute the rules using an individual patient's data and context (Figure 9.5). The Arden Syntax [30], another HL7 standard, is an example that has been broadly adopted in the healthcare field to write "if... then..." rules.

One advantage of this approach is its scalability, especially for the straightforward CDSS: New rules can relatively easily be added to the knowledge base, generally without disrupting the functioning of pre-existing rules.*[1] However, the maintenance of pre-existing rules is often challenging. When new knowledge is acquired, the old knowledge may not be obvious,

[1]*Exceptions do occur, especially when rules are chained such that the inference from one rule is used as an assertion by another rule. In practice, these links are uncommon.

Table 9.1 Summary of the publications on paradigms of knowledge integration with patient records.

Integration paradigm	Publication year	Authors	Ref. no
Knowledge agent (*n* = 5)			
	1997	Tarczy-Hornoch P, Kwan-Gett TS, Fouche L, *et al.*	[47]
	1999	Fuller SS, Ketchell DS, Tarczy-Hornoch P, *et al.*	[48]
	1999	Godin P, Hubbs R, Woods B, *et al.*	[49]
	2002	Bountis C and Kay JD	[50]
	2002	Wu R, Peters W, and Morgan MW	[51]
Expert systems (*n* = 45)			
	1995	Moser W, Bohm V, Bohmer K, *et al.*	[52]
	1996	Johansson B, Shahsavar N, Ahlfeldt H, *et al.*	[53]
	1997	Eich HP, Ohmann C, Keim E, *et al.*	[54]
	1998	Henry SB, Douglas K, Galzagorry G, *et al.*	[55]
	1999	Porcelli PJ and Lobach DF	[56]
	2000	Goldstein MK, Hoffman BB, Coleman RW, *et al.*	[57]
	2000	Müller ML, Ganslandt T, Eich HP, *et al.*	[58]
	2002	Ray HN, Boxwala AA, Anantraman V, *et al.*	[59]
	2002	Wu R, Peters W, and Morgan MW	[51]
	2003	Adams WG, Fuhlbrigge AL, Miller CW, *et al.*	[60]
	2003	Tange H, Van Der Linden H, Sas P, *et al.*	[61]
	2004	Ciccarese P, Caffi E, Boiocchi L, *et al.*	[62]
	2004	Frank L, Galanos H, Penn S, *et al.*	[63]
	2004	Goldstein MK, Coleman RW, Tu SW, *et al.*	[64]
	2005	Holbrook A, Keshavjee K, Lee H, *et al.*	[65]
	2005	Hulse NC, Del Fiol G, and Rocha RA	[66]
	2005	Shah NR, Seger AC, Seger DL, *et al.*	[67]
	2006	Conforti D, Costanzo D, Perticone F, *et al.*	[68]
	2006	Huang C, Noirot LA, Heard KM, *et al.*	[69]
	2006	Panzarasa S, Quaglini S, Cavallini A, *et al.*	[70]
	2006	Postilnik A, Palchuk MB, Vashevko M, *et al.*	[71]
	2007	Linder J, Schnipper JL, Volk LA, *et al.*	[72]
	2008	Schnipper JL, Linder JA, Palchuk MB, *et al.*	[73]
	2008	Wright A and Sittig DF	[74]
	2009	Liu D, Ajlouni M, Jin JY, *et al.*	[75]
	2010	Steurbaut K, Van Hoecke S, Colpaert K, *et al.*	[76]
	2011	Bernonille S, Nies J, Pedersen HG, *et al.*	[77]
	2011	Duke J and Bolchini D	[78]
	2011	Kam HJ, Kim JA, Cho I, *et al.*	[27]
	2012	Slonim N, Carmeli B, Goldsteen A, *et al.*	[79]
	2013	Zhang M, Velasco FT, Musser RC, *et al.*	[80]
	2014	Archer M, Proulx J, Shane-McWhorter L, *et al.*	[81]
	2014	Bouzguenda L and Turki M	[82]

Table 9.1 (Continued.)

	2014	Goldspiel BR, Flegel WA, DiPatrizio G, *et al.*	[83]
	2014	Kunhimangalam R, Ovallath S, and Joseph PK	[84]
	2014	Ninh A	[3]
	2014	Rasmussen-Torvik LJ, Stallings SC, Gordon AS, *et al.*	[85]
	2014	Serban A, Crisan-Vida M, and Stoicu-Tivadar L	[4]
	2014	Welch BM, Rodriguez-Loya S, Eilbeck K, *et al.*	[5]
	2015	Chackery DG, Keshavjee K, Mirza K, *et al.*	[86]
	2015	Moriarity AK, Klochko C, O'Brien M, *et al.*	[87]
	2016	Gaebel J, Cypko MA, and Lemke HU	[88]
	2016	Syomova II, Bologvaa EK, Kovalchuka SV, *et al*	[89]
	2017	Kopanitsa G	[90]
	2019	Seitz MW, Listl S, and Knaup P	[91]
Knowledge modeled as an ontology (n = 23)			
	2001	Müller ML, Ganslandt T, Eich HP, *et al.*	[92]
	2004	Rahman Y, Knape T, Gargan M, *et al.*	[44]
	2006	Das AK, Ahmed BA, Garten Y, *et al.*	[93]
	2007	Klimov D and Shahar Y	[94]
	2007	Tu SW, Campbell JR, Glasgow J, *et al.*	[45]
	2008	Adlassnig KP and Rappelsberger A	[95]
	2009	Groot P, Hommersom A, Lucas PJ, *et al.*	[46]
	2011	van den Branden M, Wiratunga N, Burton D, *et al.*	[24]
	2012	Jing X, Kay S, Marley T, *et al.*	[37]
	2013	Fernández-Breis JT, Maldonado JA, Marcos M, *et al.*	[96]
	2013	Garcia D, Moro CM, Cicogna PE, *et al.*	[97]
	2013	Hulse NC, Long J, and Tao C	[98]
	2013	Wang HQ, Li JS, Zhang YF, *et al.*	[43]
	2014	Jing X, Kay S, Marley T, *et al.*	[38]
	2014	Liaw ST, Taggart J, and Yu H	[99]
	2014	Lupşe OS and Stoicu-Tivadar L	[100]
	2015	Robles-Bykbaeva V, López-Noresb M, Pazos-Ariasb J, *et al.*	[42]
	2015	Shemeikka T, Bastholm-Rahmner P, Elinder CG, *et al.*	[101]
	2016	Rosier A, Mabo P, Temal L, *et al.*	[102]
	2016	Zhang YF, Gou L, Tian Y, *et al.*	[103]
	2017	Shaban-Nejad A, Lavigne M, Okhmatovskaia A, *et al.*	[104]
	2019	Chelsom J and Dogar N	[105]
	2019	Roehrs A, da Costa CA, Righi RDR, *et al.*	[106]

and the affected rules in thc rule base may not be readily identified or located. Further, the rules often contain many clauses as a collection of "if" statements

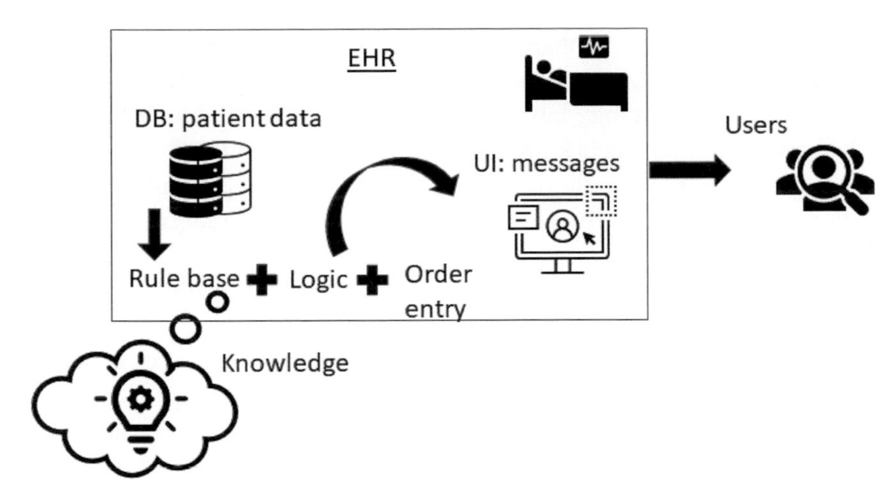

Figure 9.5 Expert systems approach: The logic is realized via a rule engine that uses rules from a rule base together with the patient's data and the user's actions (such as entering a patient order) to detect alerting conditions. When such conditions are met, alert messages are displayed to the user. The same logic and rule base can be evoked from different parts of the EHR (e.g., by data stored in the DB or an order placed by a user), and the specific rules used in each instance are usually distinct, even if they make use of the same external knowledge. EHR, electronic health record; DB, database; UI, user interface.

combined with Boolean logic ("and" and "or"). Some of these clauses may be reused in multiple rules; updating a clause in one rule does not usually ensure that it will be updated throughout the rule base. For a more complete discussion on the challenges of maintaining rule bases, refer to a paper by Zhou *et al.* [31].

Another disadvantage of an expert system is that its knowledge is not easily reused, as the rules are often closely coupled to the workflows for which the associated alerts are intended. For example, a medication dosage rule may read like this: "If Ordered-Medication is X and Order-Dose is greater than Y, send an alert that the dose may be too high." Although the typical dose range information also might be used to help detect a drug overdose, the knowledge in the former rule is not available for such an inference. For a more complete discussion on the challenges of reusing knowledge in rules, see publications by Sittig *et al.* [32, 33] and Zhou *et al.* [31].

9.3.5 Knowledge modeled as an ontology

This paradigm engages an external knowledge base in a generic format, especially in the form of an ontology. We refer to this mode as ***knowledge modeled***

as an ontology paradigm. Gruber defined ontology as "a specification of conceptualization" [34]. Ontology is an enabling technology of the semantic web. Many studies in ontology have been conducted in the biomedical field in the last two decades. Such knowledge could be selected and displayed according to a particular patient, or it could be used to support inferencing for alerts or knowledge interpretation. The logic in this paradigm is more complex, as there is a need for selection, inference, and display functions. All of these functions will be tightly coupled to the workflow of the particular EHR (Figure 9.6).

This knowledge modeled as an ontology approach has advantages in both scalability and maintenance of knowledge. The internal knowledge base *can be* a direct import of the external knowledge, requiring little human and no domain expert intervention. The external knowledge base keeps the knowledge as explicit facts in which domain concepts are related in well-defined ways [35, 36]. In addition to their use by humans, a variety of domain-independent (but workflow-specific) reasoners can use such knowledge. For example, OntoKBCF [35] is represented by the web ontology language (OWL) and has been demonstrated to provide molecular genetic and phenotypic knowledge via an EHR prototype interface [37, 38]. As early as 2009, other researchers, such as Sittig [39–41] and Welch [5], have recognized the

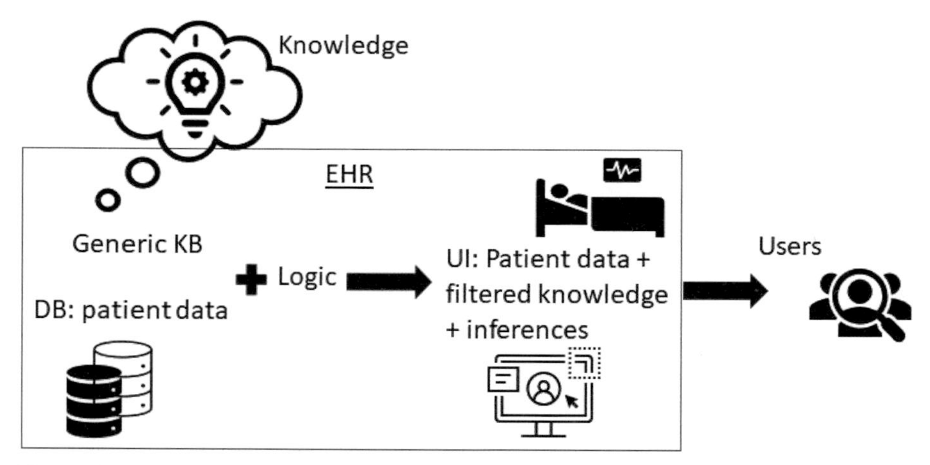

Figure 9.6 The knowledge modeled as an ontology paradigm: The external knowledge consists of a set of independent assertions (facts) about a particular domain. These facts are imported in a generic form (i.e., not tailored for any particular function) into a knowledge base in the EHR. The logic for using the knowledge is distributed throughout the EHR. EHR, electronic health record; DB, database; UI, user interface.

advantages and value of formally represented external knowledge bases. This paradigm's demonstrated and potential uses include assisting with patient data display via customizing relevant information, displaying non-duplicate patient information dynamically [37], providing patient educational supports, generating therapy plans [42] and care plans [43], and serving as a knowledge supplier to alert systems.

The advantages of the knowledge modeled as an ontology approach include the following: (1) the supporting knowledge can be customized according to the patient's characteristics, e.g., age, gender, ethnicity, and diagnoses; and (2) the content of the knowledge base can be updated and scaled up to cover additional content, without affecting the EHR to a certain extent. One disadvantage of this paradigm is that computational efficiency can be a challenge, as compared to the other paradigms.

In addition to semantic web technologies to organize the known facts, other methods include the unified modeling language (UML) [44], case-based reasoning [24], SAGE Guideline Model [45], and model checking for inconsistency [46]. Twenty-three publications are in this paradigm (Table 9.1). Table 9.1 contains a list of the publications for each paradigm.

9.4 Discussion

In this section, we summarize and compare the five paradigms, interpret the review results, and present the limitations of this scoping literature review.

9.4.1 Summary of literature review results

We have selected 73 relevant publications about the knowledge integration with patients' records from the literature search. Table 9.2 summarizes the five paradigms, including knowledge sources, mechanisms, purposes, and examples. Table 9.2 explains further the characteristics and examples of each paradigm. Table 9.3 compares different knowledge integration approaches regarding the scalability, maintainability, reusability, and explicitness of logic of the knowledge sources. Each paradigm has its strengths and weaknesses. The summarization and comparison may be useful for development of future clinical decision support.

9.4.2 Interpretation of the literature review results

We noticed that many publications lacked elaboration on knowledge integration mechanisms during our review process. With increasing EHR adoption

Table 9.2 Summary of the five paradigms of the knowledge integration with patient records.

Knowledge integration paradigm	Knowledge source/base	Mechanism of integration	Purpose/ functionality	Example
Knowledge inscription	External knowledge	Snippets get turned into permanent artifacts	Add knowledge to the raw clinical data to help readers understand and use the data	Normal ranges for lab results, a literature reference
Knowledge catalog	A list of knowledge elements	Incorporates into the EHR's internal data structure	For retrieval and display	Drug drug interactions [10]
Knowledge agent	External biomedical domain knowledge; a knowledge base concerned with information needs, resources, and the mechanisms for retrieval	It does not have to be integrated into the EHR; it can be invoked as an external and independent service	Identify appropriate knowledge resources to match information needs and pass patient information to external resources to retrieve relevant information	Infobuttons [13]
Expert system	Internal knowledge + patient information	A rule base that EHR may use; a domain-independent rules engine	Provide users with an alert or a reminder when a rule is triggered	Alerts [22], reminders [71], and diagnostic decision support system [44]
Knowledge modeled as an ontology	An outside knowledge base + internal knowledge base	Incorporates an outside knowledge base into the EHR in a generic form (internal knowledge base)	For multiple purposes, including training, education, and reminders	SAGE Guideline Model [35] and OntoKBCF [27]

rates, especially in the USA, formal approaches to knowledge integration are now more important than ever. Improving knowledge integration mechanisms will play a critical role in the effectiveness of EHRs in supporting the best practices in evidence-based patient care.

Within these paradigms, we call attention to the knowledge modeled as an ontology paradigm that we believe has potential on *scalability*, *maintainability*, and *reusability* of knowledge. Along with other authors [5, 39–41], we [107] recognize the values of formally represented knowledge bases and the challenge of constructing more comprehensive knowledge bases with finer granularity [5, 42, 43]. Meanwhile, we do recognize the limitations related to the knowledge modeled as an ontology paradigm. The implementation scale

Table 9.3 Comparison of the different paradigms of knowledge integration with patient records.

Approaches	Knowledge scalability	Knowledge maintainability	Knowledge reusability	Explicitness of logic
Knowledge inscription	Easy	Difficult	Difficult	Straightforward
Knowledge catalog	Easy	Easy	Difficult	Straightforward
Knowledge agent	It depends on external resources	It depends on external resources	Difficult	A single link to access external resources
Expert system	Difficult to scale	Difficult	Difficult	Straightforward rules
Reusable knowledge	Easy	Easy	Easy	Less straightforward distributes throughout the EHR

in the examples [42, 43, 95] is limited. Only SPELTA [42] has been deployed in five institutions in Ecuador, and the systems have been tested by clinical experts and patients successfully. Larger scale implementations are necessary to confirm the advantages of SPELTA. To meet the real-world requirements, the knowledge bases will need to include more complex and comprehensive knowledge facts. Computational efficiency and other challenges may emerge at the production level. Finally, the knowledge bases that we discussed are all manually curated, a very labor-intensive process. Meanwhile, we observe that a larger proportion of publications in this paradigm (16/23, i.e., 69.57%) than the expert systems paradigm (19/45, i.e., 42.22%) emerged in the last 10 years.

Another trend is more and more deliberate considerations about interoperability and reusability emerged, especially when knowledge facts were integrated with patients' records. Several publications in the expert system category emphasize standardization, reuse, and interoperability via using the HL7 architecture [91] and OpenEHR archetype [90].

The idea of a "learning health system" (LHS) has gained increased attention in recent years [108]. The core concept of LHS is continuous quality improvement via learning from previous operational data. The LHS provides an optimistic perspective for all aspects of the healthcare arena, including incorporating knowledge with patient records. The better maintainability and reusability of the knowledge, together with needed patients' data and better informed clinical decisions, will likely contribute to achieving LHS goals.

From a historical perspective, we include knowledge inscription and knowledge catalog as two paradigms. Their implementation is straightforward, probably explaining the absence of scientific literature describing their invention. However, these paradigms exist nonetheless and differ from the other three paradigms.

9.4.3 Limitations

There are some limitations in our literature review process: We had relatively low initial agreement rates of 53% (for relevant publications) and 70% (for irrelevant publications). However, we did come close to 100% agreement rates after revising the classification and refining inclusion/exclusion criteria iteratively. Most of the disagreements were due to a lack of sufficient detail in the publications. Although the discussions could not bring in additional details from the included publications per se, each reviewer did contribute different perspectives to collect more details from additional available resources to help make consensus decisions on classification. Since we only searched PubMed, certain publications may be missing under each paradigm. For example, Sutton and Fox have integrated clinical guidelines in formal ways [109, 110]. Our search strategies, however, missed the publications. Even within PubMed, our review only reflects the results of the search strategies we used, which inevitably cannot include all possible publications. Nevertheless, our goal was to provide preliminary coverage of paradigms as a baseline, not necessarily an exhaustive review of all publications within each paradigm.

Another limitation of the work is the comparison of different paradigms in terms of scalability, maintainability, reusability, and explicitness of their logic. The current comparison was conducted via thought experiments without quantitative measures; therefore, the results are subjective and can only be used within the context for comparison among the paradigms.

9.5 Conclusion

Clinicians, informatics researchers, and patient record developers have utilized different approaches to incorporate external knowledge into displays of patient record data. The ultimate goal of the integration is to help clinicians make more informed decisions to improve care quality. We compared five integration paradigms: knowledge inscription, knowledge catalog, knowledge agent, expert systems, and knowledge modeled as an ontology. The knowledge modeled as an ontology paradigm appears to have advantages regarding reusability, scalability, and maintainability. The application of semantic web technologies to the production level, however, is still at an early stage. Nevertheless, the summarized integration paradigms can be utilized as a benchmark for future investigation, design, and improvement.

Declarations

Consent for Publication

All co-authors agree to publish the manuscript

Data sharing statement

All raw data are available per request; please email Xia Jing to request: xjing@clemson.edu.

Acknowledgment

This study was supported partially by start-up funds for Xia Jing from the Department of Public Health Sciences, College of Behavioral, Social and Health Sciences at Clemson University.

References

[1] Greenes, R., Clinical decision support: the road to broad adoption. 2nd ed. 2014, San Diego, CA: Elsevier.
[2] Kuperman, G., et al., Medication-related clinical decision support in computerized provider order entry systems: a review. JAMIA, 2007. 14: p. 29-40.
[3] Ninh, A. DocBot: a novel clinical decision support algorithm. in Conf Proc IEEE Eng Med Biol Soc 2014. 2014.
[4] Serban, A., M. Crisan-Vida, and L. Stoicu-Tivadar. Data and Knowledge in Medical Distributed Applications. in Cross-Border Challenges in Informatics with a Focus on Disease Surveillance and Utilising Big-Data. 2014.
[5] Welch, B., et al., Clinical Decision Support for Whole Genome Sequence Information Leveraging a Service-Oriented Architecture: a Prototype, in AMIA 2014. 2014. p. 1188-1197.

[6] Ellsworth, M., et al., Point-of-Care Knowledge-Based Resource Needs of Clinicians: A Survey from a Large Academic Medical Center. App Clin Inform, 2015. 6: p. 305-317.
[7] Smith, R., What clinical information do doctors need? BMJ, 1996. 313: p. 1062-1068.
[8] Mickan, S. and D. Askew, What sort of evidence do we need in primary care? BMJ, 2006. 332: p. 619-620.
[9] Greenes, R. A., et al., Clinical decision support models and frameworks: Seeking to address research issues underlying implementation successes and failures. J Biomed Inform, 2018. 78: p. 134-143.
[10] Guise, J., L. Savitz, and C. Friedman, Mind the gap: putting evidence into practice in the era of learning health systems. . Journal of general internal medicine, 2018. 33(12): p. 2237-9.
[11] Alavi, M. and D. Leidner, Knowledge management and knowledge management systems: Conceptual foundations and research issues. MIS quarterly, 2001: p. 107-36.
[12] Floridi, L., Semantic information and the network theory of account. Synthese, 2012. 184(3): p. 431-54.
[13] Tricco, A., et al., PRISMA Extension for Scoping Reviews (PRISMAScR): Checklist and Explanation. Ann Intern Med, 2018. 169: p. 467–473.
[14] Moher, D., et al., Preferred reporting items for systematic reviews and meta-analyses: the PRISMA statement. Bmj, 2009. 339: p. b2535.
[15] Teich, J., et al., Effects of computerized physician order entry on prescribing practices. Arch Intern Med, 2000. 160: p. 2741-2747.
[16] Teich, J., J. Schmiz, and E. O'Connell, An information system to improve the safety and efficiency of chemotherapy ordering, in Proc AMIA Annu Fall Symp. 1996. p. 498-502.
[17] Cimino, J. J., An integrated approach to computer-based decision support at the point of care. Trans Am Clin Climatol Assoc, 2007. 118: p. 273-288.
[18] Rosenbloom, S., et al., Effect of CPOE user interface design on user-initiated access to educational and patient information during clinical care. J Am Med Inform Assoc, 2005. 12: p. 458-473.
[19] Maviglia, S., et al., KnowledgeLink: impact of context-sensitive information retrieval on clinicians' information needs. J Am Med Inform Assoc, 2006. 13: p. 67-73.

[20] Del Fiol, G., et al., Disseminating Context-Specific Access to Online Knowledge Resources within Electronic Health Record Systems, in MEDINFO 2013. 2013. p. 672-676.

[21] OpenInfobutton Project Team. OpenInfobutton Project. 2015 [cited 2020 Aug 29th]; Available from: https://www.openinfobutton.org/.

[22] HL7 International. HL7 Version 3 Standard: Context Aware Knowledge Retrieval Application ("Infobutton"), Knowledge Request, Release 2. [cited 2020 Dec 2nd]; Available from: https://www.hl7.org/implement/standards/product_brief.cfm?product_id=208.

[23] Wright, A., et al., Development and evaluation of a comprehensive clinical decision support taxonomy: comparison of front-end tools in commercial and internally developed electronic health record systems. J Am Med Inform Assoc, 2011. 18: p. 232-242.

[24] van den Branden, M., et al., Integrating case-based reasoning with an electronic patient record system. Artif Intell Med, 2011. 51: p. 117-123.

[25] Buchanan, B. and E. Shortliffe, Rule Based Expert Systems: The MYCIN Experiments of the Stanford Heuristic Programming Project. 1984, Reading, MA: Addison-Wesley.

[26] Gottesman, O., et al., The CLIPMERGE PGx Program: clinical implementation of personalized medicine through electronic health records and genomics-pharmacogenomics. Clin Pharmacol Ther, 2013. 94: p. 214-217.

[27] Kam, H., et al., Integration of heterogeneous clinical decision support systems and their knowledge sets: feasibility study with Drug-Drug Interaction alerts, in AMIA 2011. 2011. p. 664-673.

[28] Cho, I., et al., Design and implementation of a standards-based interoperable clinical decision support architecture in the context of the Korean EHR. Int J Med Inf, 2010. 79: p. 611-622.

[29] Berlin, A., M. Sorani, and I. Sim, A taxonomic description of computer-based clinical decision support systems. J Biomed Inform, 2006. 39: p. 656-667.

[30] Hripcsak, G., Writing Arden Syntax medical logic modules. Computers in Biology and Medicine, 1994. 24(5): p. 331-363.

[31] Zhou, L., et al., A study of diverse clinical decision support rule authoring environments and requirements for integration. BMC Med Inform Decis Mak, 2012. 12: p. 128.

[32] Sittig, D., et al., The state of the art in clinical knowledge management: An inventory of tools and techniques. Int J Med Inform, 2010. 79: p. 44-57.

[33] Sittig, D., et al., Grand challenges in clinical decision support. J Biomed Inform, 2008. 41: p. 387-392.
[34] Gruber, T., What is an ontology? Knowledge Acquisition, 1993. 5: p. 199-220.
[35] Jing, X., et al., Ontology-based knowledge base model construction-OntoKBCF. Studies in health technology and informatics, 2007. 129: p. 785-790.
[36] Jing, X., et al., Ontological Knowledge Base Model for Cystic Fibrosis –BioPortal. 2012. 2013.
[37] Jing, X., et al., Incorporating personalized gene sequence variants, molecular genetics knowledge, and health knowledge into an EHR prototype based on the Continuity of Care Record Standard. J Biomed Inform, 2012. 45: p. 82-92.
[38] Jing, X., et al., Integration of an OWL-DL knowledge base with an EHR prototype and providing customized information. J Med Syst, 2014. 38.
[39] Sittig, D., et al., Comparison of clinical knowledge management capabilities of commercially-available and leading internally-developed electronic health records. BMC Med Inform Decis Mak, 2011. 11.
[40] Wright, A., et al., Creating and sharing clinical decision support content with Web 2.0: Issues and examples. J Biomed Inform, 2009.42: p. 334-346.
[41] Sittig, D., et al., The state of the art in clinical knowledge managment: an inventory of tools and techniques. Int J Med Inform, 2010. 79: p. 44-57.
[42] Robles-Bykbaeva, V., et al., An Ecosystem of Intelligent ICT Tools for Speech-Language Therapy Based on a Formal Knowledge Model, in MEDINFO 2015. 2015. p. 50-54.
[43] Wang, H., et al., Creating personalised clinical pathways by semantic interoperability with electronic health records. Artif Intell Med, 2013. 58: p. 81-89.
[44] Rahman, Y., et al., e-clinic: an electronic triage system for the management of type 2 Diabetes Mellitus. Stud Health Technol Inform, 2004. 107: p. 246-250.
[45] Tu, S., et al., The SAGE Guideline Model: achievements and overview. J Am Med Inform Assoc, 2007. 14: p. 589-598.
[46] Groot, P., et al., Using model checking for critiquing based on clinical guidelines. Artif Intell Med, 2009. 46: p. 19-36.

[47] Tarczy-Hornoch, P., et al. Meeting clinician information needs by integrating access to the medical record and knowledge resources via the Web. in Proc AMIA Symp. 1997.

[48] Fuller, S., et al., Integrating knowledge resources at the point of care: opportunities for librarians. Bull Med Libr Assoc, 1999. 87: p. 393-403.

[49] Godin, P., et al., New paradigms for medical decision support and education: the Stanford Health Information Network for Education. Top Health Inf Manage, 1999. 20: p. 1-14.

[50] Bountis, C. and J. Kay, An integrated knowledge management system for the clinical laboratories: an initial application of an architectural model. Stud Health Technol Inform, 2002. 90: p. 562-567.

[51] Wu, R., W. Peters, and M. Morgan, The next generation of clinical decision support: linking evidence to best practice. J Healthc Inf Manag, 2002. 16: p. 50-55.

[52] Moser, W., et al., Integrated development of a knowledge-based CPR system for quality assurance in diabetes outpatient clinics, in Medinfo. 1995. p. 236-239.

[53] Johansson, B., et al., Database and knowledge base integration–a data mapping method for Arden Syntax knowledge modules. Methods Inf Med, 1996. 35: p. 302-308.

[54] Eich, H., et al., Integration of a knowledge-based system and a clinical documentation system via a data dictionary. Stud Health Technol Inform, 1997. 43: p. 431-435.

[55] Henry, S., et al., A template-based approach to support utilization of clinical practice guidelines within an electronic health record. J Am Med Inform Assoc, 1998. 5: p. 237-244.

[56] Porcelli, P. and D. Lobach, Integration of clinical decision support with on-line encounter documentation for well child care at the point of care, in Proc AMIA Symp. 1999. p. 599-603.

[57] Goldstein, M., et al., Implementing clinical practice guidelines while taking account of changing evidence: ATHENA DSS, an easily modifiable decision-support system for managing hypertension in primary care, in Proc AMIA Symp. 2000. p. 300-304.

[58] Müller, M., et al., Integrating knowledge based functionality in commercial hospital information systems. Stud Health Technol Inform, 2000. 77: p. 817-821.

[59] Ray, H., et al., Providing context-sensitive decision-support based on WHO guidelines, in Proc AMIA Symp. 2002. p. 637-641.

[60] Adams, W., et al., TLC-Asthma: an integrated information system for patient-centered monitoring, case management, and point-of-care decision support, in AMIA Annu Symp Proc. 2003. p. 1-5.
[61] Tange, H., et al., Towards a PropeR combination of patient records and protocols. Int J Med Inform, 2003. 70: p. 141-148.
[62] Ciccarese, P., et al., A guideline management system. Stud Health Technol Inform, 2004. 107: p. 28-32.
[63] Frank, L., et al., Using BPI and emerging technology to improve patient safety. J Healthc Inf Manag, 2004. 18: p. 65-71.
[64] Goldstein, M., et al., Translating research into practice: organizational issues in implementing automated decision support for hypertension in three medical centers. J Am Med Inform Assoc, 2004. 11: p. 368-376.
[65] Holbrook, A., et al., Individualized electronic decision support and reminders can improve diabetes care in the community, in AMIA Annu Symp Proc. 2005. p. 982.
[66] Hulse, N., G. Del Fiol, and R. Rocha, Development and validation of XML-based calculations within order sets, in AMIA Annu Symp Proc. 2005. p. 360-364.
[67] Shah, N., et al., Improving override rates for computerized prescribing alerts in ambulatory care, in AMIA Annu Symp Proc. 2005. p. 1110.
[68] Conforti, D., et al., HEARTFAID: A knowledge based platform of services for supporting medical-clinical management of heart failure within elderly population. Stud Health Technol Inform, 2006. 121: p. 108-125.
[69] Huang, C., et al., Implementation of virtual medical record object model for a standards-based clinical decision support rule engine, in AMIA Annu Symp Proc. 2006. p. 958.
[70] Panzarasa, S., et al., Workflow technology to enrich a computerized clinical chart with decision support facilities, in AMIA Annu Symp Proc. 2006. p. 619-623.
[71] Postilnik, A., et al., Smart Form framework as a foundation for clinical documentation platform, in AMIA Annu Symp Proc. 2006. p. 1067.
[72] Linder, J., et al., Clinical decision support to improve antibiotic prescribing for acute respiratory infections: results of a pilot study, in AMIA Annu Symp Proc. 2007. p. 468-472.
[73] Schnipper, J., et al., "Smart Forms in an Electronic Medical Record: documentation-based clinical decision support to improve disease management." J Am Med Inform Assoc, 2008. 15: p. 513-523.

[74] Wright, A. and D. Sittig, SANDS: an architecture for clinical decision support in a National Health Information Network. J Biomed Inform, 2008. 41: p. 962-981.

[75] Liu, D., et al., Analysis of outcomes in radiation oncology: an integrated computational platform. Med Phys, 2009. 36: p. 1680-1689.

[76] Steurbaut, K., et al., Use of web services for computerized medical decision support, including infection control and antibiotic management, in the intensive care unit. J Telemed Telecare, 2010. 16: p. 25-29.

[77] Bernonille, S., et al., Three different cases of exploiting decision support services for adverse drug event prevention. Stud Health Technol Inform, 2011: p. 180-188.

[78] Duke, J. and D. Bolchini, A successful model and visual design for creating context-aware drug-drug interaction alerts, in AMIA Annu Symp Proc. 2011. p. 339-348.

[79] Slonim, N., et al., Knowledge-analytics synergy in Clinical Decision Support. Stud Health Technol Inform, 2012. 180: p. 703-707.

[80] Zhang, M., et al., Enabling cross-platform clinical decision support through Web-based decision support in commercial electronic health record systems: proposal and evaluation of initial prototype implementations, in AMIA Annu Symp Proc. 2013. p. 1558-1567.

[81] Archer, M., et al., Development of an Alert System to Detect Drug Interactions with Herbal Supplements using Medical Record Data, in AMIA Annu Symp Proc. 2014. p. 249-255.

[82] Bouzguenda, L. and M. Turki, Designing an architectural style for dynamic medical Cross-Organizational Workflow management system: an approach based on agents and web services. J Med Syst, 2014. 38: p. 32.

[83] Goldspiel, B., et al., Integrating pharmacogenetic information and clinical decision support into the electronic health record. J Am Med Inform Assoc, 2014. 21: p. 522-528.

[84] Kunhimangalam, R., S. Ovallath, and P. Joseph, A clinical decision support system with an integrated EMR for diagnosis of peripheral neuropathy. J Med Syst, 2014. 38: p. 38.

[85] Rasmussen-Torvik, L., et al., Design and anticipated outcomes of the eMERGE-PGx project: a multicenter pilot for preemptive pharmacogenomics in electronic health record systems. Clin Pharmacol Ther, 2014. 96: p. 482-489.

[86] Chackery, D. G., et al., Integrating Clinical Decision Support into EMR and PHR: a Case Study Using Anticoagulation. Stud Health Technol Inform, 2015. 208: p. 98-103.
[87] Moriarity, A., et al., The effect of clinical decision support for advanced inpatient imaging. J Am Coll Radiol, 2015. 12: p. 358-363.
[88] Gaebel, J., M. Cypko, and H. Lemke, Accessing Patient Information for Probabilistic Patient Models Using Existing Standards. Stud Health Technol Inform, 2016. 223.
[89] Syomova, I., E. Bologvaa, and S. Kovalchuka. Towards Infrastructure for Knowledge-based Decision Support in Clinical Practice. in Conference on ENTERprise Information Systems / International Conference on Project MANagement / Conference on Health and Social Care Information Systems and Technologies, CENTERIS / ProjMAN / HCist 2016. 2016.
[90] Kopanitsa, G., Integration of Hospital Information and Clinical Decision Support Systems to Enable the Reuse of Electronic Health Record Data. Methods Inf Med, 2017. 56(3): p. 238-247.
[91] Seitz, M. W., S. Listl, and P. Knaup, Development of an HL7 FHIR Architecture for Implementation of a Knowledge-based Interdisciplinary EHR. Stud Health Technol Inform, 2019. 262: p. 256-259.
[92] Müller, M., et al., Towards integration of clinical decision support in commercial hospital information systems using distributed, reusable software and knowledge components. Int J Med Inform, 2001. 64: p. 369-377.
[93] Das, A., et al., Knowledge-based method for building patient decision-analytic tools, in AMIA 2006. 2006. p. 175-179.
[94] Klimov, D. and Y. Shahar, Intelligent querying and exploration of multiple time-oriented medical records. Stud Health Technol Inform, 2007. 129: p. 1314-1318.
[95] Adlassnig, K. and A. Rappelsberger, Medical Knowledge Packages and their Integration into Health-Care Information Systems and the World Wide Web. Stud Health Technol Inform, 2008. 136: p. 121-126.
[96] Fernández-Breis, J., et al., Leveraging electronic healthcare record standards and semantic web technologies for the identification of patient cohorts. J Am Med Inform Assoc, 2013. 20: p. 288-296.

[97] Garcia, D., et al., Method to integrate clinical guidelines into the electronic health record (EHR) by applying the archetypes approach. Stud Health Technol Inform, 2013. 192: p. 871-875.

[98] Hulse, N., J. Long, and C. Tao, Combining infobuttons and semantic web rules for identifying patterns and delivering highly-personalized education materials, in AMIA Annu Symp Proc. 2013. p. 638-647.

[99] Liaw, S., J. Taggart, and H. Yu, EHR-based disease registries to support integrated care in a health neighbourhood: an ontology-based methodology. Stud Health Technol Inform, 2014. 205.

[100] Lupşe, O. and L. Stoicu-Tivadar, Pediatrics prescriptions with ontologies and treatment suggestions. Stud Health Technol Inform, 2014. 205: p. 403-407.

[101] Shemeikka, T., et al., A health record integrated clinical decision support system to support prescriptions of pharmaceutical drugs in patients with reduced renal function: design, development and proof of concept. Int J Med Inform, 2015. 84: p. 387-395.

[102] Rosier, A., et al., Remote Monitoring of Cardiac Implantable Devices: Ontology Driven Classification of the Alerts. Stud Health Technol Inform, 2016. 221: p. 59-63.

[103] Zhang, Y., et al., Design and Development of a Sharable Clinical Decision Support System Based on a Semantic Web Service Framework. J Med Syst, 2016. 40: p. 118.

[104] Shaban-Nejad, A., et al., PopHR: a knowledge-based platform to support integration, analysis, and visualization of population health data. Ann N Y Acad Sci, 2017. 1387(1): p. 44-53.

[105] Chelsom, J. and N. Dogar, Linking Health Records with Knowledge Sources Using OWL and RDF. Stud Health Technol Inform, 2019. 257: p. 53-58.106. Roehrs, A., et al., Toward a Model for Personal Health Record Interoperability. IEEE J Biomed Health Inform, 2019. 23(2): p. 867-873.

[106] Roehrs, A., et al., Toward a Model for Personal Health Record Interoperability. IEEE J Biomed Health Inform, 2019. 23(2): p. 867-873.

[107] Jing, X., Constructing a bio-health knowledge base for access via a standardised EHR prototype. 2010, University of Salford: Salford, UK.

[108] Friedman, C., et al., Toward a science of learning systems: a research agenda for the high-functioning Learning Health System. J Am Med Inform Assoc, 2015. 22(1): p. 43-50.

[109] Sutton, D. R. and J. Fox, The syntax and semantics of the PROforma guideline modeling language. J Am Med Inform Assoc, 2003. 10(5): p. 433-43.
[110] Fox, J., et al., OpenClinical.net: A platform for creating and sharing knowledge and promoting best practice in healthcare. Computers in Industry, 2015. 66: p. 63-72.

Dr. Xia Jing is a health informatics scholar educated in medical informatics (MD), health informatics (Ph.D.), and clinical informatics (postdoctoral training). She is an associate professor in the Department of Public Health Sciences at Clemson University, SC, USA. Her main research interests involve exploring clinicians' research hypothesis generation process via secondary data analytic tools; she and her collaborators also explore better management and maintenance of clinical decision support rules in primary care settings, especially under-resourced settings. She teaches various undergraduate and doctoral health informatics courses. She also serves communities mainly via the American Medical Informatics Association and American Heart Association.

Dr. James Cimino is currently Distinguished Professor of Medicine and Director of the Informatics Institute at the University of Alabama-Birmingham School of Medicine. Past appointments include a Professor of Biomedical Informatics and Medicine at Columbia University and Chief of the Laboratory for Informatics Development at the NIH Clinical Center and the National Library of Medicine. He has over 35 years of experience in clinical informatics research, building information systems, teaching informatics and medicine, and caring for patients. Principle research areas in diagnostic expert systems, desiderata for controlled terminologies, mobile and Web-based clinical information systems, and context-aware decision support called "infobuttons".

Dr. Xuequn Pan is an assistant professor in the Department of Public Health at Changsha Medical University, Hunan, China. She was a clinical informatics postdoctoral fellow at the National Institutes of Health. She received her Ph.D. in Information Science from the University of North Texas. Her research interests include medical informatics, clinical decision support systems, evidence-based medicine, and research methodology in medical informatics.

Dr. Shijuan Li is an associate professor in the Department of Information Management, Peking University, Beijing, China. She holds a Ph.D. in Health Informatics from the University of Salford, UK. Her research areas include health informatics, digital reading promotion, and library information service. She is particularly interested in how innovative information and communications technologies (ICTs) influence individuals' activity and behavior and assist them in making informed decisions.

10

Semantic Checking of Information Support for Heterogeneous Resources of Train Speed Restrictions by Ontological Means

V. Shynkarenko and L. Zhuchyi

Ukrainian State University of Science and Technologies, Dnipro, Ukraine
E-mail: shinkarenko_vi@ua.fm, larisa_zhuchiy@ukr.net

Abstract

The problem of harmonization and validation of train speed restrictions represented in sources of various formats and railway transport instructions is investigated. Harmonization is carried out by ontological means. The ontology integrates information from the following sources: databases, station drawings, and natural language regulations. The ontologies of the data source and the railway model are enriched with concepts for classifying tables of speed restriction warnings and track defects. Methods and software have been developed for processing tabular format data using automated tools for data wrangling and extraction from sources of various types. Implicit knowledge in the tabular representation of speed restrictions is used to validate these tables and classify them. Semantic checking is carried out using formalized regulations for the maintenance of the track, track works, and automated issuance of speed restrictions warnings. The usage of the ontology will improve the safety of train traffic control by checking the compliance of the speed restrictions and track defects, timely cancellation of restrictions, and checking the validity of manual data entry. The method is illustrated by the integration of data sources such as railway switch and track inspection records and lists, warning records books, and forms for train drivers.

Keywords: Conceptual model, knowledge base, natural language, OWL, ontology, railway, semantic checking, speed restriction.

10.1 Introduction

Temporary speed restrictions warnings transmitted to the locomotive driver must correspond to real railway track defects and comply with the regulations of its maintenance.

Possible inconsistencies may be due to the silos of the information support of the subsystems of the railway track facilities and train traffic control. In addition, to carry out railway track maintenance work, the representative of the track maintenance must determine the speed restriction necessary for the work and submit a request manually.

One way to evolve railway information systems is to integrate them to infer new knowledge and improve safety. To do this, it is necessary to link the data with semantic relationships using a common vocabulary.

10.2 Problem Statement and Purpose

The link between the information systems of the railway track facilities and traffic control subsystems regarding the railway track defects is carried out by transmitting information manually about the defect type and speed restriction by different transportation process participants. The safety level can be increased by ontological means.

Let's Let us consider the existing sequence of processing information about the railway track defect:

- When it is identified, the maintenance worker determines the speed restriction and make records:
  - About the railway track defect in records book of railway tracks inspections PU-28 (RBRSI) according to the current maintenance of railway tracks rules (CMRTR);
  - About the speed restriction in the records book of speed restriction warnings PU-84 (RBSRW) according to the train management rules (TMR) and the corresponding databases;.

- The railway track maintenance worker transmits information about the type and location of the defect and the speed restriction to the train operations centre center worker by telephone message, according to which the train operations centre center worker makes an entry in the RBSRW of the train operations centre center according to the TMR;.
- The train operations centre center worker verifies the speed of the RBSRW and the driver handout form of speed restriction warnings (DHFSRW) before handing out to the driver according to the guidelines for automated issuance and cancellation of warnings (GAICW);).
- Train operations centre center worker gives the FVMPOS speed restriction to the train driver according to the technical operation rules of railways railways (TERR);
- When repairing a defect track maintenance, the worker makes a record about of the date of maintenance in the RBRTI and the corresponding database according to the CMRTR;
- An entry about the cancelation of the speed restriction in the RBSRW PU-84 and the corresponding database according to the TMR;
- The track maintenance worker transmits information about the cancellation of the speed restriction to the train operations centre center worker, according to which the train operations centre center worker makes a record in the RBSRW of the train operations centre center according to the TMR.

Such a sequence is complex and there may be inconsistencies are possible, because of the manual chain of information transfer, which occasionally reveals train operations centre center workers. A way to make things better is proposed.

This work aims to improve the consistency of data of various subsystems and technological chains of railway transport without changing the databases and software of the subsystems of the railway track and train traffic control.

The goal is achieved by performing a semantic checking of speed restrictions for railway infrastructure elements defects based on the integration of heterogeneous resources related to railway track defects.

10.3 Related Works

In Europe, ontologies that have transport defects concepts of various types (and not only) receive a lot of attention. Ontology development methods include the use of higher- level ontologies for the integration of heterogeneous sources [15, 20, 15] and SWRL rules [21, 12, 21].

In contrast to the approaches [12, 15, 20, 21], compositions of relations are used to develop this ontology. They have found applications in the domains of law, medicine, and public procurement. Let's Let us consider these two aspects in modern ontological research.

10.3.1 Ontological modelling in transport, taking into account defects and speed restrictions

The first exclusive railway ontology [12] was developed within the framework of the IntegRail project [24]. The use of ontology made it possible to integrate wheel impact load measurement systems and hot axle box detectors to determine the highest priority train maintenance.

Rail core ontology [20], includes such railway subsystems as infrastructure condition, fleet condition, and infrastructure information, and through the inference allows one to determine not only a defective infrastructure element but also dependent elements.

The transport disruption ontology [2] includes a description of defects of all modes of transport, including rail, to identify links with a planned trip and is based on such high-level ontologies such as Event event ontology and friend-of-a-friend ontology. By integrating information about planned trips, bus delays, accidents, and a map, it became possible to compare the route of the trip with the places of transport failure.

The system of intelligent speed adaptation (ISA) was developed in [21]. The system collects information from cameras and GPS and executes SPARQL queries to identify speeding violations by comparing vehicle speed and speed restrictions from the knowledge base. The knowledge base includes map ontology, control ontology, car ontology, and SWRL rules and was developed in Protégé.

OWL Ontologies [15] for advanced driver assistance systems and ISA systems were developed using the Protéegée ontology editor as part of the ONTOTRAFFIC project to ensure interoperability of applications (infrastructure and vehicle): vehicle security, road and traffic security, meteorological, users' users' profiles, and travel. To integrate models, top-level ontologies are used as the basis for five ontologies. Three types of speed restrictions are taken into account: permanent, temporary, and dynamic.

In the domain of the aviation industry, an ontology [11] was developed in Protégé to achieve data consistency when performing reliability analysis in the manufacture of the engines and improving the safety of aircraft operation. The reason for possible inconsistencies is the manual entry and free interpretation of the data by different participants performing the reliability analysis. Each defect is associated with a cause, an element of the engine, and can be the cause of emergencies.

10.3.2 Ontological modelling of computer, medical, and construction domains, taking into account defects

Ontology of software defects, errors, and failures [6] is based on unified foundational ontology [7] and reuses software process ontology [13]. The ontology is developed on the basis of generally accepted ISO standards as a unified scheme for representing software defects. Defects are classified by source: developer or user. Logical definitions for the concepts of "failure", "defect", "error", and others have been developed.

The ontology [3] makes it possible to prevent failures by choosing the most appropriate plan for building a structure in the building domain. The ontology is made in Protégé, which contains SWRL rules and allows for modelling the relationships between risk factors. Since some risk factors lead to others (for example, overloading a strut will lead to its destruction), the logical inference of one SWRL rule is a prerequisite for the inference of other rules.

As part of the Heartfaid [10] project, Heart heart failure ontology [22] was manually developed in Protégé, which, in addition to heart failures, also contains tests and treatment procedures and establishes a unified vocabulary for representing the heart diseases. The class of tests includes the values of the heart characteristics that lead to diseases, and the rules – the definition of a patient having a heart disease.

10.3.3 Application of the relations composition in ontology development

Lightweight ontologies include simple classes and instances connected by simple relations, while heavyweight ontologies include constructions in the form of relations compositions and logical definitions. One of the ways to develop heavy ontologies for railway transport is presented in [18] and [19]. There are some modern studies using relationship compositions in ontology development.

Rail core ontology [20] includes a property chain to represent different levels of rail network aggregation. For example, if a railway line is electrified and includes multiple nodes, then those nodes inherit the electrification characteristic. The ontology includes such aggregation levels as path, line, and network.

In [1], the ontology judicial ontology library was developed on the basis of legal texts for checking the consistency of contracts, tools for legal and judicial analysis, etc., and includes two modules: Core core Ontology ontology (linked to the Legal legal Knowledge knowledge Interchange interchange Format format LKIF-Core ontology) and Domain Ontology. The ontology can be used to search for relevant precedents, validate sentences, and derive appropriate legal norms. The texts are annotated by specialists using the Norma-Editor, and the ontology is made in the Protégé editor in OWL.

The OWL ontology linked open Tenders Electronic Daily LOTED2 represents open government data tenders in a legal context and is developed by Protégé, and which is also linked with LKIF and good relations ontologies. It is an intermediate layer for users of Tenders Electronic Daily, who may experience difficulties due to the presentation of a large number of characteristics.

One of the features of property chains is that they can be "nested". They are considered in ontologies [4] and LKIF [9]. Rinke Hoekstra developed a design pattern [24] that includes a property chain for *n*-ary relations in OWL.

"Constituent parts" of complex relations can be both relations of one and several ontologies. In [8], property chains are used to link and make inferences using the Systematized Nomenclature of Medicine—Clinical Terms (SNOMED-CT) and Functional Basis Ontology (FBO), which provide support in the early stages of medical device development within the concept ideation framework for medical device design.

10.4 Modular Railway Track Defect Ontology

A modular ontology has been developed to link regulations and data – formalized restrictions from CMRTR, track works safety arrangements rules (TWSAR), IDP, and GAICW.

In a separate module, not only rules but also vocabularies are allocated, but also vocabularies to represent the railway infrastructure and data sources. The presence of a separate vocabulary allows one to use the same concept to form different rules; for example, in this paperchapter, we check the

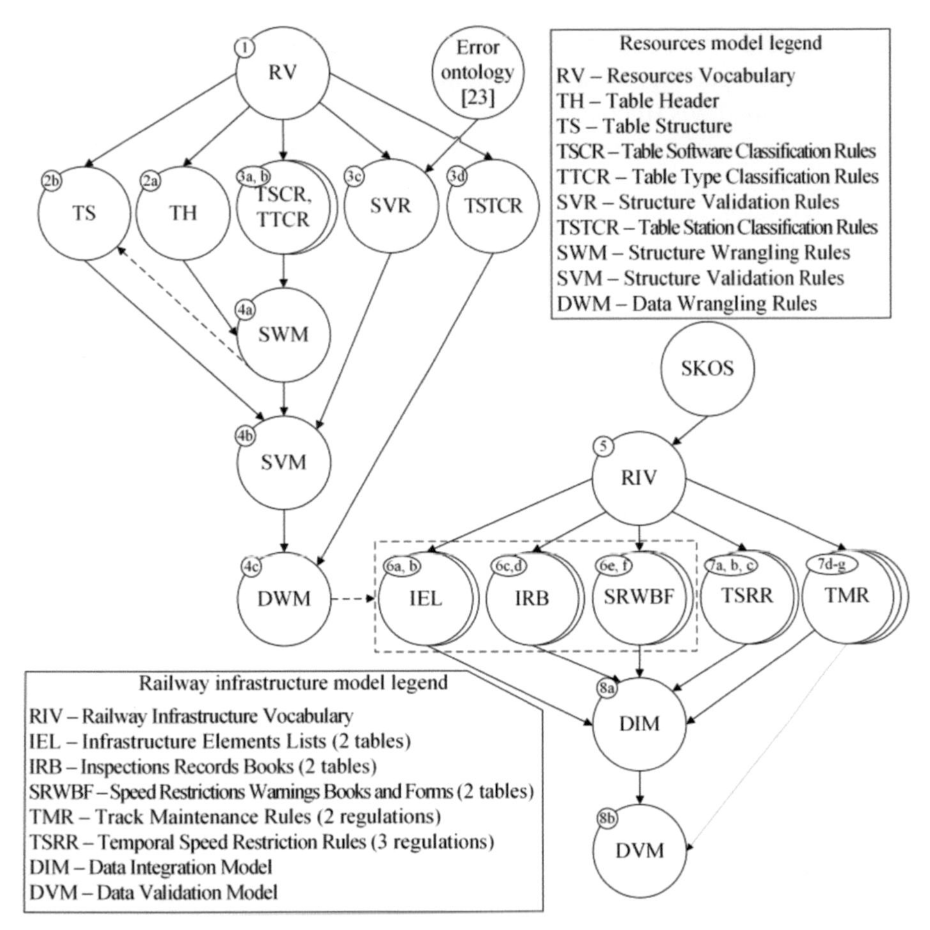

Figure 10.1 The modularity of the railway track defect ontology.

consistency of the railway trains permitted speed on the railway track and CMRTR.

Figure 10.1 shows a modular ontology of railway track defects and its formation, where each node corresponds to an ontology module.

Consider the structure of the ontology and its formation. The following models have been developed:

- The abstract resource model is a vocabulary that combines concepts to represent the structure of heterogeneous speed restriction sources;.
- A concrete resources model of the first level with data (models 2a, and 2b) is developed according to [17] and includes instances representing

the header of data source tables (model 2a) and for instances representing the structure of tables generated by data wrangling and data extraction methods (model 2b);).

- A concrete resources model of the first level with rules (models 3b - and 3c) is developed according to [16] and includes rules for classifying tables by program (model 3a), by type of table (model 3b), by station (model 3d), and data validation rules (3c model);).
- A concrete resources model of the second level (models 4a, and 4b) is used for structure validation (model 4a) and for transforming table data (model 4b);.
- The railway infrastructure abstract model is a vocabulary that combines concepts to represent these disparate sources of speed restrictions;.
- A concrete railway infrastructure model of first level with data (models 6a -–6f) includes facts from each of the following tables: RTL (model 6a), RSL (model 6b), RBRTI (models 6c), RBRSI (model 6d), DHFSRW (model 6e), and RBSRW (model 6f);).
- A concrete railway infrastructure model of the first level with rules (models 7a-–7e) is developed according to [16] and includes rules from each of the following instructions: GAICW (model 7a), TMR (model 7b), TERR (model 7c), CMRTR (model 7d), and TWSAR (model 7e);).
- A concrete railway infrastructure model of the second level (models 8a, and 8b) is intended for data integration (model 8a) and checking the consistency of data and rules (model 8b).

The ontology is formed in the following sequence:

- A vocabulary of tables and railway infrastructure is developed when performing ontologies RV, and RIV (models 1, and 5);).
- Table structure validation rules are formalized when developing SVR ontologies (the concept of classification is used in the sense of ontologies: the reasoner determines the type of an instance based on its properties) and classification rules – TSCR, TTCR, and TSTCR tables (models models 3a-–3d);).
- The table header model is developed during the making of the SWM ontology (models 4a);).
- Classification of tables according to type and program is performed by the reasoner in Protégé for the SWM ontology (model 4a);).
- The table structure data is retrieved in OpenRefine, where the TS ontology is generated (model 2b);).

- Table validation is done by reasoner in Protégé for SVM ontology (model 4b), and table classification by station – DWM ontology (model 4c););).
- Data from tables is extracted in OpenRefine, in which IEL, IRB, and SRWBF ontologies are generated;.
- The rules of the railway regulations are formalized when developing the TMR, and TSRR ontologies (models 7a-–7g););).
- Railway infrastructure instances (models 6a-–6f) of IEL, IRB, and SRWBF ontologies are loaded into a data integration ontology DIM (model 8a););).
- Instances from different tables according to the regulations of TMR (models 7d, and 7e) and TSRR (models 7a-–7c) are linked in the DIM ontology (model 8a););).
- Checking the consistency of the TMR ontology constraints (models 7f, and 7g) and the data is performed by the reasoner in Protégé for the DVM ontology.

To perform checking of the railway track maintenance technology in the railway infrastructure ontology, it is necessary to link instances from different tables:

- it is necessary to link speed from the RBSRW table and the defect value of the RBRTI and RBRSI tables to check the compliance of speed and defect;
- it is necessary to link the speed from the RBSRW and DHFSRW tables to validate of the speed restriction value which was handed out to the train driver;
- it is necessary to link cancellation from the RBRSI and the value of the repair in the table and RBRSI to validate the cancellation of the warning.

The aggregation of instructions of three levels "TERR TMR GAICW" and "TERR CMRTR and TWSAR" was done.

There are no instructions about working with DHFSRW and RBSRW and the value of defects in the TERR, since the it does not specify railway track defects and their processing. The TERR level is represented in the train sheet, where the tracks are highlighted in different colours, on which there are conditions or speed restrictions for trains. At the railways, train dispatchers' level information about the value of railway track settlement is not relevant.

There are no guidelines about checking the speed of DHFSRW and RBSRW in the TMR since it does not specify the work of all railway transport workers.

10.5 Implementation of Railway Track Defect Ontology

Consider the features of the resources and railway track defect ontologies implementation.

10.5.1 Resources ontology

An abstract ontology of resources corresponds to a vocabulary for representing table structures. Conceptualization is performed according to [17]. Vocabulary makes it possible to develop rules for the classification and validation of the RTL, RSL, RBRTI, RBRSI, DHFSRW, and RBSRW tables.

The source ontology framework is enriched with classes of these tables and station attributes. As well as relationships for linking tables to tables and tables to a station attribute.

The concrete model of resources with data is made in the form of two ontologies. Instances of ontology TH are entered into Protégé manually and represent the description of the header and programs of the tables.

Instances of the TS ontology are generated in OpenRefine and, in addition to the header and programs, also represent the values of the tables. Consider the RTL table without rail type values. In OpenRefine, a facet is performed on empty cells and an error value is assigned to empty cells.

A concrete model of resources with rules is made in the form of four models for the table classification: TSCR, TTCR, TSTCR, and SVR validation.

Tables can be classified according to three criteria, : according to the program for their development (AutoCAD), according to the type of table (RTL). and according to its belonging to the station (RTL indicative station).

Thus, a poly-hierarchy is obtained, which is an anti-pattern in the development of ontologies. In an ontology, classes are related to each other not by a strong sub-classification relationship, but by logical definitions.

The TSCR ontology contains logical class definitions:

```
drawingTable equivalentTo table and (outputOf
some autocad).
```

The TTCR ontology has logical class definitions definitions:

```
- railwaySwitchListTable Equivalent to
drawingTable and (hasPart some switchListHeader)
```

```
- switchListHeader Equivalent To tuple and
(hasElement some amountIdentifier) and (hasElement
some frogTypeIdentifier) and (hasElement some
railTypeIdentifier), and (hasElement some
sidednesIdentifier) and (hasElement some
switchTypeIdentifier).
```

The TSTCR ontology has logical class definitions:

```
- railwaySwitchListTable_railwayStation Equivalent
To railwaySwitchListTable and (hasStationAttribute
some stationAttribute_railwayStation)
- stationAttribute_railwayStation Equivalent to
stationAttribute and (isA some {"railway station"}).
```

Wherein the station attribute for tables containing information about the infrastructure elements owners is extracted from them at the stage of transforming its structure and linked to all tables associated with them using property chain and inverse relationship ń“hasStationAttributesż..”

```
isStationAttributeOf SuperPropertyOf
isStationAttributeOf o hasAssociatedTable.
```

The SVR ontology has rules like this:

```
drawingTable(?x), hasPart(?x, ?y), hasElement(?y,
?z), isA(?z, "error") -> hasError(?x, "table has
empty cell") .
```

The concrete ontology of data resources is made in the form of three models: SWM, SVM, and DWM. Consider inference path on a pair of tables containing information about the owners of infrastructure elements and RTL.

Only table instances of headers and TSCR rules are imported into SWM and TTCR and table instances are classified first as “drawingTable” because they are associated with the AutoCAD program, and then as a “railwaySwitchListTable” and “titleBlockTable,” because these tables have corresponding headers.

Using rules from SVR, table instances generated by OpenRefine are checked against their table structure model. Let’s Let us go back to the example of a corrupted table that has empty cells. Since there are no rail type values in the column, its tuple is associated with cells that have the “error” property and the table is assigned the error property “table has empty cell”.

In DWM, tables are classified by the station. A table containing information about the owners of infrastructure elements, to the class of certain station tables "titleBlockTable_railwayStation," because in the TS model, it is already linked to the station attribute "railwayStation". After linking a table containing information about the infrastructure elements owners to the RTL (RSL) table with the relationship "hasAssociatedTable", RTL is also associated with the station attribute by property chain, and it is classified by the reasoner as RTL of some station "railwaySwitchListTable_railwayStation".

10.5.2 Railway track defect ontology

The abstract model of railway infrastructure is a vocabulary that allows one to represent railway tracks and switches defects as well as speed restrictions and combines the concepts of relevant databases tables, drawings, and regulations.

The framework of the railway infrastructure ontology [16] is enriched with the concepts of railway switches and track defects, speed restrictions, as well as relations to link infrastructure elements with defects and permitted speeds.

According to the TMR, the speed restrictions are recorded in the RBSRW and the DHFSRW. These are RBSRW and DHFSRW tables, represented by classes like "book speed", "defect", "form speed", and relations like "bookSpeedCorrespondsToActualDefect", "equals", and "formSpeed".

According to the GAICW, before handing out the DHFSRW, the train operations centre center worker verifies the DHFSRW speed restrictions against the RBSRW. These are DHFSRW and RBSRW tables, with the speed value class "speed value" and relations for checking the equality "isSpeedValueOf", and "hasSpeedValue".

According to the instructions of the CMRTR and the TWSAR, the results of the inspection are recorded in the RBRTI and RBRSI. Tables of corresponding books are represented by "defectValue", "defect", and "infrastructureElement" classes and relations such as "isActualDefectOf," "isActualDefectValueOfTheElement", and "isActualDefectOf".

Classes like "P65SwitchBigRampWearingOut" and "mediumSettlement-Value" are formed to check the correspondence between speed and defect.

According to the TERR, speed restrictions or special movement conditions apply in the event of a railway track defect or maintenance work. The TERR does not define new tables, but the "isActual-SpeedRestrictionOf" relation is added, that which links railway track to the restriction.

The concrete railway infrastructure model with rules is implemented as five TSRR, and TMR models for integration, and two TMR models for checking data consistency with track facilities facility regulations.

Diagrams have been developed for processing methods of the actual and fixed defects of the railway track. The first case is divided into four levels, and the second – into three. This is Because because the TERR does not contain instructions on how to handle the cancellation of the speed restriction.

Consider the case of an actual defect and a speed restriction (Figure 10.2). According to the TERR, the presence of a speed restriction on the track means the presence of a defect. According to the CMRTR, the presence of a defect on the track means that this defect has value. According to the TMR, the presence of a speed restriction on the track means that the speed of the DHFSRW corresponds to one in the RBSRW. According to GAICW, there should be equality of in the values of these speeds.

In Figure 10.2, dashed arrows are relations compositions. The level of GAICW is highlighted in green, TMR is highlighted in orange, CMRTR and TWSAR are highlighted in blue, and TERR is highlighted in black.

Consider the case of repairing and cancelling a speed restriction warning (Figure 10.3). According to CMRTR and TWSAR, the repair of a defect means that this defect is associated with a certain value (for example, railway track settlement of 4 mm) and as a result of the repair, the entire defect (4-mm settlement) was fixed. According to the TMR, the cancellation of a

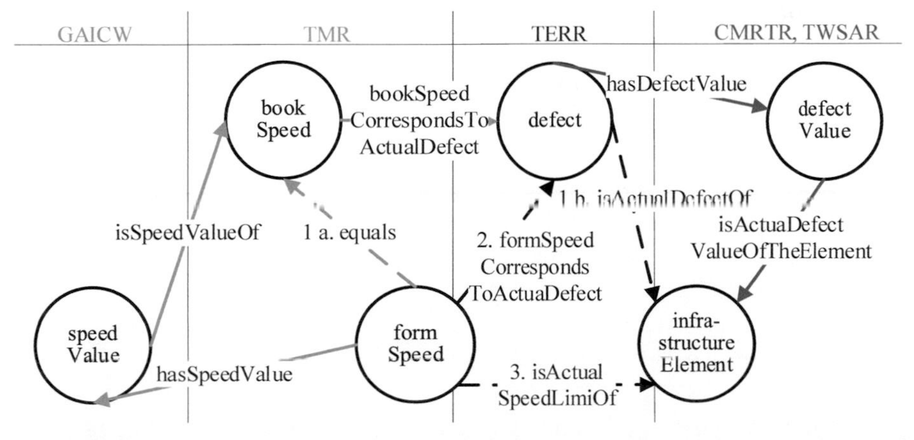

Figure 10.2 Layered aggregation of the actual defect and speed restriction relationships of the railway infrastructure.

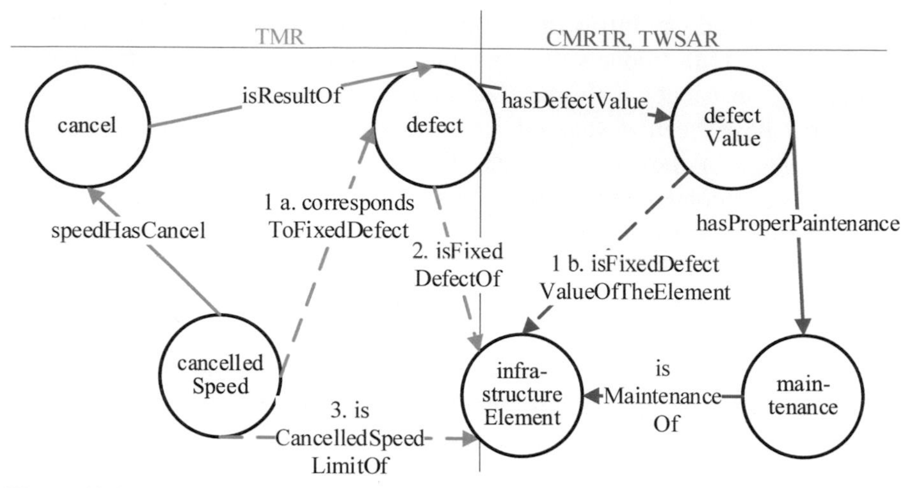

Figure 10.3 Level-by-level aggregation of relations between repair and cancellation of train speed restrictions.

restriction for railway track defect means that a request has been made to cancel the speed restriction, and the cancellation of a speed restriction for an infrastructure element means that a cancel has been given and maintenance has been made.

Compositions of relations for different levels of regulations aggregation are summarized in Table 10.1.

Logical definitions of this type have been developed to check the consistency of track failure and speed restriction.

```
   - rampWearingOut equivalent to defect and
(hasName some {"ramp wearing out"})
   - bigRampWearingOut equivalent to rampWearingOut
and hasDefectValue some bigRampWearingOutValue
   - bigRampWearingOutValue equivalent to
rampWearingOutValue and isA some xsd:int[> "10"
^^xsd:int , < "12"^^xsd:int]
   - P65SwitchBigRampWearingOut equivalent to
bigRampWearingOut and isActualDefectOf some
P65Switch or isFixedDefect of some
P65Switch
   - rampWearingOutValue equivalent to defectValue
and hasName some {"ramp wearing out"}
```

Table 10.1 Relations compositions of regulations representing railway track defects.

Regulation	Linked tables	Relations composition
GAICW	DHFSRW, and RBSRW	equals SuperProperty of hasConditionsValue o isConditionsValueOf equals SuperProperty of hasSpeedValue o isSpeedValueOf
TMR	RBSRW	correspondsToFixedDefect SuperProperty of speedHasCancel o isResultOf isCancelledSpeedRestriction of SuperProperty of correspondsToFixedDefect o isFixedDefectOf
	DHFSRW, and RBSRW	formConditionsCorrespondToActualWork SuperProperty of equals o bookConditionsCorrespondToActualWork formSpeedCorrespondsToActualDefect SuperProperty of equals o bookSpeedCorrespondsToActualDefect
CMRTR	RBSRW, and RBRTI	isActualDefectOf SuperProperty of hasDefectValue o isActualDefectValueOfTheElement isFixedDefectOf SuperProperty of hasDefectValue o isFixedDefectValueOfTheElement
	RBRTI	isFixedDefectValueOfTheElement SuperProperty of hasProperMaintenance o isMaintenanceOf
TWSAR	RBSRW	isActualWorkOf SuperProperty of hasWorkValue o isActualWorkValueOfTheElement
TERR	RBSRW, and RBRTI	isActualConditionsOf SuperProperty of formConditionsCorrespondToActualWork o isActualWorkOf isActualSpeedRestrictionOf SuperProperty of formSpeedCorrespondsToActualDefect o isActualDefectOf

- 25Speed Equivalent To BookSpeedRestriction and bookSpeedRestriction hasSpeedValue some 25SpeedValue

- 25SpeedValue Equivalent To speedValue and isA some {"25"}.

Restrictions of this type have been developed to check the consistency of track failure and speed restriction:

- 25Speed SubClass Of bookSpeedCorrespondsTo ActualDefect only P65SwitchBigRampWearingOut
- 25Speed SubClass Of correspondsToFixedDefect only P65SwitchBig RampWearingOut
- bookSpeedRestrictionHasSpeedValue Inverse Of isSpeedValueOf
- P65SwitchMediumRampWearingOut Disjoint With P65SwitchBig RampWearingOut
- bigRampWearingOutValue Disjoint With MediumRampWearingOutValue.

A concrete model of the railway infrastructure with data is made in the form of six models (ontologies) for the tables: RTL, RSL, RBRTI, RBRSI, DHFSRW, and RBSRW.

Consider the description of the tables representing the case of an actual defect and a speed restriction on the railway track.

A triple is extracted from the DHFSRW table: speed, its value, and the relation linking them. A speed is assigned a UUID to distinguish its instance from a speed value instance.

```
rail:b8b2ec24-6e5b-4f74-b916-305b16ce51d6
rail:hasSpeedValue rail:60;
rdfs:label "form speed 1" .
rail:60 rail:isA "60" .
```

A triple is extracted from the RBSRW table: the value of speed, railway track defects, and the relation linking them.

```
rail:60 rail:isSpeedValueOf rail:12ec59a9-b3ce-4fb9-aa4f-b0a1c97c3351.
rail:12ec59a9-b3ce-4fb9-aa4f-b0a1c97c3351
rail:bookSpeedCorrespondsToActualDefect
rail:settlement;
rdfs:label "book speed".
```

A triple is extracted from the RBRTI table.

```
rail:settlement rail:has defect value rail:26;
rail:has name "settlement" .
rail:26 rail:is actual defect value of the element rail:railway station pt 1;
```

```
    rail:has name "settlement";
    rail:is a "26"^^<http://www.w3.org/2001/
XMLSchema#int>.
```

The concrete railway infrastructure model is demonstrated in two knowledge bases: a track with a defect and a switch with maintenance.

Let's Let us consider the integration of these tables representing a railway track that has an actual defect and a speed restriction.

The DHFSRW table has a speed restriction value of 25 km/h. The RBSRW table has the a speed restriction value of 25 km/h and its correspondence to the railway track №1 settlement. According to property chain 1a (Figure 10.2), instances of the 25 km/h speed of DHFSRW and RBSRW are connected by the relation "equals", since both instances are linked with the value "25". According to chain 2 (Figure 10.2), the 25 km/h speed of DHFSRW is linked with the defect corresponding to it from RBSRW.

RBRTI has the defect (track settlement) related to the value of 26 mm by the relation "isA" and railway track №1. According to chain 1b (Figure 10.2), the track is linked with a defect instance.

Then, according to chain 3 (Figure 10.2), the track is linked with the speed restriction instance. This is possible if the speeds of RBSRW and DHFSRW were equal and the defect was recorded in RBSRW.

Consider checking the data of tables representing a railway track with an active defect and speed restriction.

First, classification is done according to rdfs:range and rdfs:domain. Since the speed of the DHFSRW is related to the value "60" by the relationship "hasSpeedValue" and the relation has domain and range, then the speed is classified as "formSpeedRestriction" and the value to the "speedValue" class.

The defect is linked to the defect name "track settlement" by the relation "hasName" and is classified in the "settlement" class. The railway track defect value is also named "track settlement" and is classified into a settlement value class. It is related to "26" literal by relation "isA" and is therefore further classified into the "largeSettlement" class. The "settlement" instance of defect is classified as a "bigSettlement" and, it is further classified as a "trackBigSettlement" because it is linked with track №1.

The speed value instance is linked with the literal "60" and is therefore classified as "60SpeedValue". An instance of the RBSRW speed is linked with an instance of the value 60 km/h and classified into "60 speed" class.

The settlement test instance has a value of 21 mm. If it is linked with a speed of 60 km/h, then the ontology becomes inconsistent, since the reasoner classifies the settlement as medium one because of its value and as large because of a speed restriction. The classes of large and medium railway track settlements are disjoint.

Let's Let us consider the data integration of tables representing the railway switch with the cancelled speed restriction and the repaired defect.

The RBSRW table has the "14.06 PCH Pushkin" cancel record, the speed "25", and the defect "ramp wearing out". According to chain 1a (Figure 10.3), and the defect instance "ramp wearing out" is linked with an instance of the cancelled speed restriction "60 speed".

The RBRSI table has the repair "14.06" for the railway switch №1 and the defect value of "11 mm". According to chain 1b (Figure 10.3), the instance of railway switch №1 is linked with the value of the repaired defect. RBRSI table has a defect value of "11 mm". According to chain 2 (Figure 10.3), railway switch instance №1 is linked with the repaired ramp wearing out individual. According to property chain 3 (Figure 10.3), railway switch №1 instance is linked to the cancelled speed restriction with the relation "isCancelledSpeedRestrictionOf".

10.6 Speed Restriction Checking

The checking is carried out at four levels: the structure of the tables, the correspondence of the speed restriction to the infrastructure element defect, the existence of repairs to cancel the speed restriction, and the correctness of the speed restriction handed out to the train driver. And it is implemented in three ways: rules, restrictions, and compositions of relations.

SWRL rule assigns a zero-range property to a table, and if the structure of the table does not match the model, its elements have an error property, and the ontology becomes inconsistent. This inference is necessary because data validation should be performed on tables that do not contain structural errors.

When using OWL restrictions, the speed class of 60 km/h has the restriction to correspond only to the railway "trackLargeSettlement". If the track facilities facility tables contain the value of the railway "trackMediumSettlement", and the train operations centre center tables have 60 km/h speed, the ontology becomes inconsistent, since the conditions of the CMRTR are violated. At the moment, the correspondence of restrictions and defects or maintenance work is determined by track maintenance workers; therefore, errors are possible due to the human factor.

When using OWL property chains, the railway track does not have the property of the actual or cancelled speed until the consistency conditions of the speed issued to the train driver and the repair of the railway track defect, respectively, are met.

10.7 Discussion

Approbation of the resources ontology was carried out in [17]. In this work, the ontology has been developed to process multiple tables using unique identifiers. The following criteria of table classification have been developed: a program to fill, table header, and stations (that has the infrastructure described in the table).

Approbation of the railway subsystems ontology framework was done in [16]. In this paperchapter, the station model is enriched by an aggregation of regulations that describe the defects of the railway infrastructure.

Further work is to link the ontology of speed restrictions with the "Time Ontology in OWL" because permanent speed restrictions are not taken into account here.

10.8 Conclusions

The applicability of the methods developed in [17] (tabular representation of knowledge) for ontology development and their enrichment is determined, which allow for data validation, partially eliminate the laborious process of the ontology population, and reusing reuse vocabularies and rules for various data.

Methods have been developed that allows checking the consistency of railway train speed restrictions and four regulations of railways by their aggregation that improve the level of railway transport safety, as they check tables for compliance with regulations and exclude the possibility of cancelling restrictions before repairs are completed.

References

[1] Ceci, M. and Gangemi, A. (2016). An OWL ontology library representing judicial interpretations. Semantic Web 7(3), 229-253. DOI: 10.3233/SW-140146.

[2] Corsar, D., Markovic, M., Edwards, P. and Nelson, JD (2015) "The transport disruption ontology" in International Semantic Web Conference, Springer, Cham, 329–336. DOI: 10.1007/978-3-319-25010-6 22.

[3] Ding, LY, Zhong, BT, Wu, S. and Luo, HB (2016). Construction risk knowledge management in BIM using ontology and semantic web technology. Safety science 87, 202-213. DOI: 10.1016/j.ssci.2016.04.008.

[4] Distinto, I., d'Aquin, M. and Motta, E. (2016). LOTED2: An ontology of European public procurement notices. Semantic Web 7(3), 267-293. DOI:10.3233/SW-140151.

[5] Duarte, B. B., de Castro Leal, AL, de Almeida Falbo, R., Guizzardi, G., Guizzardi, R.S. and Souza, V.E.S. (2018). Ontological foundations for software requirements with a focus on requirements at runtime. Applied Ontology 13(2), 73-105. DOI: 10.3233/AO-180197

[6] Duarte, B. B., Falbo, R. A., Guizzardi, G., Guizzardi, R. S. and Souza, V. E. (2018) "Towards an ontology of software defects, errors and defects" in International Conference on Conceptual Modeling, Springer, Cham, 349 -362.

[7] Guizzardi, G. (2005) Ontological foundations for structural conceptual models. PhD thesis, University of Twente, Enschede.

[8] Hagedorn, T. (2018) Supporting Engineering Design of Additively Manufactured Medical Devices with Knowledge Management Through Ontologies. PhD thesis, Mechanical Engineering, University of Massachusetts Amherst, Amherst. DOI: 10.7275/11326512.0.

[9] Hoekstra, R. J. (2010) Representing Social Reality in OWL 2 in Proceedings of OWL: Experiences and Directions.

[10] Jovic, A., Gamberger, D. and Krstacic, G. Heart failure ontology. Bio Algorithms Med Syst. - 2011. - T. 7. - No. 2. - S. 101-110.

[11] Lališ, A., Bolčeková, S. and Štumbauer, O. (2020). Ontology-based reliability analysis of aircraft engine lubrication system. Transportation Research Procedia 51, 37-45. DOI: 10.1016/j.trpro.2020.11.006.

[12] Lewis, R. (2015) A semantic approach to railway data integration and decision support. PhD thesis, School of Electronic, Electrical and Computer Engineering, University of Birmingham, Birmingham.

[13] de Oliveira Bringuente, A.C., de Almeida Falbo, R. and Guizzardi, G. (2011). Using a foundational ontology for reengineering a software process ontology. Journal of Information and Data Management 2(3), 511-511.

[14] Pauwels, P., Van Deursen, D., Verstraeten, R., De Roo, J., De Meyer, R., Van de Walle, R. and Van Campenhout, J. (2011). A semantic rule checking environment for building performance checking. Automation in construction 20(5), 506-518. DOI: 10.1016/j.autcon.2010.11.017

[15] Pisanelli, D. M., De Lazzari, C., Innocenti, E. B. and Zanetti, N. (2009) "An ontology supporting an on-board vehicle multimodal interaction system," in Multimodal Human Computer Interaction and Pervasive Services, IGI Global, 230-242.

[16] Shynkarenko, V. and Zhuchyi, L. (2021) "Ontological harmonization of railway transport information systems" in 5th International Conference on Computational Linguistics and Intelligent Systems, 541–554.

[17] Shynkarenko, V. and Zhuchyi, L., Ivanov, O. (2021) "Conceptualization of the tabular representation of knowledge" in IEEE 16th International Conference on Computer Sciences and Information Technologies, 248–251. DOI: 10.1109/CSIT52700.2021.9648761.

[18] Skalozub, V., Ilman, V. and Shynkarenko, V. (2017). Development of ontological support of constructive synthesizing modeling of information systems. Eastern-European Journal of Enterprise Technologies (6(4)), 58-69. DOI:10.15587/1729-4061.2017.119497.

[19] Skalozub, V., Ilman, V. and Shynkarenko, V. (2018). Ontological support formation for constructive-synthesizing modeling of information systems development processes. Eastern-European Journal of Enterprise Technologies 5(4(95)), 55–63. DOI: 10.15587/1729-4061.2018.143968.

[20] Tutcher, J. (2016) Development of semantic data models to support data interoperability in the rail industry. PhD thesis, Electronic, Electrical, and Systems Engineering, University of Birmingham, Birmingham.

[21] Zhao, L., Ichise, R., Mita, S., and Sasaki, Y. (2014) "An ontology-based intelligent speed adaptation system for autonomous cars" in Joint International Semantic Technology Conference, Springer, Cham, 397–413. DOI: 10.1007/978-3-319-15615-6 30,

[22] A knowledge based platform of services for supporting medical-clinical management of heart failure withing eldery population. Available at: https://cordis.europa.eu/project/id/027107 [accessed March 2, 2022].

[23] Error Ontology. Available at: https://github.com/SPAROntologies/error [accessed March 2, 2022].

[24] N-Ary Relation Pattern (OWL 2). Available at: http://ontologydesignpa tterns.org/wiki/Submissions:N-Ary Relation Pattern (OWL 2) [accessed March 2, 2022].

[25] The InteGRail project. Available at: http://www.integrail.eu/dedefect.ht m [accessed March 2, 2022].

11

A Tool for Automatic Anomaly Identification in OWL Ontologies

João Paulo Orlando[1] and Dilvan A. Moreira[2]

[1]Federal Institute of São Paulo, IFSP, Matão, Brazil
[2]University of São Paulo, ICMC, São Carlos, Brazil
E-mail: orlando@ifsp.edu.br; dilvan@icmc.usp.br

Abstract

Ontologies form the basis of the semantic web. They represent a formal specification of concepts and relationships in a domain of interest. As the semantic web develops, there is a growing number of ontologies publicly available online. This fast growth of ontologies creates difficulties in using, understanding, and managing them. In addition, problems (also called anomalies) in the representation of ontology structures, such as circularities, redundancies, inconsistencies, and deficiencies, may occur during the construction or integration of these ontologies. There is a need for techniques that help users to evaluate ontology quality and uncover such anomalies. To tackle this problem, we present an ontology to describe ontology structures, called MetaFOR, and a system that uses it to spot anomaly occurrences, called ONTO-Analyst. The system generates a structural description of an ontology, using MetaFOR, and then queries this description, using SPARQL queries, to find anomalies. The ONTO-Analyst system was tested using a representative set of 18 anomaly types, taken from the literature, in a set of 608 OWL ontologies from four major public repositories. The tool found more than 3 million occurrences of 12 anomaly types: three circularities, five redundancies, one inconsistency, and three deficiencies. The testing results demonstrate that the ONTO-Analyst method is capable of automatically identifying many kinds of anomalies and that even widely used ontologies have some anomaly issues.

Keywords: Anomaly identification, MetaFOR, ontology, ontology engineering, semantic web..

11.1 Introduction

The semantic web explores how meaning can be associated with content on the web, so this information is machine-processable. For this reason, it is necessary to have inference methods and a structured collection of information (ontologies) [1]. Ontologies are the basis of building the semantic web and represent a shared knowledge of a domain of interest. To define and instantiate ontologies on the Web, the World Wide Web Consortium (W3C) recommends the web ontology language (OWL). OWL expressiveness is not sufficient to model all problem types. For example, some problems need Horn-like rules. Moreover, ontologies can use rules to facilitate the definition of logical deductions, for the development of semantic applications [2].

In 2004, W3C first considered the use of the semantic web rule language (SWRL) to represent rules. SWRL complements OWL because it includes a high-level abstract syntax for Horn-like rules [3]. SWRL rules can be added to an OWL file as valid OWL. Even though SWRL is not yet a W3C recommendation (standard), it is a popular language with support in many tools, such as Protégé, Pellet[1] and HermiT.[2]

As the semantic web becomes more popular, there are more and more ontologies publicly available online, creating difficulties to find appropriate ones [4]. In addition, problems may occur when these different ontologies are integrated [2]. These problems, or anomalies, can be divided into four types: circularity, redundancy, inconsistency, and deficiency [2]. A circularity occurs when a class is defined as a specialization or generalization of itself. A redundancy is generated with the addition of an already defined assertion or one that can be inferred by another. Inconsistency is contradictory knowledge. Deficiency is an anomaly that only affects the design of an ontology.

To help solve these problems, new techniques are needed to help users identify anomalies in ontologies. With that in mind, we created a technique

[1]Pellet: https://github.com/complexible/pellet
[2]HermiT: http://www.hermit-reasoner.com/

that uses an ontology about ontology structures to identify such anomalies. This ontology is called *metadata description for ontologies/rules* (MetaFOR) [5]. MetaFOR instances represent classes, properties, and rules (entities) of other ontologies. MetaFOR properties (linking MetaFOR instances) represent the relationship, between these entities. The following code

```
mf:URI1_hasBondWith
a mf:ObjectProperty;
a mf:SymmetricObjectProperty;
hasInstancesNumber 0;
usedInRestrictions false;
usedInRules false;
isLeaf false;
hasDomain mf:URI1_Atom;
hasRange mf:URI1_Atom;
isEquivalent mf:URI1_hasBondWith;
hasDirectSub mf:URI1_hasDoubleBondWith;
hasSub mf:URI1_hasDoubleBondWith.
```

shows a property, `hasBondWith`, being represented as a MetaFOR instance (the URI1 prefix was added to avoid name clashes).

Using the MetaFOR representation (or format) for an ontology, it is possible to write SPARQL queries to identify anomalies in its structure. We also created a system, called ONTO-Analyst to read an ontology, convert its entities and relations to MetaFOR format, and to run a set of SPARQL queries to find anomalies. Ontology users can use predefined query sets, to find common ontology anomalies, or write their own queries to enforce agreed design constraints.

To show that MetaFOR and ONTO-Analyst are able to identify anomalies in ontologies, we tested them using a set of representative ontology anomalies, described in the scientific literature. In our tests, we used SPARQL queries to test 18 anomaly types (three circularities, seven redundancies, five inconsistencies, and three deficiencies) in 608 ontologies from four well-established public repositories (and two papers). These ontologies are a representative set of publicly available ontologies (in OWL). Many of these ontologies belong to important projects (such as the Gene Ontology). ONTO-Analyst found more than 4 million anomaly occurrences divided in 12 types: three circularities, five redundancies, one inconsistency, and three deficiencies. The results showed that inconsistencies are the least common anomalies.

The test results show the importance of the ONTO-Analyst system. The system was able to detect more than 4 million anomaly occurrences in the ontology set, even in the most popular ontologies. Another contribution is the identification of the most common anomalies in open OWL ontologies. Two deficiencies, *lazy class or property* and *chains of inheritance*, account for more than 98% of the anomaly occurrences. Circularities and redundancies are also common in ontologies and deserve more attention in the development of new techniques for anomaly detection. Many related works describe ontology anomalies or techniques to identify them, but they do not show the occurrence of these anomalies in a large set of representative OWL ontologies.

This chapter is divided into seven sections: Related works (Section 11.2), ONTO-Analyst presentation (Section 11.3), Experiments (Section 11.5) (Section 11.5), Discussion (Section 11.6), and Conclusion (Section 11.7). The most important section, Experiments, is divided into five subsections representing the five steps taken for the validation of the ONTO-Analyst system: (i) selection of ontology repositories (subsection 11.5.1), (ii) ontologies download (Subsection 11.5.2), (iii) conversion of the chosen OWL ontologies to MetaFOR format (Subsection 11.5.3), and (iv) analysis of the identification results (subsection 11.5.4).

11.2 Related Work

There are several papers describing and formalizing problems or anomalies that occur in ontologies [6]–[9]. Based on this information, other authors developed techniques to automatically identify anomalies in ontologies [2, 10]–[14]. They identify, using different techniques, the occurrences of, at least, one of these anomaly types: redundancies, inconsistencies, circularities, or deficiencies.

Feng *et al.* [10] developed a method to discover three redundancies, a circularity, and three inconsistencies using constraint logic programming techniques and reasoners. The authors implemented a prototype to verify their method and validated it with a case study in one ontology. Hui *et al.* [14] presented a technique to identify and eliminate cycles in an ontology's taxonomy. The authors validated the technique by applying it to one ontology. Hussain *et al.* [13] detected inconsistencies by finding contradicting derivations produced under a given logic program. The authors validated this approach detecting all inconsistencies in an ontology and generating all possible minimal inconsistent resolve candidates (MIRCs) for extracting a maximal consistent ontology. Sun *et al.* [11] created a tool to identify four redundancy types in SWRL rules. It uses an algorithm for identifying and eliminating implication redundancy.

All of these methods work well, but they only detect a small anomaly set and were tested using only one ontology. Poveda-Villallón [12] shows problem occurrences in an ontology set. A catalog of 41 pitfalls (term similar to anomalies) and a system, called OOPS!, are used. The system allows ontology developers to (semi) automatically diagnose ontologies. But the system validation is based on qualitative metrics other than quantitative. ONTO-Analyst allows quantitative analysis of anomaly occurrences. It also detects anomalies in SWRL rules found in ontologies; not all systems implement this feature [12]–[14].

Baumeister *et al.* [2] present the only work we found that allows users to define new anomalies (in addition to a limited set of predefined ones). Since the verification of OWL ontologies with rule extensions is not tractable in general, they propose to verify ontologies at the symbolic level by using a declarative approach using the DATALOG* language. This language is used to easily specify and test known anomalies in a compact manner. These tests run in a system, called DisLog Developers' Kit, and introduce supplements to existing verification techniques to support the design of ontologies with rule enhancements. These tests focus on the detection of anomalies that especially occur due to the combined use of rules and ontological definitions. They can identify circularities, inconsistencies, redundancies, and deficiencies in ontologies. Users must know the DATALOG* language and some system to run it in order to define new anomalies. This language is not very popular and can be difficult to understand for ontology developers.

To define new anomalies, ONTO-Analyst uses SPARQL queries. SPARQL is a query language inspired by SQL and is well-known by the semantic web community.

The MetaFOR ontology, used by ONTO-Analyst, was presented in a previous work [5]. In it, we showed that it is possible to automatically identify certain ontology anomalies using MetaFOR. We used 11 anomalies: nine taken from the literature [2] (a circularity, five contradictions, and three redundancies) and two user-defined. They were manually inserted in a test ontology (family relationships ontology[3]). Rules were defined for each anomaly type, using SWRL. The test ontology was converted to the MetaFOR format and we were able to identify all inserted anomalies. All rules were written using SWRL, but, since then, we decided to switch to SPARQL (since it is more popular). This previous work shares some of the same limitations found in the literature, such as the use of a small set of anomalies and being tested using only one ontology.

---

[3]http://protegewiki.stanford.edu/wiki/Protege_Ontology_Library

11.3 ONTO-Analyst System

The ONTO-Analyst system performs the conversion of OWL ontologies to the MetaFOR format and identifies anomaly occurrences. These OWL ontologies can also have SWRL rules. MetaFOR [5] is an ontology to represent ontology structures. Figure 11.1 shows its class hierarchy.

ONTO-Analyst uses instances of the MetaFOR classes (Classes, ObjectProperty, DataProperty, and Rules), as shown in Figure 11.1, to represent the classes, properties, and rules, of other ontologies and the relations between them [5].

The system was built in Java and uses the OWL API (a popular Java API to work with OWL ontologies) to read ontology files and provide the basic functionality to manipulate ontology entities. The system works in three stages. The first one is responsible to load the information necessary for the ontology conversion and has four components: OntologyLoader, EntityLoader, RelationLoader, and URIManager (Figure 11.2). The second stage performs the conversion to the MetaFOR format using the Converter component and its sub-components. The third stage executes the SPARQL queries to identify anomaly occurrences (using the AnomalyFinder component).

Figure 11.2 shows the component diagram[4] for ONTO-Analyst with their provided and required interfaces in each conversion stage.

11.3.1 First stage

In the first stage, the ONTO-Analyst system loads all information needed for the conversion. The stage starts with the OntologyLoader component. It encapsulates and isolates the OWL API from the rest of the system. This separation allows the substitution of the OWL API by other APIs, if necessary.

The OWL API (version 3.5.2) supports the creation, manipulation, and serialization of OWL ontologies [15]. ONTO-Analyst uses it to manipulate ontologies in all stages of the conversion, via the OntologyLoader component. Presently, the OWL API can load ontologies in six different file formats (or serializations): RDF/XML, OWL/XML, OWL Functional Syntax, Turtle, KRSS, and OBO Flat. Therefore, all these file formats are supported by ONTO-Analyst.

[4]A component diagram is a UML diagram (unified modeling language).

Ontologics can usc many URIs, especially when integrating multiple ontologies. The URIManager component manages these URIs. It renames URIs with shorter names to be used as entity names in MetaFOR format.

The components EntityLoader and RelationLoader load java classes that implement the EntityInterface and RelationInterface interfaces. Classes that implement these interfaces are used to create entities, in MetaFOR format, to represent the entities of the ontology being analyzed. Users can easily

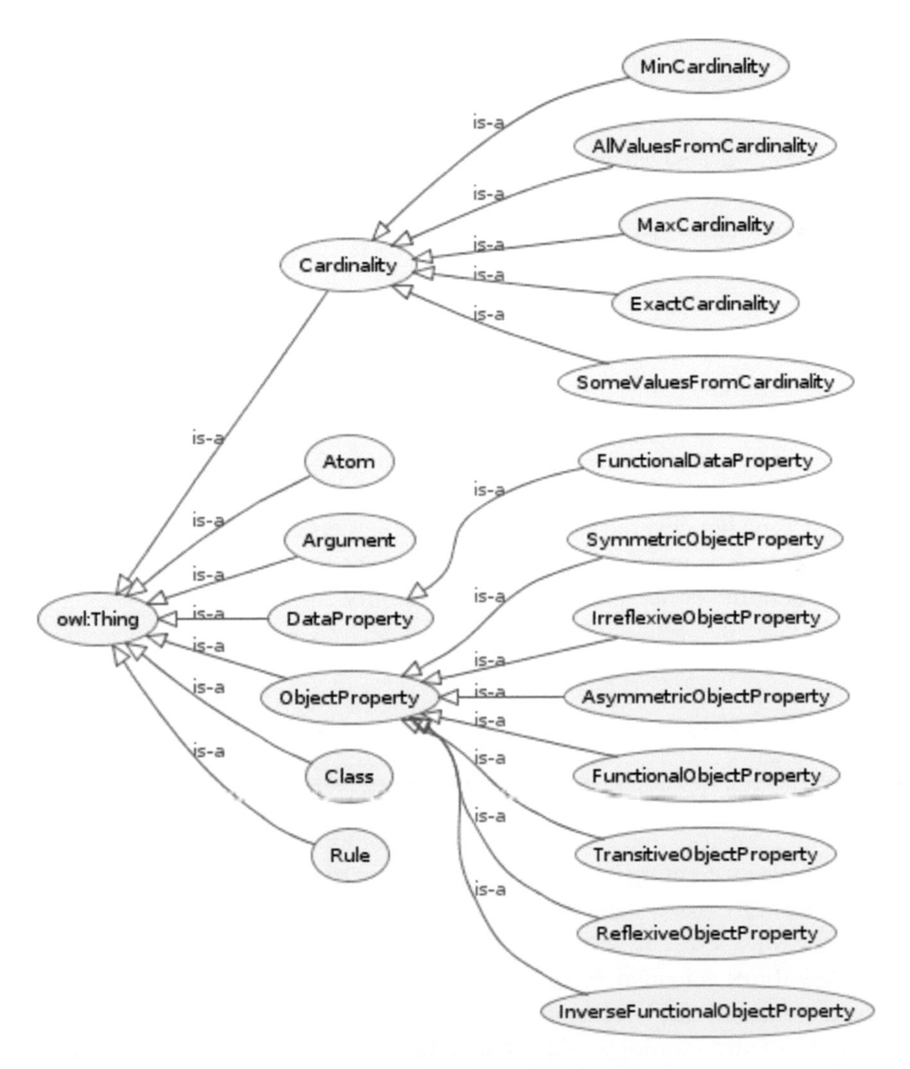

Figure 11.1 MetaFOR class hierarchy.

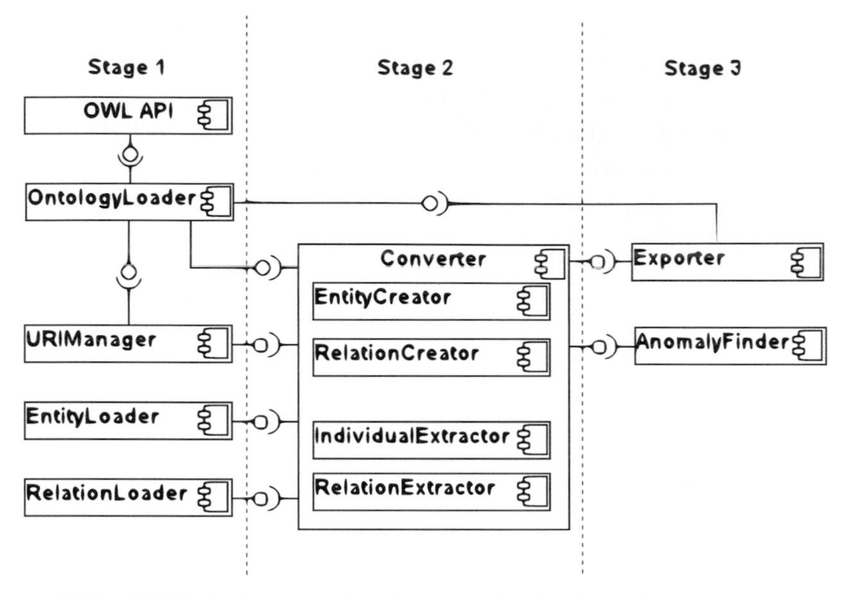

Figure 11.2 ONTO-Analyst staged architecture showing its software components and interfaces.

substitute the implementations provided. For instance, users may need to create more detailed representations, for some ontology entities, to detect a new anomaly type not covered by the current implementation. EntityInterface and RelationInterface interface implementations are loaded, from JAR files, using Java's default service-provider loading facility.

11.3.2 Second stage

The second stage implements the actual ontology conversion to the MetaFOR format.

The EntityCreator and RelationCreator components instantiate the MetaFOR ontology creating its classes, object properties, and data properties. They use the Java classes loaded by the EntitiesLoader and RelationLoader components.

After MetaFOR is loaded, the conversion begins. The IndividualExtractor component creates MetaFOR instances representing classes, properties, rules, and atoms from the ontology being converted. For instance, consider that there is an ontology to be converted, which has two data properties: *hasHeight* and *hasWeight*. This component will create one instance of the *DataProperty* MetaFOR class for each property: *URI1_hasHeight* and *URI1_hasWeight* (these are examples of the shorter names created by the URIManager component).

The RelationExtractor component creates MetaFOR instances representing relations (using object or data properties) between entities from the ontology being converted. For example, consider that an ontology has an object property *hasParent* with a domain restriction for class *Person* (only instances of *Person* can have this object property). A MetaFOR relation *hasDomain* will be defined between the instances *URI1_hasParent* (of type *ObjectProperty*) and *URI1_Person* (of type *Classes*). Both these instances were created by the IndividualExtractor component.

11.3.3 Third stage

After the second stage converts an ontology, the Exporter component saves the resulting MetaFOR description. This description can be examined using an ontology editor, such as Protégé.

The AnomalyFinder component loads and executes SPARQL queries to identify anomaly occurrences in the MetaFOR description. It also allows users to edit these queries, changing the way an anomaly is found or creating new queries to identify new anomalies.

The next section shows the set of anomalies used to test the ONTO-Analyst system.

11.4 Anomalies to be Identified

To test the ONTO-Analyst system, we need to choose a representative set of anomalies to be detected and to create the SPARQL queries to find them. Four anomaly types were chosen from the scientific literature. The criteria to choose the anomalies were to get different types that occur in different ontology entities (classes, properties, cardinalities, and rules).

The following anomaly types were chosen:

- Circularity: Circular anomalies happen when classes are defined as specializations or generalizations of themselves [7, 2]. Circular anomalies have a strong impact on reasoner performance [2].
- Redundancy: Redundant knowledge is generated by the addition of an assertion already existing or that can be inferred by another [7]. It is a hard-to-find anomaly as it does not affect the inference results in an

ontology. But it indicates flaws in the knowledge base design that may become real issues as the base evolves over time.

- Inconsistency: Contradictory knowledge is another important anomaly kind for ontologies [2]. It can occur in two ways: inconsistent assertions are created in the ontology and assertions are never created because of contradictions.
- Deficiency: This anomaly type is more subtle than others because it is considered a problem that affects ontology design [2]. It does not cause errors in the reasoning but may be an indicator of problematic and/or badly designed areas in an ontology [16]. The removal of this kind of anomalies improves the ontology usability and maintainability [17].

From these four types, 18 specific anomalies were chosen (Table 11.1). Of these anomalies, six occur in classes, four in object properties, nine in SWRL rules, and two in cardinalities. Three of these anomalies occur in classes and properties at the same time. About the type, three anomalies are circularities, seven are redundancies, five are inconsistencies, and three are deficiencies.

ONTO-Analyst executes SPARQL queries to identify occurrences of these anomalies in the MetaFOR description of an ontology. In the next subsection, each anomaly type is explained.

Circularities

11.4.1 Exact circularity in taxonomy

This circularity occurs when two classes, C1 and C2, are equivalents, wherein C1 has subclass S1, C2 is a subclass of S2, and S1 is equivalent of S2. This last incorrect assertion creates an unwanted circularity in the taxonomy [2]. This is a simple circularity example. It uses direct subclasses. But a SPARQL query, using hasSub and hasSuper properties, can identify all subclasses or superclasses in the hierarchy and identify more complex cycles. Furthermore, the same query works also with circularity in properties.

11.4.2 Circular properties

Circular property anomalies occur in ontologies in which there are two inverse properties that have the same or equivalent domains and the same or equivalent ranges.

For example, there are two ontologies, A and B, with the following two equivalent classes: *A:Lecture* is equivalent to *B:Course* and *A:Professor* is

Table 11.1 Chosen anomalies with occurrence and type information.

Anomaly name	Occurrence in	Anomaly type
Exact circularity in taxonomy	Classes and Properties	Circularity
Circular properties	Object properties	Circularity
Circularity between rules and taxonomy	Rule	Circularity
Redundancy by repetitive taxonomic definition	Classes and Properties	Redundancy
Redundant cardinalities	Cardinalities	Redundancy
Redundant derivation in the antecedent	Rule	Redundancy
Redundant implication	Rule	Redundancy
Redundant implication of transitivity or symmetry	Rule	Redundancy
Redundant use of transitivity and symmetry	Rule	Redundancy
Rule subsumption	Rule	Redundancy
Partition error in taxonomy	Classes	Inconsistency
Multiple functional properties	Cardinalities	Inconsistency
Contradicting rules	Rule	Inconsistency
Incompatible rule antecedent	Rule	Inconsistency
Self-contradicting rule	Rule	Inconsistency
Chains of inheritance	Classes	Deficiency
Lonely disjoint classes	Classes	Deficiency
Lazy class or property	Classes and properties	Deficiency

equivalent to *B:Professor*. Further, in these ontologies, there are the following defined properties: *A:lectures* with domain *A:Professor* and range *A:Lecture* and *B:teaches* with domain *B:Professor* and range *B:Course*. If *A:lectures* is defined as the inverse of *B:teaches*, then a circular property anomaly is created.

11.4.3 Circularity between rules and taxonomy

This circularity happens when a consequent atom implies some antecedent atom, leading to a circularity [2]. For example, consider an ontology that has a class Father that is a *subClassOf* Person and the following rule:

```
Person(?x), hasChild(?x, ?y) -> Father(?x)
```

The rule should be considered as a restricted *subClassOf* relation between the consequent atom and antecedent atom (Person and Father). It leads to a circularity [2]. Such a restriction would be better represented in the taxonomy.

Inconsistencies

11.4.4 Partition error in taxonomy

This kind of error happens when a wrong combination of disjoint and derived relations occur. For example, if two disjoint classes have the same subclass. If these are direct subclasses, the inconsistencies are easier to find, but it is much more difficult to find them if they appear down on the hierarchy.

Given that MetaFOR describes an ontology's structure, a great deal of the SPARQL queries used to detect anomalies can be simple. For example, for this anomaly:

```
SELECT DISTINCT ?cf
WHERE { ?c1 metaFor:isDisjoint ?c2 .
        ?c1 metaFor:hasSub ?cf .
        ?c2 metaFor:hasSub ?cf }
```

11.4.5 Multiple functional properties

This inconsistency type happens because functional properties must have only one value for each instance [2], but users can define for them minimum cardinality restrictions greater than 1. This situation generates contradictory knowledge.

11.4.6 Contradicting rules

These inconsistencies occur when two rules have equivalent or the same antecedent atoms and two of their consequent atoms are disjoint (one in each rule). If a rule would fire, then also the other rule would fire, which derives contradicting conclusions.

11.4.7 Incompatible rule antecedent

These inconsistencies occur when there is an incompatibility among antecedent atoms [2]. The cause of this anomaly is a disjoint relationship between two predicate atoms of the same rule antecedent, as long as these atoms have the same variables. For instance, an ontology with two disjoint classes Student and Teacher has the following rule:

```
Student(?x), Teacher(?x) -> TeachingAssistant(?x)
```

There can be no x that is a Student and a Teacher at the same time. In order to identify this anomaly, it is necessary to find rules that have disjoint class or property predicates in their antecedent atoms.

11.4.8 Self-contradicting rule

Similar to the last inconsistency (Incompatible Rule Antecedent – Section 11.4.7), in the self-contradicting anomaly, there is a disjointness incompatibility between two atoms, but one is in the antecedent and the other in the consequent. In this case, there is a predicate in the antecedent atom that is disjoint with a predicate in the consequent atom. Distinctively, from the last anomaly, this one can create inconsistent results if not fixed. For instance, an ontology with two disjoint classes Student and Teacher has the following rule:

```
Student(?x), Lecture(?y), teaches(?x, ?y) -> Teacher(?x)
```

In order to identify this anomaly, it was necessary to create two SPARQL queries: one to treat disjoint classes and another to treat disjoint properties.

Redundancies

11.4.9 Redundancy by repetitive taxonomic definition

This anomaly is a redundant definition of *subClassOf* or *subPropertyOf* in a hierarchy of classes or properties [7]. This anomaly can occur in two situations:

- Direct repetition: This occurs when two or more definitions of sub (*subClassOf* or *subPropertyOf*) are defined for the same classes or properties. Using the OWL API, the system is free of this anomaly, because the OWL API do not load repeated definitions. So, this situation does not need to be verified by ONTO-Analyst.
- Indirect repetition: This occurs if there are the following definitions: *subClassOf(C, B)*, *subClassOf(B, A)* and *subClassOf(C, A)*. This situation also can occur with properties.

11.4.10 Redundant cardinalities

Cardinalities can be redundant in some types of situations. In Baumeister and Seipel [2], the authors cite two situations that can lead to this redundancy:

- define a property with the minimum cardinality restriction greater than or equal to zero;
- define a property that has a maximum cardinality equal to 1 and a functional super property.

11.4.11 Redundant implication

This redundancy occurs when an antecedent atom implies some consequent atom [2]. In this case, a predicate found in the rule consequent is *superClassOf* a predicate found in the antecedent. For example, in an ontology, where a Father class is defined as a *subClassOf* a Person, the following rule creates a redundancy (as Father is already a Person):

```
Father(?x), hasChild(?x, ?y) -> Person(?x)
```

This redundancy is similar to the circularity between rules and taxonomy (Section 11.4.3).

11.4.12 Redundant implication of transitivity or symmetry

This redundancy can be divided into two situations: transitivity and symmetry.

The transitivity (first situation) occurs if a transitive property *P* is defined, and *P(x, y) and P(y, z)* is enough for a reasoner to infer *P(x, z)*. This situation happens when a rule has at least three properties, nominated here as P1, P2, and P3, which are transitive and equivalent, and it uses P1 and P2 to assert P3: $P1(?x, ?y), P2(?y, ?z), \cdot \rightarrow P3(?x, ?z)$. Despite not being wrong, this rule is unnecessary as reasoners can infer P3.

The second situation occurs with symmetry. This occurs when a symmetry property *P* defines that if there is *P(x, y)*, the reasoner can infer *P(y, x)*. A redundant implication of symmetry occurs when a rule uses two properties P1 and P2, which are equivalent and symmetric, to assert: $P1(?\mathrm{x}, ?\mathrm{y}), \cdot - > \mathrm{P2}(?\mathrm{y}, ?\mathrm{x})$. The reasoner can already infer this consequent.

11.4.13 Redundant use of transitivity and symmetry

This anomaly is similar to the one previously seen in Section 11.4.12 and can also be divided into two situations: transitivity and symmetry.

A redundant use of transitivity occurs when a rule has, at least, three properties, named as P1, P2, and P3, which are equivalent and transitive, and the rule uses P1, P2, and P3 in the antecedent $P1(?\mathrm{x}, ?\mathrm{y}), \mathrm{P2}(?\mathrm{y}, ?\mathrm{z}), \mathrm{P3}(?\mathrm{x}, ?\mathrm{z}), ... - > \cdot$. Despite not being wrong, the antecedent atom *P3(?x, ?z)* is not necessary because this statement can be inferred by the reasoner.

A redundant symmetry use happens when a rule uses two properties, P1 and P2, which are equivalent and symmetric, and it also uses P1 and P2 in the antecedent: $P1(?\mathrm{x}, ?\mathrm{y}), \mathrm{P2}(?\mathrm{y}, ?\mathrm{x}), ... - > ....$ The atom $P2(?\mathrm{y}, ?\mathrm{x})$ is not necessary because this atom will always be true when the $P1(?\mathrm{x}, ?\mathrm{y})$ was defined.

11.4.14 Redundant derivation in the antecedent

This redundancy occurs when an antecedent atom *X* implies other antecedent atom *Y* [2]. The atom *Y* is redundant in the rule antecedent and can be removed without changing the rule results. For example, the following rule shows this anomaly because Father is *subClassOf* Person (so Father is not needed).

```
Person(?x), Father(?x), ... -> ...
```

11.4.15 Rule subsumption

This redundancy happens when the antecedent of a rule can be fully mapped to another one and their consequents are the same. The subsumed rule can be removed without modifying the ontology result.

The next section shows experimental results obtained using the ONTO-Analyst system to detect anomalies in a large set of ontologies from public repositories.

Deficiencies

11.4.16 Chains of inheritance

Ontologies can have as many class levels, created by *subClassOf* declarations, as their creators want [17]. During the creation of this class hierarchy or during the integration of ontologies, subclass cascades can be created. If the classes, in these cascades, are not used anywhere (by instances, restrictions, or rules), a chain of inheritance is formed. Baumeister and Seipel [2] define this deficiency as a class that has just a few instances or none at all.

The query, designed to detect this anomaly, tests only for classes that have no instances, but it is simple and easy to change this query to include classes with few instances: the user needs to add a FILTER clause to test if the number of instances is less than what is required. Furthermore, the user may change the query to test for more classes in cascade, instead of only three. This is an example of a major advantage of this approach: users can easily change their SPARQL queries to identify anomalies based on their requirements [5].

11.4.17 Lonely disjoint classes

This deficiency occurs when a class is not disjoint with any of its siblings, but it is disjoint with, at least, two classes in another branch of the class hierarchy [17, 2]. A lonely disjoint class can be created when a class is moved to another branch of the taxonomy, but the attached disjointedness descriptions are not changed accordingly or by incorrect alignment during an integration task. The existence of a lonely disjoint class can cause unintended reasoning results or even errors [2].

11.4.18 Lazy class or property

Baumeister and Seipel [2] define an ontology entity as possibly lazy when it is never or rarely used in real-world applications. They define four conditions to find a lazy entity:

- The entity is a leaf in the hierarchy.
- The entity is not used by any rule.
- The entity does not have instances.
- The entity is not used by other entities (for instance, on restrictions).

This characterization can be useful as a good indicator for entity utility, if entities and instances are defined in the same place. In our experiments, instances for many ontologies were not available. For instance, we did not read data annotated with the GO ontology. In these cases, this anomaly only constitutes a "deficiency" in the context of a rich ontology, where one expects to see restrictions that define the entities. But it can still be used to measure how similar to a pure taxonomy an ontology is. An ontology with few lazy classes/properties is more likely to be a richer ontology and not just a taxonomy.

11.5 Experiments

The experimental goals are to investigate if ONTO-Analyst and MetaFOR are able to identify a set of representative anomalies in a large number of ontologies, found in widely used public repositories. Given that almost all ontologies came from bio-ontology repositories, the experiments also identify the most common anomalies found in such types of ontologies. The experiments were divided into five stages:

1. selection of ontology repositories (subsection 11.5.1);
2. download of the chosen ontologies (subsection 11.5.2);
3. conversion of the chosen ontologies to MetaFOR format (subsection 11.5.3);
4. selection of anomalies to be identified (subsection 11.4);
5. identification and analysis of the anomalies (subsection 11.5.4).

11.5.1 Ontology repository selection

To choose ontology repositories, we first created a list using all repositories from a comprehensive survey of 11 ontology repositories [18] and from the University of Manchester – OWL Research Group website[5]. Then, we applied the following criteria to each site in this list (18 sites):

- ontologies available in a format that is supported by the OWL API;
- easy downloading of ontologies (with no need for human intervention).

Using these criteria, we chose the following ontology repositories:

- BioPortal repository: It is a repository for biomedical ontologies that was developed by the National Center for Biomedical Ontology (NCBO). BioPortal allows researchers to evaluate and contribute to the development of ontological content. Subsequently mentioned as BIOPORTAL.
- Dumontier Laboratory repository: It is a repository for bio-ontologies that represent biological and scientific concepts and relations. Subsequently, mentioned as DUMONT.
- Tones ontology repository: It is a web service with ontologies used mainly for testing purposes. It has a range of different ontologies ranging in expressiveness and size. It is subsequently mentioned as TONES.

Beyond the ontologies in these repositories, two ontologies with SWRL rules from papers were chosen. They will be represented by the acronym PAPERS.

[5]OWL research group:
http://owl.cs.manchester.ac.uk/tools/repositories/

All in all, the experiments used 546 ontologies. From this total, only eight ontologies had SWRL rules, with a total of 452 SWRL rules among them. The usage of a class or property is often a good indicator for its actual utility.

To solve this lack of ontologies with rules in the experiments, other ontology repositories (with ontologies with SWRL rules) were sought but with no success. For that reason, the GitHub[6] repository was included. GitHub is a well-known source code repository that has thousands of projects. It had many ontologies with SWRL rules. The criteria to select these ontologies were the same as for the other repositories, with the addition of a third criterion: Ontologies had to be verified with Pellet (version 2.3.1) and/or Hermit (version 1.3.8) reasoners and have no errors. That criterion was added to exclude ontologies with deliberate errors on them. Such ontologies are used to test programs, like Pellet. Eighty-two ontologies with rules from GitHub were used. These will be represented by the acronym GIT.

11.5.2 Ontologies download

The places where the ontologies were found are the following:

BIOPORTAL: bioportal.bioontology.org/ontologies

DUMONT: dumontierlab.com/index.php?page=ontologies

TONES: rpc295.cs.man.ac.uk:8080/repository/browser[7]

PAPERS: From the papers of Hastings *et al.* [19] and Young *et al.* [20].

GIT: github.com/search?q=extension%3Aowl+ swrl+imp

For the DUMONT, TONES, and BIOPORTAL, we created a simple tool to extract all URLs with some file extensions (".owl", ".xrdf", ".ttl", and ".obo") and download the ontologies. Some ontologies have more than one file in different extensions. In this case, the ".owl" extension was given priority. Table 11.2 shows that 71 (DUMONT), 219 (TONES), and 254 (BIOPORTAL) ontologies were downloaded in April 2015. In the GIT, ontologies were downloaded and selected manually on May 7th and 8th 2015. In total, this experiment used 628 ontologies.

11.5.3 Conversion to MetaFOR format

Six hundred and eight ontologies were converted to the MetaFOR format. The specifications of the computer used for the conversions are:

[6]GitHub: https://github.com/

[7]This web service is now deprecated, the ontologies are stored at https://zenodo.org/record/32717

Table 11.2 Repositories with download date and number of ontologies.

Name	Download date	Number of ontologies
DUMONT	13th April 2015	71
TONES	13th April 2015	219
BIOPORTAL	26th April 2015	254
PAPERS		2
GIT	7th/8th May 2015	82
		628

Table 11.3 Conversion results to metaFor format.

Repository name	Number of ontologies	Converted and passed reasoner	Converted but failed reasoner	Not converted
DUMONT	71	69	0	2
TONES	219	199	14	6
BIOPORTAL	254	225	17	12
PAPERS	2	2	0	0
GIT	82	82	–[8]	–
Totals	628	577	31	20

- Intel Core™ i7-990X with six cores.
- 16 GB of RAM memory.
- Windows 8.

Table 11.3 shows the conversion results. All ontologies were verified with Pellet (version 2.3.1) or Hermit (version 1.3.8) reasoners before conversion. Two reasoners were used because some ontologies had assertions that the Hermit reasoner did not support (SWRL built-ins) and Pellet was slow in some cases. Hermit was used first, and Pellet was only used if Hermit failed. Thirty-one ontologies caused both reasoners to crash or not finish processing (after 24 hours); in these cases, we considered that errors had been detected.

The third column of Table 11.3 shows that 577 ontologies (91.9%) did not have errors detected by our reasoner setup. To have a high percentage here is important because it shows the basic quality of the ontologies involved in the experiments. The fourth column shows that 31 ontologies had errors detected (more details in Table 11.4).

The system converted 608 ontologies to the MetaFOR format, from a total of 628 (the OWL API could not open 20 ontologies). The conversion efficacy was 96.8%. Finally, the fifth column (Table 11.3) shows the number of ontologies not converted (more details in Table 11.5).

[8]Only ontologies that were verified using the reasoner were considered.

Table 11.4 Ontologies converted to MetaFOR format that failed reasoner verification.

Repository name	Problems found by reasoner	Reasoner stopped	Out of memory
DUMONT	0	0	0
TONES	7	7	0
BIOPORTAL	9	4	4
Totals	16	11	4

Table 11.5 Ontologies with problems in the MetaFOR conversion.

Repository name	With imports	With the OWL API
DUMONT	1	1
TONES	6	0
BIOPORTAL	1	11
Total	8	12

Converted ontologies that failed reasoner verification (column 4, Table 11.3) are divided into three groups (Table 11.4): Ontologies with inconsistencies found by the reasoners (column 2); ontologies that ran in the reasoners for more than 24 hours, our time limit (column 3); ontologies that caused out of memory errors in the reasoners (column 4). In the end, 31 ontologies that were converted to MetaFOR failed reasoner verification but were not discarded.

Table 11.5 shows ontologies with problems in the MetaFOR conversion. The following problems were found: some imports no longer exist in their URLs (column 2) and the OWL API could not read the file format (column 3). Both problems are not problems with the ONTO-Analyst converter; in other words, the converter would probably convert these ontologies, if the OWL API could load them.

11.5.3.1 Data summary of the structures of the analyzed ontologies

The summary, in Table 11.6, shows the vast amount of data that were analyzed in the experiments.

Ontologies have lots of classes, on average ∼7,000 per ontology. Object properties were found less often than classes, ∼38 on average. But all ontologies have at least one object property and one class. The experiments

Table 11.6 Data summary of the entities in the 608 ontologies analyzed.

	Classes	Object properties	Rules	Cardinalities
Sum	4,377,570	23,593	979	7
Number of ontologies with this entity	608	608	90	4
Avg per ontology	∼ 7,199	∼ 38	∼ 1.6	∼ 0.01
Max	403,295	950	159	4

analyzed 979 rules from 90 ontologies. We found a very small number of cardinality declarations, which was surprising. Cardinalities should have been an important kind of restriction in ontologies, but, in the 608 ontologies analyzed, only seven occurrences were found.

11.5.4 Identification and analysis of the anomalies

With the 608 ontologies converted to MetaFOR format and the SPARQL queries to find the 18 chosen anomalies, the AnomalyFinder component (of the ONTO-Analyst tool) was used to find the anomalies. A general overview of the results obtained is presented in Section 11.5.4.1, followed by a detailed view of each anomaly that was detected in the ontologies (Section 11.5.4.2). Finally, the results for some widely used ontologies are shown (Section 11.5.4.3).

11.5.4.1 General overview

The general results are divided by entity type in which the anomalies occur: classes, properties, rules, and cardinality statements. In the next three tables, the results for classes, properties, and rules are shown. The results for cardinality statements are not shown because the ONTO-Analyst did not find any anomaly in cardinalities.

Table 11.7 shows seven anomalies regarding classes. The "Lazy class or property," "Redundancy by repetitive taxonomic definition" and "Exact circularity in taxonomy," anomalies, shown in Table 11.7, can also occur in properties.

The first line in Table 11.7 (Exact circularity in taxonomy) is a circularity that has occurred 583 times in classes. It occurred less often than other anomalies. One should remember that more than 4 million classes were analyzed. This anomaly was distributed over just 70 ontologies and one ontology had 79 occurrences in it.

Table 11.7 General results for anomalies in classes.

Anomalies	Total occurrences	Ontologies with anomalies	Max occurrences in one ontology
Exact circularity in taxonomy	583	70	79
Partition error in taxonomy	3289	52	2028
Redundancy by repetitive taxonomic definition	45,894	265	10,825
Lonely disjoint classes	4325	116	932
Chains of inheritance	811,726	487	106,293
Lazy class or property	3,556,207	592	375,836

```
SELECT DISTINCT ?cf
WHERE { ?c1 metaFor:isDisjoint ?c2 .
        ?c1 metaFor:hasSub ?cf .
        ?c2 metaFor:hasSub ?cf }
```

The second line (Partition error in taxonomy) is an inconsistency that occurs 3289 times. It is important to emphasize that more than 61% of these occurrences (2028) were found in only one ontology. In the next subsection, this situation will be discussed in more detail. This anomaly was the one that occurred less often in the analyzed ontologies. Only 52 ontologies had at least one occurrence, which may be due to the fact that reasoners can easily identify inconsistencies.

The third line (Redundancy by repetitive taxonomic definition) is a redundancy in the subclass or subproperty definitions that occur in ~1% of all classes and properties analyzed. However, one-fourth of them belongs to one ontology.

The other three lines are deficiencies (anomalies in the ontology design). "Lonely disjoint classes" occurred more than 4 thousand times, "Chains of inheritance" occurred more than 800 thousand times and finally, the "Lazy class or property" occurred more than 3 million times and had the highest occurrence (for all anomalies in entity types).

Table 11.8 shows the results for anomalies that occurred only in object properties. The only anomaly in it is "Circular properties," which is a circularity. In total, more than 23,000 object properties were analyzed and only 28 object properties had this anomaly (in five ontologies).

Table 11.9 presents the anomalies that occurred in rules. In total, 979 rules were analyzed. The first anomaly is one type of circularity, the next three are inconsistencies, and the rest are redundancies. The first anomaly occurred 114

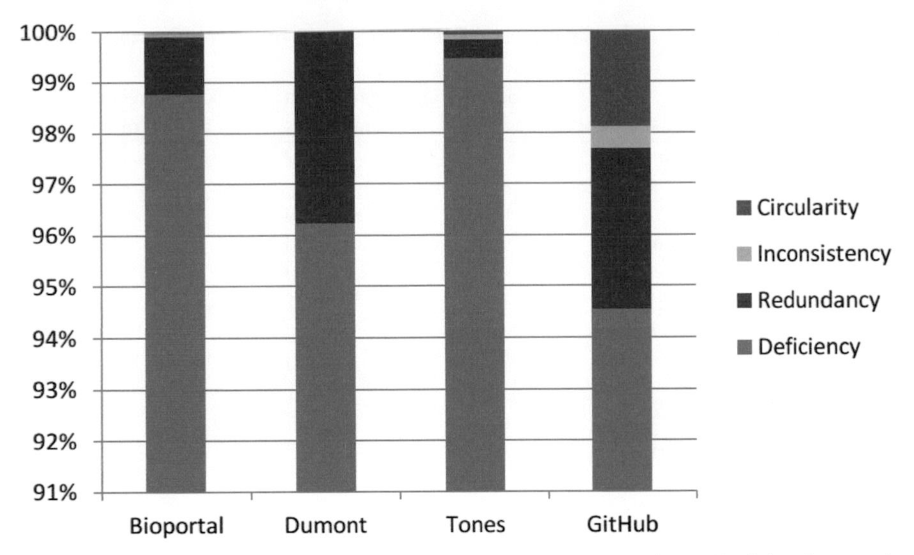

Figure 11.3 The concentration of anomaly types in each repository. Deficiencies are the most common anomaly type. The usage of a class or property is often a good indicator for its actual utility.

times in 19 ontologies and had the highest occurrence for rule anomalies. No anomalies were found for the three inconsistencies (in Table 11.9), which are probably due to the fact that reasoners can easily identify such anomalies. Regarding redundancy anomalies, only "Redundant implication" did not occur, whereas the others occur a few times. It is important to remember that the number of ontologies with SWRL rules is limited.

The influence of the ontology repositories in the results of the experiments is presented in two charts. Figure 11.3 shows the influence of each repository in the occurrence of each anomaly type. Each bar corresponds to 100% of the occurrences in each site and each color (in the bars) correspond to an anomaly type. Deficiencies are by far the most common anomaly type in all repositories, followed by redundancies. Circularity appears in a more representative number only in the GitHub repository.

Table 11.8 General results for anomalies in object properties (from a total of 4.3 million anomaly occurrences).

Anomalies	Total occurrences	Ontologies with anomalies	Max occurrences in one ontology
Circular properties	28	5	10

Table 11.9 General results for anomalies in rules (from a total of 4.3 million anomaly occurrences).

Anomalies	Total occurrences	Ontologies with anomalies	Max occurrences in one ontology
Circularity between rules and taxonomy	114	19	20
Contradicting rules	0	0	0
Incompatible rule antecedent	0	0	0
Self-contradicting rule	0	0	0
Redundant derivation in the antecedent	13	1	13
Redundant implication	0	0	0
Redundant implication of transitivity or symmetry	1	1	1
Redundant use of transitivity and symmetry	4	2	2
Rule subsumption	4	2	2

Figure 11.4 shows the influence of each repository in the anomaly occurrences for each entity type. Each bar corresponds to 100% of the anomaly occurrences in each repository, and each color, in the bars, corresponds to an entity type. The figure shows that anomalies are overwhelmingly more common in classes, almost 100% of the cases. The GitHub repository shows a relevant number of rule anomalies, but only ontologies with rules were chosen from GitHub creating a bias towards rule anomalies.

11.5.4.2 Analysis of some specific anomalies

In this subsection, some pertinent situations are analyzed for some anomaly types.

11.5.4.3 Most relevant ontologies

This subsection shows results for the top 10 ontologies in BioPortal (a popular bio ontology repository) and the 2 ontologies found in scientific papers. The top 10 ontologies of BioPortal were selected by checking the number of projects that used them in BioPortal. Thus, the 10 ontologies with more projects in the BioPortal were selected on April 30th 2015. Table 11.10 shows the list of selected relevant ontologies with their acronyms. Selected ontologies had a total 323,935 classes, 1493 data properties, 512 object properties, and 253 rules.

Table 11.11 shows the anomalies found in the selected ontologies. Of the 18 anomalies described in this chapter, nine were found in the selected ontologies. This situation shows that even popular heavily used ontologies may contain important anomalies. Table 11.11 shows the occurrences of three deficiencies (lines 1, 5, and 6), three circularities (lines 2, 3, and 4), two redundancies (lines 8 and 9), and one inconsistency (line 7) in these ontologies.

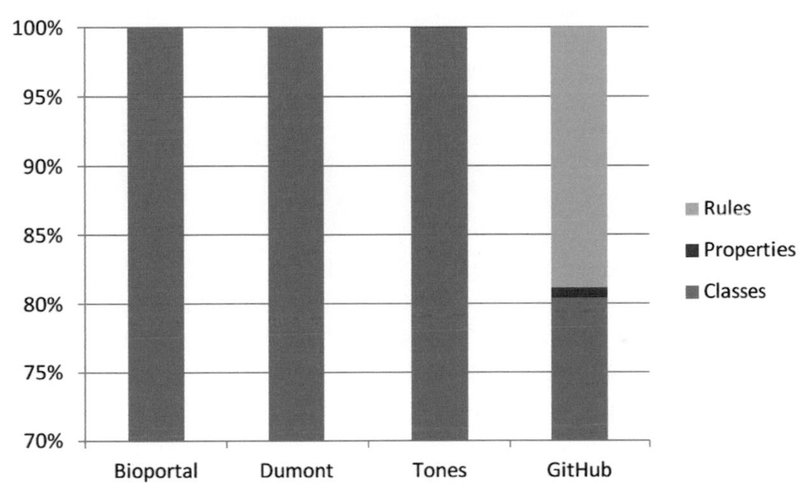

Figure 11.4 The concentration of anomalies by entity types in each repository. Anomalies are overwhelmingly more common in classes, almost 100% of the cases for all repositories but GitHub (where only ontologies with rules were chosen, creating a bias towards rule anomalies).

Table 11.10 List of relevant ontologies from Bioportal.

Ontology name	Acronym
Gene ontology	GO
Ontology for biomedical investigations	OBI
Basic formal ontology	BFO
Systematized nomenclature of medicine – clinical terms	SNOMEDCT
Open biological and biomedical ontologies relationship types	OBOREL
National Cancer Institute thesaurus	NCIT
Sequence types and features ontology	SO
Phenotypic quality ontology	PATO
Foundational model of anatomy	FMA
Chemical entities of biological interest ontology	CHEBI
Autism phenologue [20]	AUT
Chemistry ontology [19]	CHO

Table 11.11 Results from Bioportal's most relevant ontologies.

Anomalies	Total occurrences
Chains of inheritance	297,297
Circularity between rules and taxonomy	3
Circular properties	8
Exact circularity in taxonomy	2
Lazy class or property	240,983
Lonely disjoint classes	253
Partition error in taxonomy	460
Redundancy by repetitive taxonomic definition	1788
Rule subsumption	2

11.6 Discussion

In the experiments, three circularities, seven redundancies, five inconsistencies, and three deficiency types were chosen from the scientific literature to be detected by ONTO-Analyst. It found 12 types of anomalies: three circularities, five redundancies, one inconsistency, and three deficiencies. Some of the results of the experiments were unexpected and demand further discussion.[9]

11.6.1 Too many anomalies?

In the experiments, ONTO-Analyst found more than 4 million anomalies in 608 ontologies. At first glance, this number seems too high. Most ontologies are created in tools, such as Protégé, that constrain entries to a formal grammar and can validate ontologies using reasoners. How can they still have anomalies?

Some anomalies happened a lot more than others. We had 3,556,207 occurrences of *Lazy classes or property* vs. 811,726 *Chains of inheritance* vs. 54,255 from all other kinds of anomalies.

Lazy classes or property deficiency is the most common kind of anomaly, with 3.5 million occurrences vs. 866 thousand from all other kinds. Figure 11.5 shows the frequency that this deficiency occurs in ontology classes or properties. In the pie chart, all ontologies are distributed in the percentage bands. The highest occurrence was in the range of 60% – 80% (245). The

[9]SNOMEDCT: Not included because it is not available for download, only for online access.

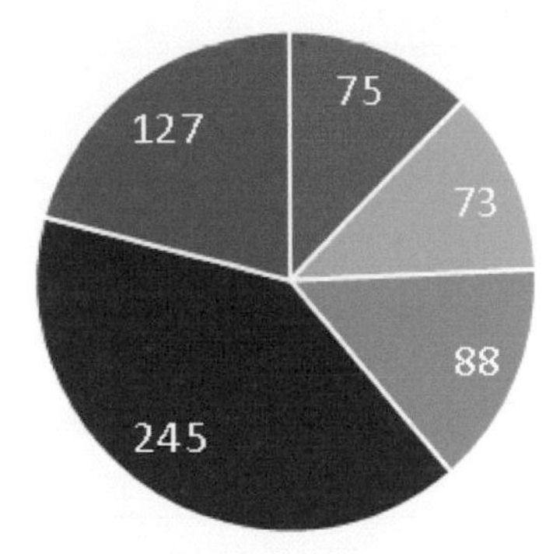

Figure 11.5 Distribution of the occurrence percentage of "Lazy class or property" in the total of classes and properties for each ontology.

sum of the ranges 60% – 80% and 80% – 100% represents more than half of this deficiency.

In another chart (Figure 11.6), a scatter plot is presented in which the gray circles represent each ontology and the black line represents the linear trend line in the relationship between vertical and horizontal axes. A linear trend line is a best-fit straight line that is used with simple linear data sets. It usually shows that something is increasing or decreasing at a steady rate. The chart shows the relation between *Lazy class or property* occurrences (the vertical axis) and the total of classes and properties in each ontology (the horizontal axis). This deficiency is fully related to the number of classes and properties in each ontology.

Given that a lazy class is a class with no instances that are not used in rules or axioms, what this result shows us is that a large part of the ontologies tested (61%) are basically just taxonomies (have more than 60% of lazy classes or properties). They have lots of classes without instances, restrictions, or rules using them. Given the number and scope of the ontologies used, this is an indication of the lack of description power and sophistication of open bio ontologies in general. If classes, from these ontologies, are used as metadata for experiment databases, they will say very little about their instances, apart from their taxonomic classification.

If the goal of an ontology is being a taxonomy, then this deficiency is not an issue. But this anomaly is still useful to allow users to determine how similar to a taxonomy an ontology is.

Chains of inheritance are the next most common anomaly with 811,726 occurrences. It refers to cascades of subclasses that are not used elsewhere

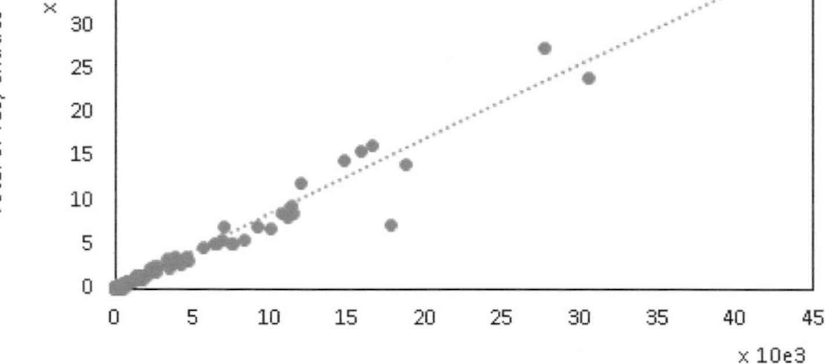

Figure 11.6 The relation between *Lazy class or property* occurrence and the total of classes and properties in each ontology.

(by axioms, rules, or instances). They are very similar to the *Lazy class or property* anomalies. Indeed, many classes share both anomalies.

Given the fact that the top two anomalies are just informing us that we are dealing with taxonomies, we can discount them and end up with 54,255 occurrences of other anomalies – a number much more reasonable for the 608 ontologies analyzed.

Finally, reasoners, used by Protégé and other tools to validate ontologies, verify well inconsistencies. ONTO-Analyst found only a small amount of inconsistency (460) of only one type (of the five types tested). Other anomaly types are more common because the tools that are used by ontology developers do not focus on their identification.

11.6.2 Anomaly detection

In this section, we analyze some pertinent situations to show how useful the detection of the anomaly types can be.

Redundancy by repetitive taxonomic definition is the third most common anomaly with 45,894 occurrences. Analyzing this redundancy, it was discovered that ontologies, with the inconsistency "Partition error in taxonomy," also had this redundancy, because reasoners created many "subClassOf" relations in the associated classes with this inconsistency. To eliminate this influence from the other anomaly, ontologies that also had "Partition error

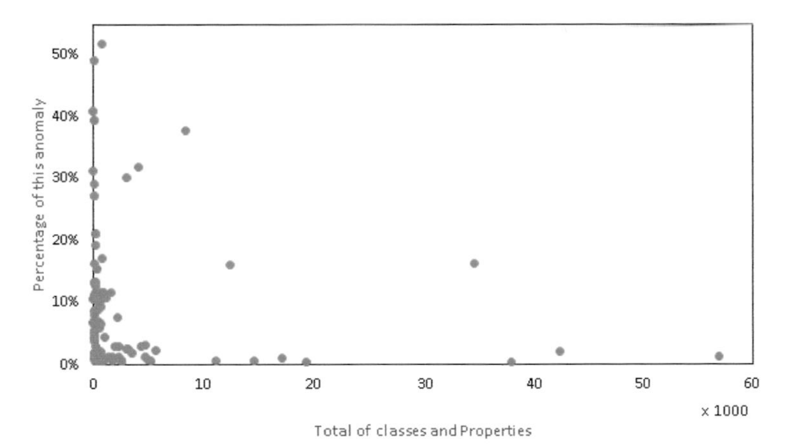

Figure 11.7 Relation of the occurrence percentage of "Redundancy by repetitive taxonomic definition" anomalies to the total of classes and properties in each ontology (each dot represents an ontology).

in taxonomy" were discarded when analyzing this redundancy. Therefore, 46 ontologies, from a total of 265, were discarded. They contained 25,530 occurrences of this redundancy from a total of 45,894 occurrences in all analyzed ontologies. A scatter plot was created, Figure 11.7, with the remaining occurrences. It presents the occurrence percentage of this redundancy, on the vertical axis, and the total of classes and properties, on the horizontal axis. Each point represents an ontology. The occurrence percentage is calculated by the division of total entities per occurrence total. The scatter plot shows that the occurrence percentage is higher in smaller ontologies.

Compared to the top two anomalies, this one may indicate a more serious issue. Given that only indirect repetition of subclass or subproperties is allowed by the OWL API, this anomaly occurs if there are the following definitions: *subClassOf(C, B)*, *subClassOf(B, A)*, and *subClassOf(C, A)*. This makes classes A, B, and C equivalent, probably not the intention of the ontology creators. The same is valid for properties.

The remaining 8380 anomalies are distributed in nine types.

The **Partition error in taxonomy** was the only inconsistency anomaly found in the analyzed ontologies. For example, in the NIFSTD[10] ontology, this inconsistency occurred 2028 times. Analyzing the occurrences, two classes, *Continuant,* and *Occurrent*, are responsible for almost all the

[10]NIFSTD: http://bioportal.bioontology.org/ontologies/NIFSTD

occurrences found. These classes are disjoint and all classes that are subclasses of both have this inconsistency. So, probably, a problem must have occurred during the construction of this ontology because no subclass of both *Continuant* and *Occurrent* can have instances. In the same ontology, there is another pertinent occurrence of this inconsistency: the class *BSPO_0000126* is disjoint with itself. So, the two subclasses of *BSPO_0000126* cannot have instances. These two examples cover all different occurrences of this inconsistency type in all analyzed ontologies.

Exact circularity in taxonomy: - For this circularity, three types of occurrences were found, each showing a different situation in the analyzed ontology. The first situation is when a class/property has itself as a subclass/subproperty. This situation was found, for instance, in the software ontology[11] from BIOPORTAL. In this ontology, there is a class, called *ObsoleteClass*, that is a subclass of itself. This ontology has 4068 classes and this class was the only one in this situation. So, probably, this assertion was created by mistake.

The second situation is when a class/property has a subclass/subproperty that is also equivalent. For example, the FBbi ontology [12], from BIOPORTAL, has one occurrence of this circularity in *FBbi_00000013* and *FBbi_00000419* classes. *FBbi_00000419* has *FBbi_00000013* as a superclass and equivalent class.

The third situation occurs when four different entities create a circularity. An ontology that relates concepts and terminologies used for human nutrition in a clinical and biomedical setting, called BNO[13], presents an occurrence with this situation. The classes FAT, Poly, Saturated, and Unsaturated form a circularity, wherein each class is a subclass of each of the other three. If class *X* is a subclass of *Y* and *Y* is a subclass of *X*, then all instances of *X* are instances of *Y* and all instances of *Y* are instances of *X*. So, *X* and *Y* are equivalents. With this, it is possible to conclude that the classes FAT, Poly2, Saturated, and Unsaturated are equivalents and that creates a circularity in this ontology. Given the classes names, that was not what the ontology authors intended.

Finding this situation can be complicated in an ontology that has many classes and properties, but the ONTO-Analyst system allows users to easily find situations like these. This is a big advantage as it allows users to

[11]Software: http://bioportal.bioontology.org/ontologies/SWO
[12]FBbi: http://bioportal.bioontology.org/ontologies/FBbi
[13]BNO: http://bioportal.bioontology.org/ontologies/BNO

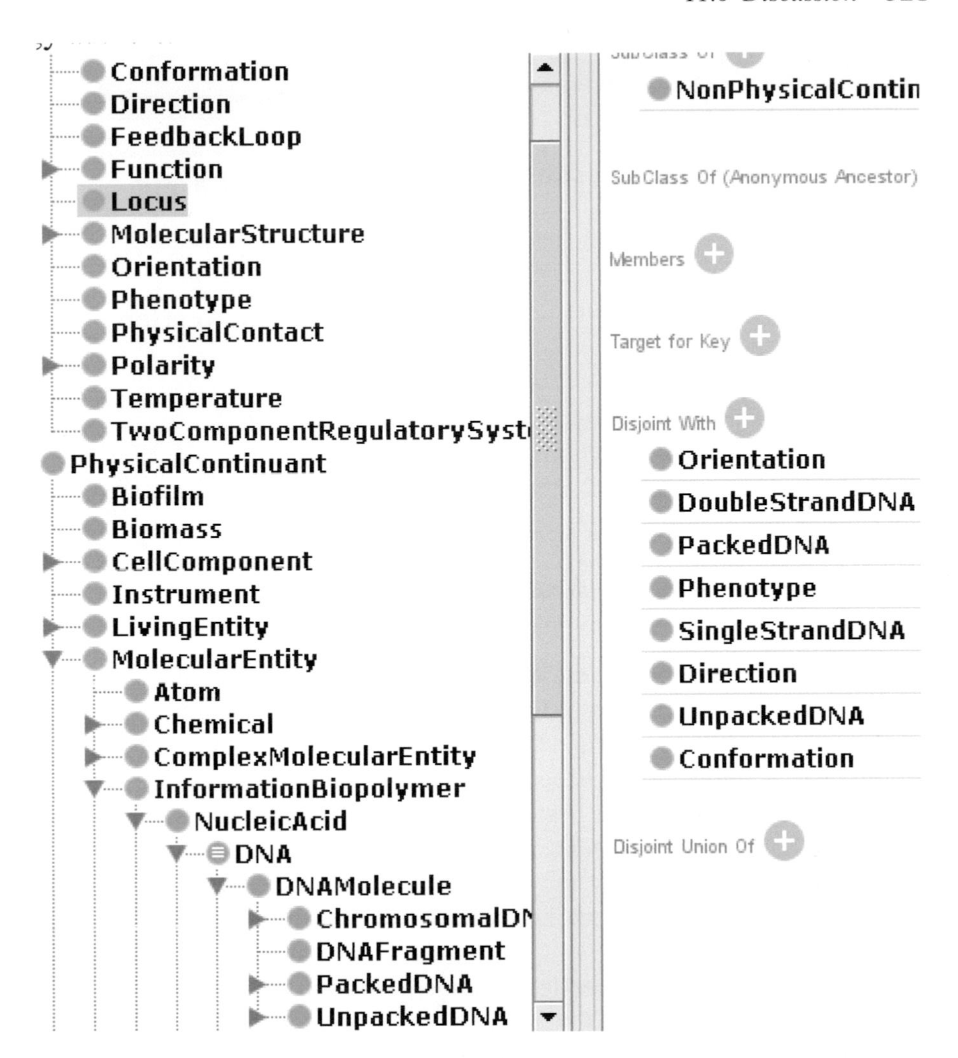

Figure 11.8 Lonely disjoint class deficiency: the *Disjoint With* field (right side) shows the classes disjoint with the Locus class (left side). Locus is disjoint from classes from two different branches of the class hierarchy. This is a situation that may require the attention of the ontology authors.

search for something more specific or for something more generic, like in these examples where the system identified three different circularity occurrences

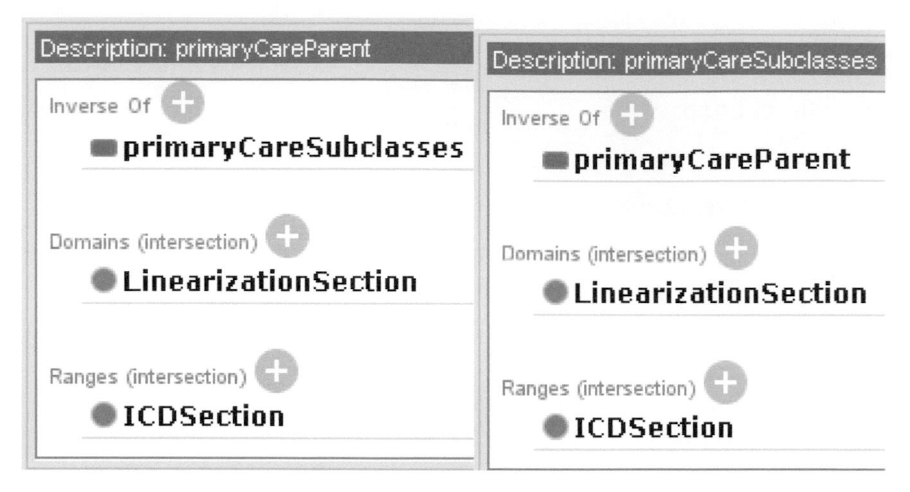

Figure 11.9 An occurrence of circular properties in the ICF ontology. Left: *primaryCare-Parent* property definition. Right: *primaryCareSubclasses* property definition. They are the inverse of each other but have the same domain and range.

To show a case of the **Lonely disjoint class** deficiency, we used the ontology GRO[14] from TONES. Figure 11.8 shows a slice of its class hierarchy: the Locus class (left side) is selected and the image shows a list of disjoint classes (right side). Locus is disjoint from Conformation, Direction, Orientation, and Phenotype classes, from the same branch of the class hierarchy, and is disjoint from PackedDNA and UnpackedDNA classes, from another branch. This is an example of a situation that may require the attention of the ontology authors. In some situations, the ontology can be redesigned to fix these badly designed areas.

To show the **Circular properties** anomaly, we used the ontology ICF (international classification of functioning, disability, and health), from BIOPORTAL. We found six object properties with this circularity in it. Figure 11.9 shows the occurrence of this anomaly in two object properties that are the inverse of each other but have the same domain and range: *primaryCareParent* (in the left of the figure) and *primaryCareSubClasses* (in the right of the figure). The consequence of this anomaly is that all instances that have these properties will be classified in both classes LinearizationSection and ICDSection.[15]

[14]GRO: https://goo.gl/VNy8ak
[15]ICF: http://bioportal.bioontology.org/ontologies/ICF

11.6.3 Rules are not used in ontologies

SWRL rules are not common in our ontology set. The few rule sets found were usually small and did not have many problems. To be able to test rules, we had to go after ontologies with rule sets and include them in the experiments. For anomalies related to SWRL rules, five of the nine verified anomaly types were found. Bellow, we show three anomaly examples detected by ONTO-Analyst:

Circularity between rules and taxonomy: - An example of this circularity is found in the BCGO[16] ontology form BIOPORTAL. This ontology is used to assign a grade to a breast cancer tumor. The ontology contains the following rule that creates a circularity between *Nucleus* and its subclass *Mitosis*. The rule should be considered as a restricted *subClassOf* relation between *Nucleus* and *Mitosis*.

```
 Nucleus(?x), hasEccentricity(?x, ?value),
 greaterThan(?value, 0), lessThan(?value, 1)
->
 Mitosis(?x)
```

Redundant derivation in the antecedent:- This anomaly was found 13 times in an ontology from GIT, called contextOntology[17]. The following rule shows an occurrence of this redundancy. The redundancy occurs because the class *face-recognition-sensor* is a subclass of *context-element*. In other words, the atom *context-element(?faceRef)* is not necessary in the rule because all instances of *face-recognition-sensor* are instances of *context-element*.

```
 alarm-state-sensor(?alarmRef),context-element(?alarmRef),
 context-element(?faceRef),
 face-recognition-sensor(?faceRef),
 has-value-of-service(?alarmRef, ?alarmValue),
 has-value-of-service(?faceRef, ?faceValue),
 equal(?alarmValue, 0), equal(?faceValue, 1)
->
 AcceptableSensorValue(AlarmStateSensorI, true)}
```

[16]BCGO: http://bioportal.bioontology.org/ontologies/BCGO
[17]ContextOntology: https://goo.gl/5NidDu

Rule subsumption: - To demonstrate this anomaly, two rules found in an ontology of PAPERS are shown. The ontology was developed so that researchers can share and integrate data on autism [20]. The two rules that have this redundancy are the following:

```
   Rule1:
     AUTISMC000000(?a),
     ADI_2003_acqorlossoflang_aphrase(?a, ?phraseage),
     SubjectKey(?a, ?subjectID),
     createOWLThing(?phenotype, ?subjectID),
     greaterThan(?phraseage, 33)
    ->
     AUTISMC1000033(?phenotype),
     subject_has_quality_or_disposition(?subjectID,?phenotype)
Rule2:
  AUTISMC000000(?a),
  ADI_2003_acqorlossoflang_aphrase(?a, ?phraseage),
  SubjectKey(?a, ?subjectID),
  createOWLThing(?phenotype, ?subjectID),
  greaterThan(?phraseage, 36)
 ->
  AUTISMC1000033(?phenotype),
  subject_has_quality_or_disposition(?subjectID,?phenotype)
```

The only difference between the two rules is the atom with *greaterThan* in which the variable *?phraseage* is tested as greater than 33; in the first rule, and, in the second, the variable is tested as greater than 36. Therefore, it is possible to conclude that rule 2 subsumes rule 1. This ontology has 152 rules and all rules have many atoms. The management of such rule set and the ontology can be quite complex without techniques that help users identify anomalies in it.

Finally, it is important to highlight that ONTO-Analyst can find problems in rules without having to fire them. Reasoners can only check rules that fire. If a rule with a problem does not satisfy the requirements to be fired, it does not generate inferences in the ontology, and consequently, it is not checked by reasoners. That can be problematic given the fact that ontologies may lack instances that trigger problematic rules, or lack any instances, or even have rules that are never fired (due to errors in them).

11.6.4 Top 10 bioportal ontologies

When compared to the whole ontology set, the top 10 had a similar result profile. The top 2 anomalies were the same: *Lazy class or property* and *Chains of inheritance*. They point out the fact that even the top 10 ontologies are mostly formed by taxonomic structures.

Another interesting point is that ONTO-Analyst was able to find other kinds of anomalies in the top 10 set. It shows that even well used better quality ontologies still have problems in their structures. They can benefit from using ONTO-Analyst.

11.7 Conclusion

This chapter describes a system, called ONTO-Analyst, used to identify anomalies in OWL ontologies (with or without SWRL rules). ONTO-Analyst can identify certain patterns/problems/anomalies using SPARQL queries in ontologies converted to the MetaFOR format (based on an ontology to describe the structure of ontologies). Furthermore, this chapter described experiments performed to identify problems, using the ONTO-Analyst tool, in ontologies from 4 well-established ontology repositories.

In total, 628 ontologies were downloaded and 608 were analyzed. Our results show that the ONTO-Analyst system was able to automatically identify a set of representative anomalies, described in the scientific literature, in these ontologies. Given the importance of the four ontology repositories used and the broad scope covered by the 608 ontologies, the ONTO-Analyst results represent also an analysis of the field of public OWL ontologies covering some of the most used and relevant OWL public ontologies today.

The results showed that inconsistencies are less common than other anomalies. They also showed that most ontologies are mainly taxonomies. No anomaly occurrences were found in cardinalities, mainly because only seven cardinality declarations were found in the 608 analyzed ontologies. Clearly, these ontologies do not use most of the modeling capabilities offered by the OWL language. The same can be said for the top 10 ontologies. The reasons why that is happening deserve some further research. However, the presence of anomalies may not make an ontology difficult to use in practice, but it certainly indicates areas where ontologies should be examined and improved.

Ontologies can be hard to administer without tools to help in the identification of problems. The majority of the 18 anomaly types used in the experiments have been described since 2001 [7]. Given its user extensibility using SPARQL, ONTO-Analyst is well suited to this task.

Acknowledgments

This work was financed in part by the Coordenação de Aperfeičoamento de Pessoal de Nível Superior - Brasil (CAPES) - Finance Code 001. Additional support was provided by grant U54 HG004028 from the U.S. National Institutes of Health.

References

[1] Tim Berners-Lee, James Hendler, and Ora Lassila. The Semantic Web. *Scientific American*, pages 29–37, May 2001.

[2] Joachim Baumeister and Dietmar Seipel. Anomalies in ontologies with rules. *Web Semantics: Science, Services and Agents on the World Wide Web*, 8(1):55–68, March 2010.

[3] Ian Horrocks, Peter F. Patel-Schneider, Harold Boley, Said Tabet, Benjamin Grosof, and Mike Dean. SWRL: A Semantic Web Rule Language Combining OWL and RuleML, May 2004.

[4] Vijayan Sugumaran and Jon Atle Gulla, editors. *Applied Semantic Web Technologies*. Auerbach Publications, Boca Raton, FL, 1 edition edition, August 2011.

[5] João Paulo Orlando, Mark A. Musen, and Dilvan A. Moreira. User Extensible System to Identify Problems in OWL Ontologies and SWRL Rules. In *Rule Technologies: Foundations, Tools, and Applications*, number 9202 in Lecture Notes in Computer Science, pages 112–126. Springer International Publishing, August 2015. DOI: 10.1007/978-3-319-21542-6_8.

[6] M. Fahad, A. Qadir, and M. W. Noshairwan. Semantic Inconsistency Errors in Ontology. In *IEEE International Conference on Granular Computing, 2007. GRC 2007*, pages 283–283, November 2007.

[7] Asunción Gómez-Pérez. Evaluation of ontologies. *International Journal of Intelligent Systems*, 16(3):391–409, March 2001.

[8] Du Zhang. Fixpoint semantics for rule-base anomalies. In *Fourth IEEE Conference on Cognitive Informatics, 2005. (ICCI 2005).*, pages 10–17, August 2005.

[9] L. Zhou, H. Huang, G. Qi, Y. Ma, Z. Huang, and Y. Qu. Measuring Inconsistency in DL-Lite Ontologies. In *IEEE/WIC/ACM International Joint Conferences on Web Intelligence and Intelligent Agent Technologies, 2009. WI-IAT '09*, volume 1, pages 349–356, September 2009.

[10] Yuzhang Feng, Yang Liu, Yuan-Fang Li, and Daqing Zhang. Discovering Anomalies in Semantic Web Rules. In *2010 Fourth International Conference on Secure Software Integration and Reliability Improvement (SSIRI)*, pages 33–42, June 2010.

[11] Yunchuan Sun, Junsheng Zhang, Wei Zhao, and Yingjie Tian. Managing and Refining Rule Set for SWRL. In *4th International Conference on Wireless Communications, Networking and Mobile Computing, 2008. WiCOM '08*, pages 1–5, October 2008.

[12] María Poveda-Villalón. *Ontology Evaluation: a pitfall-based approach to ontology diagnosis*. Ph.D. Thesis, Universidad Politécnica de Madrid, Madrid, 2016.

[13] S. Hussain, J. D. Roo, A. Daniyal, and S. S. R. Abidi. Detecting and Resolving Inconsistencies in Ontologies Using Contradiction Derivations. In *2011 IEEE 35th Annual Computer Software and Applications Conference*, pages 556–561, July 2011.

[14] Z. Hui, Z. Zhen, and Z. Junwu. A Method of Eliminating Direct Super-Class to Detect Cycle of Concepts. In *International Forum on Information Technology and Applications, 2009. IFITA '09*, volume 3, pages 7–10, May 2009.

[15] Matthew Horridge and Sean Bechhofer. The OWL API: A Java API for OWL Ontologies. *Semant. web*, 2(1):11–21, January 2011.

[16] Muhammad Fahad, Muhammad Abdul Qadir, and Syed Adnan Hussain Shah. Evaluation of Ontologies and DL Reasoners. In Zhongzhi Shi, E. Mercier-Laurent, and D. Leake, editors, *Intelligent Information Processing IV*, number 288 in IFIP – The International Federation for Information Processing, pages 17–27. Springer US, October 2008. DOI: 10.1007/978-0-387-87685-6_5.

[17] Muhammad Fahad and Muhammad Abdul Qadir. A Framework for Ontology Evaluation. In *Proceedings of the 16th International Conference on Conceptual Structures*, volume 354, pages 149–158, Toulouse, France, 2008.

[18] Mathieu d'Aquin and Natalya F. Noy. Where to publish and find ontologies? A survey of ontology libraries. *Web Semantics: Science, Services and Agents on the World Wide Web*, 11:96–111, March 2012.

[19] Janna Hastings, Michel Dumontier, Duncan Hull, Matthew Horridge, Christoph Steinbeck, Ulrike Sattler, Robert Stevens, Tertia Hörne, and Katarina Britz. Representing Chemicals using OWL, Description Graphs and Rules. In *Proceedings of the 7th International Workshop on OWL: Experiences and Directions (OWLED 2010)*, pages 1–10, 2010.

[20] L. Young, S.W. Tu, L. Tennakoon, D. Vismer, V. Astakhov, A. Gupta, J.S. Grethe, M.E. Martone, A.K. Das, and M.J. McAuliffe. Ontology driven data integration for autism research. In *22nd IEEE International Symposium on Computer-Based Medical Systems, 2009. CBMS 2009*, pages 1–7, August 2009.

12

Ontological Modeling for the Personalization of Learning Environment of the University

Tatiana Avdeenko, Marina Murtazina, and Natalia Pustovalova

Novosibirsk State Technical University, Russia
E-mail: tavdeenko@mail.ru; murtazina@corp.nstu.ru; nvpustovalova@gmail.com

Abstract

This chapter is devoted to the study of the possibilities of using ontological models for knowledge representation to analyze the relationship between psychometric and neurophysiological data and a student's performance. Here, we discuss the issues of managing the process of forming professional skills through the digital environment of educational organizations. Management of the process of forming professional skills is based on the personalization of the issuance of educational content. At the center of educational activity is the student with his individual psychological and neurophysiological characteristics, which are reflected in the development of the personality at all stages of the professional path, the degree of formation, and manifestation of professional abilities. Therefore, our study is focused on representing the results of diagnosing psychological and neurophysiological student's characteristics, academic performance, and relationships between them as a learner model.

Keywords: Cognitive function, educational content, Ontology, personal learning environment, learner model, psychometric characteristic.

12.1 Introduction

This chapter discusses the issues of building a conceptual model of knowledge representation in relation to solving the problem of creating a personal educational environment for learners that take into account their individual characteristics when building educational trajectories and choosing the type of educational content. Within the framework of the currently widespread concept of life-long learning (learning throughout life), the learning process can affect students of all ages, from preschoolers to the elderly. At the center of educational activity is the learners with their individual psychological and neurophysiological characteristics, which are reflected in the development of the individuals at all stages of their professional path, the degree of formation, and manifestation of professional abilities. In this regard, our study is focused on integrating into the learner model (in this chapter, we consider university students as learners) the results of diagnosing individual psychological and neurophysiological characteristics, the results of academic performance, and their relationships.

Ontologies were chosen as a model of knowledge representation as a modern tool that successfully combines the most attractive features of existing approaches. Ontologies provide not only a visual way of representing the concept hierarchy with a built-in property inheritance mechanism, as well as a way of linking concepts through various types of associative relations (semantic network), but also the ability to implement the so-called axioms, being the knowledge of a more complex type, presented in the form of logical rules, with the possibility of subsequent application of logical inference to obtain new facts and knowledge (a limited logical model of knowledge representation, for example, in the form of Horn clauses, as in the swi-prolog language).

In the last decade, ontologies have become widespread due to a significant expansion of their functionality. Indeed, due to the fact that in recent years tools and languages for knowledge representation and processing have been actively developed, in particular OWL, Turtle, the SPARQL query language, RDF notation, and much more, it has become possible to expand the typical capabilities of ontologies in the field of knowledge representation by processing data based on semantics and advanced semantic search [1]. Also, due to the development of the next generations of XML and HTML, it became possible to enrich ontological models using web technologies, namely, annotation [2], tagging [3], etc. Thanks to this, ontologies, as a means of representing knowledge, have acquired additional possibilities for representing semantic and structural relationships for describing application domains.

Thanks to the development of relatively easy-to-learn and easy-to-use ontology creation tools (Protégé, Ontolingua, OntoSaurus, WebOnto, OntoEdit, etc.), a lot of knowledge description projects for various application domains have appeared. Many research teams not only use ontologies in their work but also publish them for use by other researchers. Therefore, ontologies are becoming an increasingly popular tool for formalizing knowledge and exchanging knowledge to solve specific applied, often interdisciplinary, tasks that require semantic data analysis or knowledge inference.

The major result of the present work is the system of ontologies implemented in the Protégé 5.5 environment, which combines ontologies for information support of the process of forming professional skills and an ontology of cognitive functions. The chapter has the following structure. Section 12.2 gives a brief paper overview in the development of ontological models for the purposes of managing the process of formation and development of professional skills based on psychodiagnostic data, knowledge about the systemic organization of cognitive functions, and their connection with the learning cycle. Section 12.3 presents a formal model that combines ontologies for information support of the process of forming professional skills and an ontology of cognitive functions. An outstanding feature of the suggested approach from others presented in the literature is the mechanism for integrating knowledge in the field of theory and practice of digital educational content management, practical psychodiagnostics, and neurocognitive science. Section 12.4 describes the implementation of the proposed system of ontologies in the Protégé 5.5 environment. In Section 12.5, we describe the proposed application of the constructed ontology to create a personal learning environment (PLE). Section 12.6 contains conclusions on the chapter.

12.2 Application of an Ontological Approach to Managing the Process of Forming Professional Skills based on Psychodiagnostics

Psychodiagnostics, one of the key branches of psychology, is engaged in a comprehensive study of the issues of measuring, evaluating, and analyzing individual psychological and neurophysiological characteristics of a person. In the field of view of practical psychodiagnostics are the diagnostics of cognitive mental processes, intelligence, mental states, and individual personality traits. At present, psychodiagnostics has already firmly become a part of psychological knowledge in various fields of professional activity, including education, healthcare, and medicine. A scientific knowledge content-rich accumulation about the relationship between brain activity patterns and mental states, processes, and intelligence has occurred in the early 21st century. In the last few years, scientific research trends are aimed at identifying relationships between psychometric and neurophysiological data based on machine learning methods and ontological models of knowledge representation [4–6], as well as the study of the possibilities of integrating knowledge about the management of professional activities and the results of psychodiagnostics to improve feedback mechanisms in control systems [7]. In our work, we suppose that professional skill development is based on the personalization of the issuance of educational content.

12.2.1 Application of Ontologies to Represent Knowledge About the Study of Cognitive Functions

By the beginning of the 21st century, many studies of human cognition and intelligence were carried out, aimed at explaining the structure of intelligence, the correlation of individual characteristics when performing a variety of tasks, and describing cognitive functions and methods for measuring them. In order to systematize the accumulated knowledge and reasonably combine them, it became necessary to create appropriate information resources. One of the first such resources was the Phenowiki and the Cognitive Atlas [8, 9]. The first project contains wiki-like descriptions of cognitive tasks and cognitive concepts. As part of the second project, a formal ontology of mental concepts, mental tasks, and mental processes was created. A unique resource in the field of analysis and visualization of electroencephalogram signals is the Cloudwave project. This project uses the epilepsy and seizure ontology (EpSO). The ontology contains class sets for annotating EEG signals, as well as classes for presenting knowledge of brain anatomy and placing EEG electrodes on the scalp. Ontology allows accumulating knowledge about pathological and normal EEG patterns [10].

The first works marked the beginning of active research in this direction, which continues to this day. Thus, in [11], it is argued that in order to understand the correspondence between brain function and mental processing, it is necessary to move from brain mapping strategies to the search for selective associations. To solve this problem, it is advisable to use formal ontologies that describe the structure of mental processes. In [12], a neuropsychological testing ontology is presented. The class structure of the ontology is designed to accumulate the results of neuropsychological tests and indicate their connection with the state of cognitive functioning of the subjects. In [13], it is proposed to use a formal ontological model to standardize the terminology required when performing EEG mapping. The work [14] discusses the potential role of cognitive individual differences in the context of developing ontologies for personalizing user characteristics. In [15], a psychologically oriented ontology CCOnto is presented for identifying state markers and character traits of university students. In [16], a neurocognitive integrated ontology was proposed, which combines knowledge related to neuropsychological tests, mild cognitive impairment in Alzheimer's disease, cognitive functions, and brain regions. The work [16] discusses the joint use of ontological models of knowledge representation and data mining algorithms.

Despite the large number of works in this area, it should be noted that the above works are focused on separate aspects of the representation of knowledge about the methods of studying the brain, and the relationship between the psychological and neurophysiological results of individual diagnostics. Therefore, the construction of an integrated ontological model that will accumulate knowledge about base cognitive functions and methods for their study from the neuropsychology and neurophysiology standpoint is relevant.

12.2.2 Ontologies as a Mechanism for Implementing a Personalized Approach in Professional Activities

An ontology can be determined as a set of domain objects, which includes a glossary of terms, semantic relationships from which new information can be obtained through "the inference," and intelligent queries for any specific domain [17, 18]. A separate scientific area deals with the application of ontologies in the implementation of a personalized approach in various areas of professional activity: in healthcare [19], in corporate governance [20], in the activities of educational organizations, and in the field related to providing educational services [21–25].

Ontologies are also used to manage knowledge, skills, and competencies at a university [26]. In the face of the 2020 pandemic, educational organizations were forced to accelerate the introduction and improvement of digital tools and information systems to support remote interaction between actors in the educational environment, as a result of which they faced new challenges and requests. Today, personalization is one of the most significant trends in education at all levels. Universities are actively creating a personalized educational environment for students to develop their professional competencies.

Employers use personalized solutions to monitor and control employee skill levels. At the same time, not only the IT architecture of educational organizations is undergoing active changes, the requirements for the educational process, its structure, the composition of participants (actors), as well as the means used are changing. Therefore, intellectual technologies, including ontologies, are most actively used in education [27]. This allows not only to obtain additional knowledge based on existing facts but also to reduce the amount of memory used by storing functions that describe dependencies (that is, knowledge), and not the data itself. Therefore, ontologies are used for the following purposes:

- to present the experience of experts in specific disciplines (mathematical analysis, English, programming, etc.); thus, those knowledge and skills that need to be taught (domain model or module-teacher) are determined [29];
- to represent the features and characteristics of various kinds of learners, mainly students and schoolchildren, which have a significant impact on the educational result (learner model) [30];
- to represent the knowledge about the educational process and its results, for example, about how an educational strategy can be built and implemented to achieve the required educational result or about what resources need to be used (tutor model) [31].

Developers of information systems for educational institutions face the question of what data about students should be collected. Indeed, incorrectly formulated requirements for input information can lead to the fact that the learner model will not correspond to reality, and the system will operate with large amounts of data, which leads to an increase in costs, and at the same time will not provide adequate information for making managerial decisions. Therefore, it is very important to carefully design the input data for building a learner model, that is, to select those individual characteristics of students that will be taken into account by the university when building PLE in order to improve the quality of education.

According to [28], the architecture of intelligent learning systems is based on a four-component model. Its structure, which ensures the integration of the main types of knowledge necessary for organizing the learning process (knowledge about strategies and methods of teaching, about the application domain, and about the learner [32–35]), includes a module containing data about the learner, a module-teacher, a module containing data about application domain (expert), and interface. The implementation of individual modules and interfaces between them are examples of private university architectures. At the same time, artificial intelligence tools, namely, ontologies, are quite often used in modeling and implementing certain components of such an architecture [27].

The above analysis shows the relevance of developing mechanisms for managing the process of forming professional skills based on ontological models for representing knowledge about the results of psychodiagnostics, their relationship with academic performance, systemic organization of cognitive functions, and the learning cycle.

12.3 Formal Ontological Model for Managing the Process of Forming Professional Skills based on Individual Psychometric and Neurophysiological Assessment

A system of ontological models for the process of forming professional skills based on psychometric and neurophysiological assessment of individual cognitive abilities can be presented as a tuple:

$$Z=\left\langle \mathrm{Onto}^{\mathrm{Learner}}, \mathrm{Onto}^{\mathrm{EduContent}}, \mathrm{Onto}^{\mathrm{CognitFunction}}, R \right\rangle, \quad (12.1)$$

where $\mathrm{Onto}^{\mathrm{Learner}}$ is the learner ontology, which presents individual results of psychodiagnostics and academic performance.

$\mathrm{Onto}^{\mathrm{EduContent}}$ is the ontology of educational content, where knowledge on the theory and practice of managing digital educational content is accumulated.

$\mathrm{Onto}^{\mathrm{CognitFunction}}$ is the ontology of cognitive functions.

$R=\{r_i | i=\overline{1,k}\}$ is a finite set of binary relations, which is a union of sets $R=R_{\mathrm{EduContent}}^{\mathrm{Learner}} \cup R_{\mathrm{CognitiveFunction}}^{\mathrm{EduContent}} \cup R_{\mathrm{Learner}}^{\mathrm{CognitiveFunction}}$, $R_{\mathrm{EduContent}}^{\mathrm{Learner}}$ is a finite set of binary relations defined on classes, $R_{\mathrm{EduContent}}^{Learner} \subseteq C^{\mathrm{Learner}} \times C^{\mathrm{EduContent}}$, combining two subsets of relations $R_{\mathrm{EduContent}}^{\mathrm{Learner}} = R_{\mathrm{ISA}}^{\overset{\mathrm{Learner}}{\mathrm{EduContent}}} \cup R_{\mathrm{ASS}}^{\overset{\mathrm{Learner}}{\mathrm{EduContent}}}$, $R_{\mathrm{ISA}}^{\overset{\mathrm{Learner}}{\mathrm{EduContent}}}$ $R_{\mathrm{ISA}}^{\overset{\mathrm{Learner}}{\mathrm{EduContent}}}$ is a set of transitive, antisymmetric, and irreflexive hierarchy relations between classes, $R_{\text{ASS}}^{\overset{\text{Learner}}{\text{EduContent}}}$ is a set of associative relations defined on ontology classes, $R_{\text{CognitiveFunction}}^{\text{EduContent}}$ is a finite set of binary relations defined on classes, $R_{\text{CognitiveFunction}}^{\text{EduContent}} \subseteq C^{\text{EduContent}} \times C^{\text{CognitiveFunction}}$, combining two subsets of relations $R_{\text{CognitiveFunction}}^{\text{EduContent}} = R_{\text{ISA}}^{\overset{\text{EduContent}}{\text{CognitiveFunction}}} \cup R_{\text{ASS}}^{\overset{\text{EduContent}}{\text{CognitiveFunction}}}$, and $R_{\text{Learner}}^{\text{CognitiveFunction}}$ is a binary relation finite set defined on classes, $R_{\text{Learner}}^{\text{CognitiveFunction}} \subseteq C^{\text{CognitiveFunction}} \times C^{\text{Learner}}$, combining two subsets of relations $R_{\text{Learner}}^{\text{CognitiveFunction}} = R_{\text{ISA}}^{\overset{\text{CognitiveFunction}}{\text{Learner}}} \cup R_{\text{ASS}}^{\overset{\text{CognitiveFunction}}{\text{Learner}}}$.

The learner ontology $\text{Onto}^{\text{Learner}}$ can be represented by a tuple:

$$\text{Onto}^{\text{Learner}} = \left\langle C^{\text{Learner}}, R^{\text{Learner}}, S^{\text{Learner}}, D, A^{\text{Learner}}, E^{\text{Learner}} \right\rangle, \quad (12.2)$$

where $C^{\text{Learner}} = \{c_i^{\text{Learner}} | i = \overline{1,n}\}$ is a finite non-empty set of classes that make up the basic concepts of the learner model, which is a union of sets $C^{\text{Learner}} = C_{\text{AP}}^{\text{Learner}} \cup C_{L}^{\text{Learner}} \cup C_{P}^{\text{Learner}} \cup C_{\text{PLT}}^{\text{Learner}} \cup C_{\text{PLS}}^{\text{Learner}} \cup C_{\text{PPC}}^{\text{Learner}} \cup C_{\text{TT}}^{\text{Learner}}$; $C_{\text{AP}}^{\text{Learner}}$ is a set of classes for representing knowledge about the academic performance and achievements of the student, C_{L}^{Learner} is a class describing the identification characteristics of students, C_{P}^{Learner} is a set of classes necessary for categorization and storage of personal, motivational, and cognitive characteristics of students, $C_{\text{PLT}}^{\text{Learner}}$ is a set of classes describing personal learning trajectory of the student,$C_{\text{PLS}}^{\text{Learner}}$ is a set of classes describing personal learning styles, C_{PPC}^{Learner} is a set of classes describing types of information perception, and $C_{\text{TT}}^{\text{Learner}}$ is a set of classes describing types of psychometric tests.

$R^{\text{Learner}} = \{r_i^{\text{Learner}} | i = \overline{1,t}\}$ is a binary relation finite set defined on classes, $R^{\text{Learner}} \subseteq C^{\text{Learner}} \times C^{\text{Learner}}$, combining two subsets of relations $R^{\text{Learner}} = R_{\text{ISA}}^{\text{Learner}} \cup R_{\text{ASS}}^{\text{Learner}}$.

$S^{\text{Learner}} = \{s_i^{\text{Learner}} | i = \overline{1,w}\}$ is a data properties finite set defined for class instances as a relation $S^{\text{Learner}} \subseteq C^{\text{Learner}} \times D$.

$D = \{d_i | i = \overline{1,s}\}$ is a set of XML schema data types.

$A^{\text{Learner}} = \{a_i^{\text{Learner}} | i = \overline{1,p}\}$ is a logical axiom finite set of the learner ontology $\text{Onto}^{\text{Learner}}$, which is a union of sets $A^{\text{Learner}} = A_{\text{SubClass}}^{\text{Learner}} \cup A_{\text{Equivalent}}^{\text{Learner}} \cup A_{\text{Disjoint}}^{\text{Learner}} \cup A_{\text{Enumeration}}^{\text{Learner}} \cup A_{\text{RelationProperties}}^{\text{Learner}} \cup A_{\text{Chain}}^{\text{Learner}}$, $A_{\text{SubClass}}^{\text{Learner}}$ is a set of inheritance axioms, $A_{\text{Equivalent}}^{\text{Learner}}$ is a set of equivalence axioms, $A_{\text{Disjoint}}^{\text{Learner}}$ is a set of class disjoint axioms, $A_{\text{Enumeration}}^{\text{Learner}}$ is a set of class enumeration axioms, $A_{\text{RelationProperties}}^{\text{Learner}}$ is a set of axioms for specifying the characteristics of binary relations, such as transitivity, symmetry, asymmetry, etc., $A_{\mathrm{Chain}}^{\mathrm{Learner}}$ is a set of axioms for establishing relationship properties between class instances, and $E^{\mathrm{Learner}}=\{e_i^{\mathrm{Learner}}|i=\overline{1,a}\}$ is a class instance set.

The educational content ontology $\mathrm{Onto}^{\mathrm{EduContent}}$ can be represented as a tuple:

$$\mathrm{Onto}^{\mathrm{EduContent}}=\langle C^{\mathrm{EduContent}}, R^{\mathrm{EduContent}}, S^{\mathrm{EduContent}}, D, A^{\mathrm{EduContent}}, E^{\mathrm{EduContent}}\rangle, \quad (12.3)$$

where $C^{\mathrm{EduContent}}=\{c_i^{\mathrm{EduContent}}|i=\overline{1,m}\}$ is a finite non-empty set of classes that describe the basic concepts of the application domain "Educational content," which is a union of sets $C^{\mathrm{EduContent}}=C_{\mathrm{CP}}^{\mathrm{EduContent}}\cup C_{\mathrm{ICC}}^{\mathrm{EduContent}}\cup C_{\mathrm{LCA}}^{\mathrm{EduContent}}\cup C_{\mathrm{LCF}}^{\mathrm{EduContent}}\cup C_{\mathrm{LCC}}^{\mathrm{EduContent}}$, $C_{\mathrm{CP}}^{\mathrm{EduContent}}$ is a set of classes to represent knowledge about the components of educational content, $C_{\mathrm{ICC}}^{\mathrm{EduContent}}$ is a set of classes to represent the learning process according to the Kolb's cycle, $C_{\mathrm{LCA}}^{\mathrm{EduContent}}$ is a set of classes that describe the actions available to participants in the educational process, $C_{\mathrm{LCF}}^{\mathrm{EduContent}}$ is a set of classes describing the purpose of educational content objects, $C_{\mathrm{LCC}}^{\mathrm{EduContent}}$ is a set of classes describing types of educational content objects, and $R^{\mathrm{EduContent}}=\{r_i^{\mathrm{EduContent}}|i=\overline{1,u}\}$ is a binary relation finite set defined on classes, $R^{\mathrm{EduContent}}\subseteq C^{\mathrm{EduContent}}\times C^{\mathrm{EduContent}}$, including two subsets of relations $R^{\mathrm{EduContent}}=R_{\mathrm{ISA}}^{\mathrm{EduContent}}\cup R_{\mathrm{ASS}}^{\mathrm{EduContent}}$, $S^{\mathrm{EduContent}}=\{s_i^{\mathrm{EduContent}}|i=\overline{1,h}\}$ is a data properties finite set defined for class instances as a relation $S^{\mathrm{EduContent}}\subseteq C^{\mathrm{EduContent}}\times D$, $A^{\mathrm{EduContent}}=\{a_i^{\mathrm{EduContent}}|i=\overline{1,q}\}$ is a logical axiom finite set of educational content ontology $\mathrm{Onto}^{\mathrm{EduContent}}$, which is a union of sets $A^{\mathrm{EduContent}}=A_{\mathrm{SubClass}}^{\mathrm{EduContent}}\cup A_{\mathrm{Equivalent}}^{\mathrm{EduContent}}\cup A_{Disjoint}^{\mathrm{EduContent}}\cup A_{\mathrm{RelationProperties}}^{\mathrm{EduContent}}\cup A_{\mathrm{Chain}}^{\mathrm{EduContent}}$, and $E^{\mathrm{EduContent}}=\{e_i^{\mathrm{EduContent}}|i=\overline{1,f}\}$ is a class instance set.

The ontology $\mathrm{Onto}^{\mathrm{CognitFunction}}$ can be represented as a tuple:

$$\mathrm{Onto}^{\mathrm{CognitFunction}}=\langle C^{\mathrm{CognitFunction}}, R^{\mathrm{CognitFunction}}, S^{\mathrm{CognitFunction}}, D, A^{\mathrm{CognitFunction}}, E^{\mathrm{CognitFunction}}\rangle, \quad (12.4)$$

where $C^{\mathrm{CognitFunction}}=\{c_i^{\mathrm{CognitFunction}}|i=\overline{1,y}\}$ is a finite non-empty set of classes that describe the main concepts of the subject area "Cognitive functions," which is a union of sets $C^{\mathrm{CognitFunction}}=C_{\mathrm{BN}}^{\mathrm{CognitFunction}}\cup$

$C_{\text{EEG}}^{\text{CognitFunction}} \cup C_{\text{CFAP}}^{\text{CognitFunction}} \cup C_{\text{PP}}^{\text{CognitFunction}}$, $C_{\text{BN}}^{\text{CognitFunction}}$ is a set of classes for representing knowledge about large-scale brain networks, $C_{\text{EEG}}^{\text{CognitFunction}}$ is a set of classes for representing knowledge about the EEG as a method for studying the unconscious components of mental states, $C_{\text{CFAP}}^{\text{CognitFunction}}$ is a set of classes describing cognitive functions, abilities, and processes, $C_{\text{PP}}^{\text{CognitFunction}}$ is a set of classes describing methods of psychodiagnostics, $R^{\text{CognitFunction}}=\{r_i^{\text{CognitFunction}} | i=\overline{1,x}\}$ is a binary relation finite set defined on classes, $R^{\text{CognitFunction}} \subseteq C^{\text{CognitFunction}} \times C^{\text{CognitFunction}}$, combining two subsets of relations $R^{\text{CognitFunction}}=R_{\text{ISA}}^{\text{CognitFunction}} \cup R_{\text{ASS}}^{\text{CognitFunction}}$, $S^{\text{CognitFunction}}=\{s_i^{\text{CognitFunction}} | i=\overline{1,o}\}$ is a data property finite set defined for class instances as a relation $S^{\text{CognitFunction}} \subseteq C^{\text{CognitFunction}} \times D$, $A^{\text{CognitFunction}} = \{a_i^{\text{CognitFunction}} | i = \overline{1,c}\}$ is a logical axioms finite set of the ontology of cognitive functions $\text{Onto}^{\text{CognitFunction}}$, which is a union of sets $A^{\text{CognitFunction}} = A_{\text{SubClass}}^{\text{CognitFunction}} \cup A_{\text{Equivalent}}^{\text{CognitFunction}} \cup A_{\text{Disjoint}}^{\text{CognitFunction}} \cup A_{\text{RelationProperties}}^{\text{CognitFunction}} \cup A_{\text{Chain}}^{\text{CognitFunction}}$, and $E^{\text{CognitFunction}} = \{e_i^{\text{CognitFunction}} | i = \overline{1,e}\}$ is a class instance set.

12.4 Implementation of Ontology Models in Protégé 5.5

The Protégé 5.5 editor was used to build the OWL ontologies described in the previous section. Each top-level class of OWL ontologies is a child class of the predefined owl:Thing class, which includes all domain objects. The ontology system is implemented as four owl files: Main.owl, LearnerModel.owl, EducationalContent.owl, and CognitiveFunction.owl. The Main.owl file enables combining other files and manipulate them as a single ontology. The LearnerModel.owl file is designed to represent knowledge about the student, and the EducationalContent.owl file is about educational content and pedagogical strategies for its issuance to the student, depending on the personal characteristics, determined by the results of psychodiagnostics. The CognitiveFunction.owl file is for accumulating interdisciplinary knowledge necessary for research cognitive functions of the human brain.

12.4.1 Learner Ontology

The learner ontology underlies the personalized digital educational environment. It is the relationship structure of this model that determines what output data will be collected and available for further processing and analysis. At the same time, researchers should take into account that each student has a relatively stable set of characteristics (cognitive, personal, and motivational) [36], which determines his/her capabilities in the educational process. But these relatively stable parameters can be influenced by the emotional state of the student (fear, grief, happiness, fun, etc.).

Although emotions certainly play an important role in the learning process, their influence is difficult to capture and evaluate. However, there are some studies about this question; for example, [37], from which we can conclude that personalized environment can not only take into account the influence of emotions but also purposefully use them to improve the educational process with proper technical and methodological equipment. An example is the computer games industry, where user experience design (UX – user experience) is a separate field of activity. There are attempts to transfer experience of using various stimuli (light, sound, content, etc.) to create a certain mood in other areas, in particular, in the development of educational content [38, 39].

To build the learner ontology, which integrates a set of student's characteristics, that affect the ability to perceive and assimilate information, as well as the relationships between these characteristics and learning goals, we used the methods of multivariate data analysis. Thus, the results of the Amthauer and Raven tests are supposed to be used as input data for assessing cognitive characteristics. To assess personal characteristics, we used the results of big five and Bachard tests. Motivational characteristics were assessed using Gray's short questionnaire. Also, the model must take into account academic performance, data on which is usually stored in the learning management system of an educational organization.

The choice of psychometric characteristics for testing students implies compliance with certain requirements. The test materials used should have a scientific basis, be standardized, and be adapted for the language of a particular audience. In our implementation of the learner model, the test materials indicated in Table 12.1 were used [34, 40].

Table 12.1 Description of the test materials used in building the learner model.

Stimulus material	Field of application
Intellect structure test by R. Amthauer (IST)	Analytical intelligence
Progressive matrices by J. Raven	Fluid intelligence
Big five, 5PFQ	Personality
Gray–Wilson personality questionnaire, GWPQ	Motivation
Emotional intelligence scales, by K. Bachard	Emotional intelligence

Table 12.2 The examples of object properties for learner ontology.

Name	Domain	Range	Description
hasPersonality	Learner	Personality	Determines the set of student's characteristics
learnerPerformDependsOn	LearnerPerformance	Personality	Presents the dependence between the academic performance and student's characteristics
prefLearnStyleDependsOn	PrefLearningStyle	Personality	Allows to set the relation between the learning style and the student's characteristics
prefPersChanDependsOn	PrefPerseptionChan	Personality	Allows to set the relation between the style of perception of educational content and the student's characteristics
decMakeStyle DependsOn	DecisionMakeStyle	Personality	Allows to set the relation between the decision-making style and the student's characteristics
describedBy TestsType	Personality	TestsType	Allows to set the relation between the types of tests and the student's characteristics
corresponds	PersonalLearningTrajectory	Personality	Allows to set the relation between the personal educational trajectory and the student's characteristics
dependsOnDecMakeStyle	PersonalLearningTrajectory	DecisionMakeStyle	Allows to specify the dependence of a personal educational trajectory on the style of decision making
dependsOn PrefLearStyle	PersonalLearningTrajectory	PrefLearningStyle	Allows to specify the dependence of the personal educational trajectory on the preferred learning style
dependsOn PrefPersChan	PersonalLearningTrajectory	PrefPerseptionChan	Allows to specify the dependence of a personal educational trajectory on the modality

The learner model implemented as the OWL ontology. We present the structure of ontology classes in Figure 12.1. Table 12.2 contains the examples of relations for learner ontology.

This OWL ontology includes knowledge about characteristics that can be directly measured through psychological tests or special hardware and software systems (for EEG recording).

We designed the learner class to store learner's identification data about each student, due to data properties of an instance of the learner class.

The LearnerPerformance class contains performance information that reflects academic (AcademicPerformance), sports (SportAchiev), and scientific (ScientificAchiev) achievements, as well as the project activities (ProjectAchiev). This structure makes it possible to take into account the different work-orientation of each person. Due to this structure, we propose to diagnose students' preferences and subsequent counseling to build an individual educational strategy or choose a direction. Second, by comparing

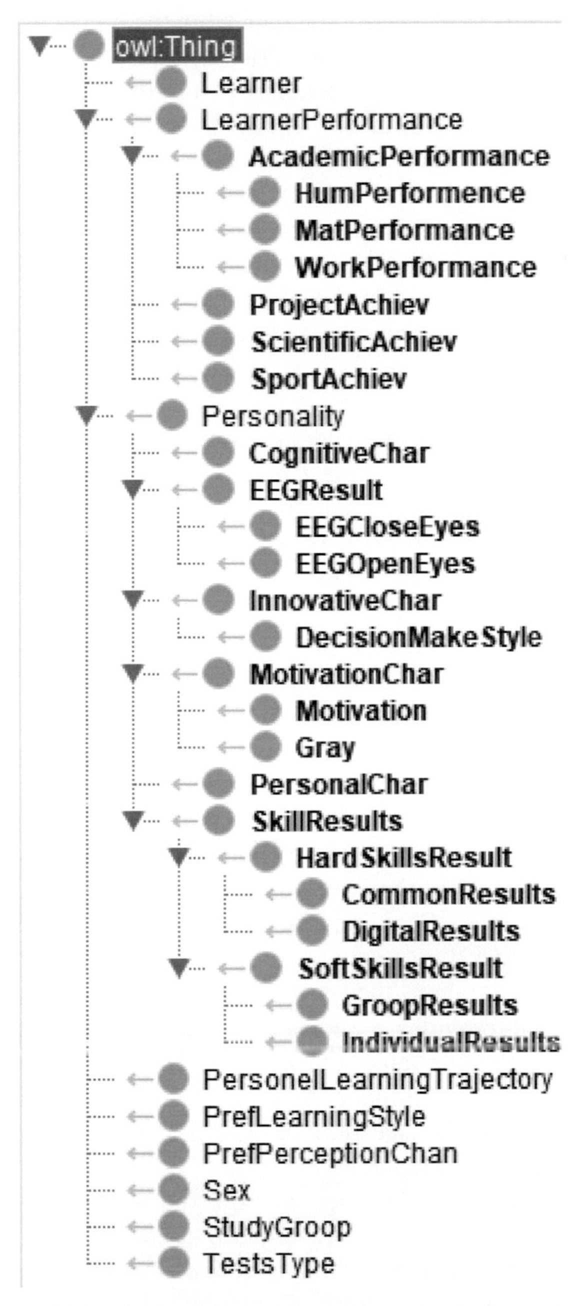

Figure 12.1 Class structure of learner ontology.

the level of student achievement with the data from the model of an expert in the field of professional activity, it is possible to make a decision about the compliance of students (students, specialists of the organization, etc.) with specific qualifications or requirements. Today, there are blocks of questions devoted to a person's sports and artistic preferences in diagnostic surveys on educational platforms. When interviewing, recruiters are sure to determine what interests the jobseeker or employee has, and whether he or she has achievements in these areas. All these aspects are essential for characterizing a person from the point of view of his professional development. Data values stored in instances of the class and subclasses of LearnerPerformance must be normalized according to the methodology for calculating the student's rating (for example, used in NSTU).

The personality class contains subclasses for categorizing and storing individual characteristics, including test results (cognitive, personality, and motivational characteristics) and results of measuring of instrumental tools (EEG) [37]. For the EEGResults class, two subclasses are provided for storing the parameters of electroencephalograms of individuals with closed and open eyes (EEGCloseEyes and EEGOpenEyes). For each data instance (for one test) from the EEGResults class, instances of the coherence assessment for all electrodes are added to the ontology. Through the data properties for the data instance from the CoherenceEstimation class, references to the data array of coherence estimates for different frequency intervals and the result of visualization of coherence estimates are established. The CoqnitiveChar subclass is designed to store data on the results of Amthauer tests (specialized by subtests) and Raven. Subclass MotivationChar contains the results of testing BIS (behavioral inhibition system) and BAS (behavioral activation system) by GWPQ. PersonalChar contains the results of the Bachard test (by subtests) and the big five (by categories). The DecisionMakeStyle subclass is designed to store data about the propensity of individuals to innovative thinking. Currently, it provides a subclass of InnovativChar, instances of which can take one of the five values that characterize the style of responding to changes depending on the results of testing. This element of the ontology is in the progress, since the question of which characteristics determine "innovativeness" requires further study.

The Skill Result class is designed to organize a connection with the ontology of the subject area (an expert model that enables defining a set of competencies, knowledge, and skills that a student should have for a particular subject area/educational program). Communication is realized through top-level ontology. According to modern pedagogical, methodological, and didactic paradigms, all skills are divided into Hard and Soft skills to determine the specific skills of an individual in the subject area and general ones, including cultural and social ones. In the HardSkillsResults class, two subclasses are distinguished (CommonResults and DigitalResults) to determine the level of subject skills that form the basis of professional competencies in the subject area, as well as to determine the level of digital technology proficiency and the general level of digital literacy. The SoftSkillsResults class contains subclasses for storing values that characterize the individual level of common cultural (IndividualResults) and social (GroopResults) competencies. Values for each person, stored as class instances, are calculated as a function of general academic performance (GPA) for the subject categories defined above in the AcademicPerformance class (MatPerformance – includes basic mathematical disciplines, HumPerformance – basic humanities disciplines, and WorkRelatPerformance – disciplines important from the point of view of future occupation).

PersonelLearningTrajectory is a class designed for connection with the educational content ontology. Based on the knowledge inference mechanism and the rules embedded in it, the student should be provided with educational content that takes into account the peculiarities of his or her perception and needs. Furthermore, the description of possible educational outcomes is also stored in the educational content ontology.

PrefLearningStyle allows to store information about what style of cognition is specific for a student in accordance with the D. Kolb cycle. Information about learning styles is presented in the cognitive function ontology, and the preferred style is determined due to the top-level ontology.

PrefPerceptionChan is a class for describing the preferred perception channel (audio, video, and kinesthetic), which is essential for forming an individual trajectory and recommendations for the issuance of educational content. This information is defined using the educational content ontology.

The TestsType class is required to store information about which tests are available to describe a learner model. In addition, this class is used to communicate with the cognitive function ontology.

The Sex and StudyCroup classes are needed to store information about students' sex, field of occupation, and study group.

12.4.2 Educational Content Ontology

Previous attempts to describe modern educational content have focused mainly on electronic educational resources (EERs) and similar forms of presentation. Now, with the development of technologies such as, for example, learning record store (LRS), when information systems try to take into account the offline activities of students, it makes sense to present a model of educational content that includes all possible forms of its organization, presentation, and use. Now we must take into account new aspects (for example, user experience) for designing of educational content, as well as implementing a learner model for analysis using information systems and technologies (digital diagnostics and digital footprint).

A personalized approach involves the issuance of content that meets the needs and abilities of students. Methodically, it is necessary to classify students into categories depending on patterns of content perception. For such classification designing, we propose an additional construct – the learning style. The learning style is understood as individually peculiar ways of assimilation of information in the learning process. The learning style influences the style of interaction with educational content. Therefore, when developing a model of educational content, it is worth considering the relationship between the structure and forms of its presentation, as well as those characteristics of students that affect the perception of content. Developing an educational content ontology, we will suppose that the learning style is determined by psychometric testing data. Analyzing a digital footprint allows us to track the results of issuing educational content according to the learning style (see Figure 12.2).

The class structure of the educational content ontology is shown in Figure 12.3. The examples of properties–relations are given in Table 12.3.

ContentPlan defines the components of educational content in terms of content's objects, its applications, and efficiency. It is also used to interact with the PersonelLearningTrajectory, PrefLearningStyle, and PrefPerceptionChan classes of the learner ontology. The preferred learning style and the

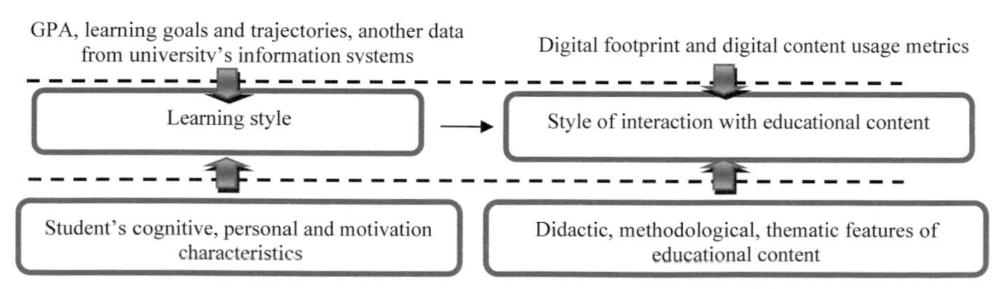

Figure 12.2 Relations between models of learning and educational content.

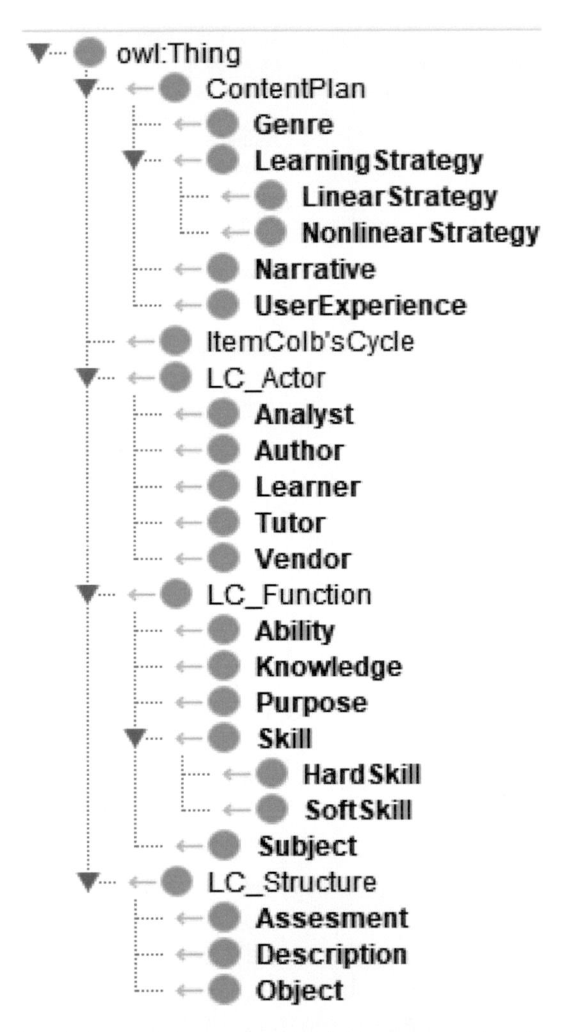

Figure 12.3 Relations between models of learning and educational content.

preferred channel of perception, together with data on individual characteristics, are necessary to determine the learning strategy within which the content plan is formed.

The Genre subclass contains information about the types of content organization (textbook, manual, quest, educational movie, etc.). The LearningStrategy subclass is necessary to store information about possible educational strategies, precisely defining linear and non-linear content organization

Table 12.3 The examples of object–properties for educational content ontology.

Name	Domain	Range	Description
contains	LC_Structure	LC_Function	Allows to set the dependence of the structure of educational content on the functions available to actors when working with it
createObject	Author	Object	Allows to set the relationship between the object and the author
create-Assessment	Author	Assessment	Allows to set the relationship between assessment materials and the author
isTutorEstimation	Tutor	Assessment	Allows to set the tutor's assessment of the quality of assessment materials
isAnalystEstimation	Analyst	Assessment	Allows to set the analyst's assessment of the quality of assessment materials
hasPlanned-Skills	Learner	Skill	Allows to specify the competencies that the student should receive in the learning process

(LinearStrategy and NonlinearStrategy, respectively) depending on the level of independence, educational goals, and the level of training of the student. The Narrative subclass is intended to describe stories if the content type implies them. The UserExperience subclass is intended to describe the emotional and psychological impact types to enhance the effect of educational materials if the genre suggests it.

The ItemKolb'sCycle class is necessary to connect with the cognitive function ontology and the learner ontology. The learning process is presented as a cycle and at its different stages, the content must perform the appropriate functions.

The LC_Actor class is needed to determine the actions available on educational content objects (Object, see the LC_Structure class) for various participants in the educational process. In addition, each of the subclasses Analyst, Author, Learner, Tutor, and Vendor as instances can store information about access rights to objects.

The LC_Function class defines the functional purpose of educational content objects. They can serve to form Ability, Knowledge, or Skill, taking into account the division into Hard and Soft skills, defining educational goals (Purpose) and topics (Subject).

The LC_Structure class defines types of educational content objects. These can be directly semantic objects (Object), metadata (Description), and assessment objects (Assessment), for example, quizzes, tests, a list of questions, etc.

12.4.3 Cognitive Function Ontology

The developed OWL ontology offers a conceptual framework for studying cognitive functions by establishing relations between concepts that reflect knowledge about brain activity patterns and human cognitive abilities. The ontology includes concepts that reflect the terminology used in the study of cognitive functions from the standpoint of the analysis of psychometric and neurophysiological data. The ontology also includes knowledge about the patterns of electrode application, the correspondence of the application patterns to each other, areas of the brain, and their connection with the learning cycle. We define some logical axioms of property–objects relations. These axioms are designed to trace the connections of brain activities with the help of the EEG.

The class structure of a cognitive function ontology, part of the relation-properties and data properties of the ontology, instances of the BasicCognitiveFunction class, and examples of axioms are shown in Figure 12.4.

The MentalProcess class corresponds to the concept of the subject area "Mental processes," which is a class of the most rapidly flowing, short-term mental phenomena. A subclass of mental processes is cognitive processes. This domain concept is denoted by the CognitiveProcess class.

The PsychodiagnosticMethod class is designed to describe psychodiagnostic methods. In this class, subclasses are distinguished that display specific methods for studying cognitive processes.

The classes EEGSignalFrequency, Electrode, ElectrodeLocation, Electrode Placement-System, Lobe, LocationType, and Rhythm are used to represent knowledge about the EEG as a method for studying unconscious components of mental states in the ontology.

The ElectrodePlacementSystem class has two subclasses: ElectroencephalographSystem and NeuroheadsetSystem. The first one will contain copies of data about the circuits used by electroencephalographs, and the second one by neuroheadsets.

The CerebralCortexRegion class is used to describe areas of the cerebral cortex. The BrodmannArea class is designed to represent the Brodmann atlas. With the help of properties-relationships in the ontology, knowledge about the location of the Brodmann fields concerning the electrodes placed according to the International 10-10 system is presented. The ontology includes 47 Brodmann cytoarchitectonic areas according to the map of the cerebral cortex compiled at the Institute of the Brain of the USSR Academy of Medical

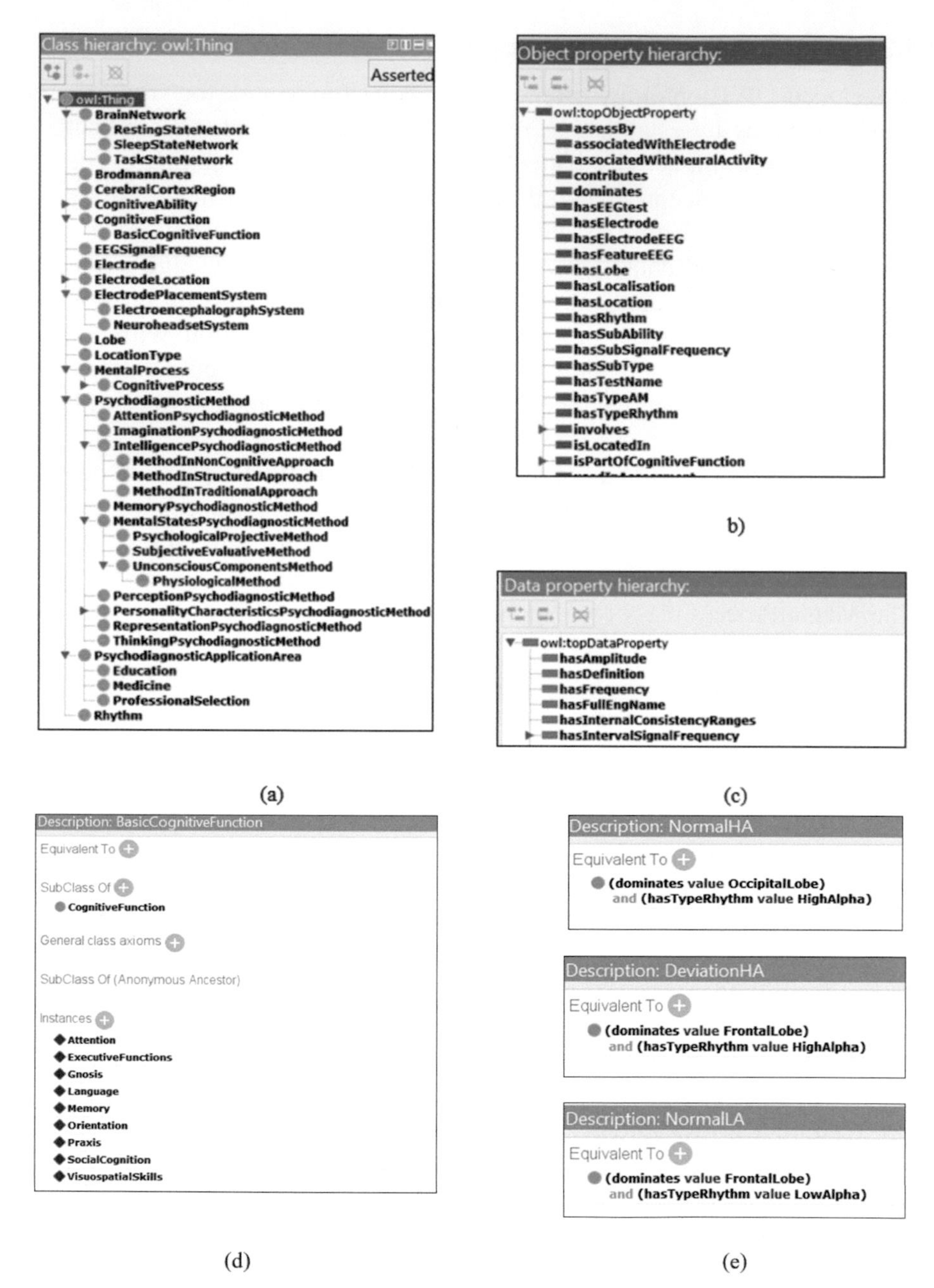

Figure 12.4 (a) Class structure of cognitive functions ontology. (b) Fragment of properties-relationships. (c) Fragment of a property-data. (d) Instances of the BasicCognitiveFunction class. (e) Examples of ontology axioms.

Sciences. Compiled by domestic scientists in the middle of the last century, the edition of the atlas of cytoarchitectonic areas by the German scientist K. Brodmann is actively used in modern studies of cognitive functions. Connections between Brodmann areas and electrodes are established according to the atlas presented on the resource [41].

The Psychodiagnostic ApplicationArea class is designed to accumulate knowledge about areas of professional activity in which psychodiagnostic methods can be applied.

Properties-relationships have been added to the top-level ontology, which allows linking brain regions with the D. Kolb cycle according to the J. Zull model [42].

The BrainNetwork class contains subclasses corresponding to large-scale brain networks RestingStateNetwork, SleepStateNetwork, and TaskStateNetwork. The RestingStateNetwork subclass is intended to describe resting-state networks corresponding to resting state (RS). Instances of this class are various types of resting networks, among which the following are best known: the default mode network, the sensorimotor network, the visual network, the executive control network, the lateralized frontoparietal network, auditory temporal network (auditory network), and temporoparietal network (temporoparietal network). The SleepStateNetwork subclass is intended to describe brain networks corresponding to sleep stages. Subclass of TaskStateNetwork – for performing certain types of tasks.

12.5 Methodological Basis for Building a Personalized Digital Educational Environment based on Ontological Models

The modern educational process pays more and more attention to the student's personality, characteristics, and needs. PLE is a set of objects of the educational process, obtained from the information and communication educational environment by adapting under the goals, content, and planned learning outcomes, the needs and abilities of the student, and acting as a means of personalization of the individual. The personalization of the digital educational environment is taking place in parallel with digitalization, e-learning development, and virtual technologies. Such an environment is important not only for students but also for analysts and researchers. Data about students' characteristics is used in the pedagogical design of educational programs and results, for content management and motivation of students to complete courses. Personalization of the educational environment through the integration of software tools that automate educational activities (LMC, LCMS, RS, LRS, etc.) with the proposed ontological models will improve the educational process and expand its functionality, for example, through the use of digital portraits [43].

Figure 12.5 shows the stages we have identified for building a personalized digital educational environment based on ontological models.

At the first stage, before building ontological models, information about the subject area contained in various publications was studied, and the available experience of experts was summarized, as a result of which the concept of description and an approximate list of characteristics that should be used in the process of personalizing the digital educational environment (cognitive, personal, and motivational) were determined. It was also decided to take into account the influence of the emotional component.

Before designing ontologies in the Protégé environment, the concepts of learner models, educational content, and cognitive functions were defined. Table 12.4 provides an example of describing the relationships between such

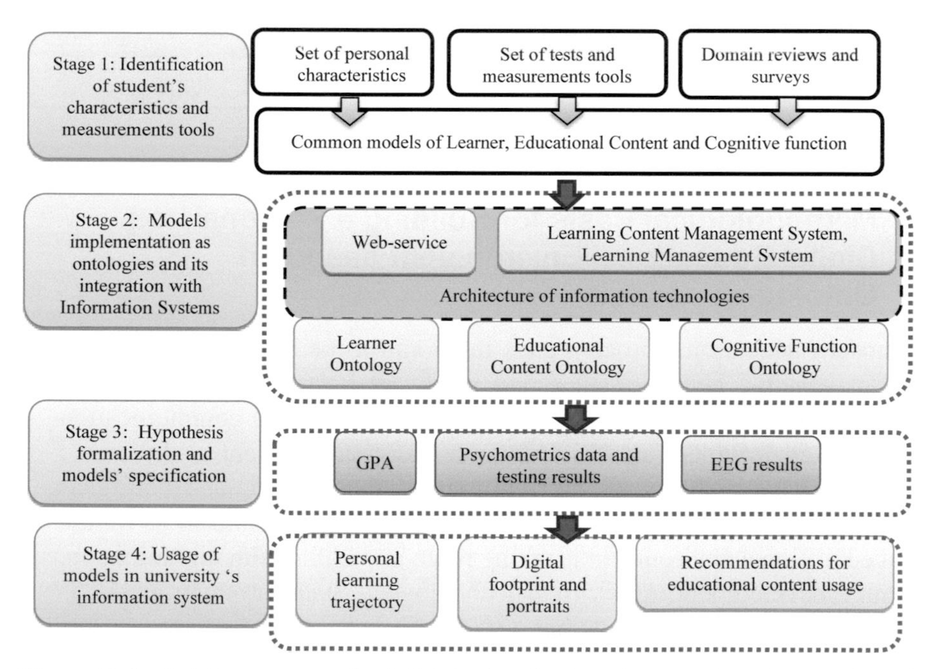

Figure 12.5 Building a personalized digital educational environment based on ontological models.

Table 12.4 Relations between the educational content and learner models in the context of personalization.

Learner characteristics	Cognitive	Personal	Motivational
Interpretation	Preferred modality is audio/video/text/formulas/graphics. Preferred learning style is about theory/practice	Better results by self-facilitated/with a tutor/in a team Preferred learning style linear/non-linearly organized material	Driving force is the motivation to achieve/to avoid failure Preferred content in classic/gamified or game form
Indicators	Cognitive characteristics (determine the types of stimuli that are perceived better) – audio, video, text, etc.	The ability to set educational goals and achieve them	Susceptibility and response to different types of challenges
Tests and measuring tools	Intellect structure test (IST), Raven progressive matrices, and GPA from university's information systems	Big five (5PFQ), emotional intelligence scales, and GPA from university's information systems	Gray–Wilson personality questionnaire (GWPQ)

conceptual models. Test materials were also identified to obtain psychometric data about students.

At the second stage, the models were implemented in the Protégé environment. During the design, input data from the information environment of the university was used for subsequent integration and development of services to provide recommendations.

At the third stage, we clarified the hypotheses and specified presented ontological models. For this purpose, we used statistical methods, in particular regression, factor analysis, as well as structural equation modeling. As a result, the existence of relationships between predictors and factors that have a significant impact on academic performance was confirmed. We obtained additional information about the semantic content of such relationships. All this information at the third stage is reflected in the presented ontologies.

During the fourth stage, using the inference mechanism and a system of axioms, it is supposed to form recommendations, primarily on the selection of educational content for students, based on a digital "portrait."

12.6 Conclusion

This chapter presented the results of a system of ontological models' development. This system is designed to monitor the relationship between psychometric and neurophysiological data and their influence on success in professional activities. The area of the research is the learning process. The presented system of ontologies integrates the learner model, the model of educational content, and the knowledge about the methods of psychodiagnostics.

The basic element of the designed system is the learner ontology, which includes a set of domain concepts that reflect the student's characteristics. These characteristics affect the student's ability to perceive educational content. Due to designing the learner ontology, we faced a large number of different indicators, describing several unobservable structures that include observable variables, for example, the concept of "intelligence." Those indicators cannot be measured directly; instead, researchers develop a hypothesis about intelligence and create tools for measurement in the form of questions or tests. We propose a set of tests for the study of different types of mental activity in terms of the educational process. The included tests must have an adaptation for the Russian-speaking audience. Then, using different types of statistical analysis, we refined the set of tests and included the results of psychodiagnostics into the ontological model. In addition, for a more objective assessment of the student's cognitive abilities, it is possible to include the results of the EEG in the model as a specific method for studying the unconscious components of mental states. These results are intended to be used together with knowledge from the cognitive function ontology for the organization of educational activities for a better perception of educational content, taking into account students' personalities.

As part of the study, educational content ontology was also developed. That ontology meets modern ideas and requirements of the educational process. In the near future, we plan to integrate an educational content ontology with a learner ontology at the architecture of the digital educational environment. Based on the constructed ontological models, the stages of implementation of a personalized educational environment were identified.

References

[1] Zhou, X., Chen, J., Wu, B., and Jin, Q.(2014) "Discovery of action patterns and user correlations in task-oriented processes for goal-driven learning recommendation" *in Proceedings of the IEEE Transactions on Learning Technologies.* 7(3), 231-245. DOI: 10.1109/TLT.2013.2297701.

[2] Uren, V., Cimiano, P., Iria, J., Handschuh, S., Vargas-Vera, M., Motta, E., et al (2006) Semantic annotation for knowledge management: Requirements and a survey of the state of the art. *J. Web Semant.* 4, 14-28.

[3] Makris, K., Gioldasis, N., Bikakis, N., and Christodoulakis, S. (2010) "Ontology Mapping and SPARQL rewriting for querying federated RDF data sources" *in Proceedings of the OTM Confederated International Conferences*, 1108–1117.

[4] Batrancourt, B., Dojat, M., Gibaud, B., and Kassel, G. (2015) A multilayer ontology of instruments for neurological, behavioral and cognitive assessments. *Neuroinformatics* 13(1), 93-110. DOI: 10.1007/s12021-014-9244-3.

[5] Kristanto, D., Liu, M., Liu, X., Sommer, W., and Zhou, C. (2020) Predicting reading ability from brain anatomy and function: From areas to connections. *NeuroImage* 218, 116966. DOI: 10.1016/j.neuroimage.2020.116966

[6] Kadam, S. T., Dhaimodker, V., Patil, M. M., Reddy Edla, D., amd Kuppili, V. (2019). EIQ: EEG based IQ test using wavelet packet transform and hierarchical extreme learning machine. *Journal of neuroscience methods* 322, 71–82. DOI: 10.1016/j.jneumeth.2019.04.008

[7] Gnedykh, D. S., Krasilnikov, A. M., and Shchur, A. D. (2020) The specifics of computer-based psychodiagnostic situation in education. *EpSBS* 87, 87–96. DOI: 10.15405/epsbs.2020.08.02.11.

[8] Bilder, R. M., Sabb, F. W., Parker, D. S., Kalar, D., Chu, W. W., Fox, J., Freimer, N. B., and Poldrack, R. A. (2009) Cognitive ontologies for neuropsychiatric phenomics research. *Cognitive neuropsychiatry* 14(4-5), 419–450. DOI: 10.1080/13546800902787180.

[9] Poldrack, R. A., Kittur, A., Kalar, D., et al. (2011) The cognitive atlas: toward a knowledge foundation for cognitive neuroscience. *Frontiers in Neuroinformatics* 5, 17. DOI: 10.3389/fninf.2011.00017.

[10] Jayapandian, C., Wei, A., Ramesh, P., et al. (2015) A scalable neuroinformatics data flow for electrophysiological signals using MapReduce. *Frontiers in Neuroinformatics* 9, 4. DOI: 10.3389/fninf.2015.00004.

[11] Poldrack, R. A. (2010) Mapping mental function to brain structure: how can cognitive neuroimaging succeed? *Perspectives on psychological science: a journal of the Association for Psychological Science* 5(5), 753–761. DOI: 10.1177/1745691610388777.

[12] Cox, A. P., Jensen, M., Ruttenberg, A., Szigeti, K., and Diehl, A. D. (2013) "Measuring cognitive functions: hurdles in the development of the neuropsychological testing ontology" *in Proceedings of the 4th International Conference on Biomedical Ontology*. Canada, Montreal: CEUR Workshop Proceedings.

[13] Podsiadly-Marczykowska, T., and Goszczynska, H. (2012) Ontology of EEG Mapping – Preliminary Research. *Lecture Notes in Computer Science*. 7339, 183-198. DOI: 10.1007/978-3-642-31196-3_19

[14] Tsianos, N., Germanakos, P., Belk, M., Lekkas, Z., Samaras, G., and Mourlas, C. (2013) An Individual Differences Approach in Designing Ontologies for Efficient Personalization / I. Anagnostopoulos, M. Bieliková, P. Mylonas, N. Tsapatsoulis (eds). *Semantic Hyper Multimedia Adaptation. Studies in Computational Intelligence* 418, 3-21. DOI: 10.1007/978-3-642-28977-4-1.

[15] Bolock, A. E., Abdennadher, S., and Herbert, C. (2021). An Ontology-Based Framework for Psychological Monitoring in Education During the COVID-19 Pandemic. *Frontiers in psychology*, 12, 673586. DOI: 10.3389/fpsyg.2021.673586.

[16] Gomez-Valades, A., Martinez-Tomas, R., and Rincon, M. (2021) Integrative base ontology for the research analysis of Alzheimer's disease-related mild cognitive impairment. *Front. Neuroinform* 15, 561691. DOI: 10.3389/fninf.2021.561691

[17] Razmerita, L. (2011) "An Ontology-based framework for modeling user behavior —A case study in knowledge management" *in Proceedings of the IEEE Transactions on Systems, Man, and Cybernetics - Part A: Systems and Humans* 41(4), 772-783. DOI: 10.1109/TSMCA.2011.2132712.

[18] Ma, Y., Zhang, X., Jin, B., and Lu, K. (2014) A generic implementation framework for measuring ontology-based information. *International Journal of Computational Intelligence Systems* 7(1), 136–146. DOI:10.1080/18756891.2013.856173

[19] Kuziemsky, C. E., and Lau, F. (2010) A four stage approach for ontology-based health information system design. *Artif. Intell. Med.* 50(3), 133-48. DOI: 10.1016/j.artmed.2010.04.012.

[20] Katifori, A., Vassilakis, C., Daradimos, I., Lepouras, G., Ioannidis, Y, Dix, A., Poggi, A., and Catarci, T. (2008) Personal ontology creation and visualization for a personal interaction management system. *Workshop on The Disappearing Desktop: Personal Information Management*, CHI2008, Florence, 2008.

[21] Razmerita, L., and Lytras, M. (2008) Ontology-based user modelling personalization: Analyzing the requirements of a semantic learning portal. *Emerging Technologies and Information Systems for the Knowledge Society* 5288, 354-363. DOI:10.14236/ewic/EL2005.10.

[22] Henze, N., Dolog, P., and Nejdl, W. (2004) Reasoning and ontologies for personalized e-learning in the Semantic Web. *Educ. Technol. Soc.* 7(4), 82-97.

[23] Dolog, P., and Nejdl, W. (2007)Semantic Web technologies for the adaptive Web. *The Adaptive Web*, Germany, Berlin:Springer-Verlag, 697-719.

[24] Sicilia, M. A., Lytras, M. D., Rodrguez, E., and Garca-Barriocanal, E. (2006) Integrating descriptions of knowledge management learning activities into large ontological structures: A case study. *Data Knowl. Eng.* 57(2), 111-121. DOI: 10.1016/j.datak.2005.04.001.

[25] Luna, V., Quintero, R., Torres, M., Moreno-Ibarra, M., Guzmán, G., and Escamilla, I. (2015) An ontology-based approach for representing the interaction process between user profile and its context for collaborative learning environments. *Computers in Human Behavior* 51(Part B), 1387–1394. DOI: 10.1016/j.chb.2014.10.004.

[26] Vargas-Vera, M., and Lytras, M. D. (2008) Exploiting Semantic Web and ontologies for personalised learning services: Towards Semantic Web-enabled learning portals for real learning experiences. *Int. J. Knowl. Learn* 4(1), 1-17. DOI:10.1504/IJKL.2008.019734.

[27] Rani, M., Nayak, R., and Vyas, O. P. (2015) An ontology-based adaptive personalized e-learning system, assisted by software agents on cloud storage. *Knowledge-Based Systems* 90, 33–48. DOI: 10.1016/j.knosys.2015.10.002.

[28] Gilmore, P., and Self, J. (1988) The application of machine learning to intelligent tutoring systems. Artificial Intelligence and Human Learning. Intelligent Computer-aided Instruction. Chapman and Hall, London, 179-196.

[29] Kholief, M., Nada, N., and Khedr, W. (2012) Ontology-oriented inference-based learning content management system. *International Journal of Web & Semantic Technology (IJWesT)* 3(3), 131-142.

[30] Chen, W., and Mizoguchi, R. (2004) Leaner model ontology and leaner model agent.

[31] Chang, M., D'Aniello, G., Gaeta, M., Orciuoli, F., Sampson, D., and Simonelli, C. (2020) Building ontology-driven tutoring models for intelligent tutoring systems using data mining. *IEEE Access* 8, 48151-48162. DOI:10.1109/ACCESS.2020.2979281.

[32] Zhou, Y., and Evens, M. (1999) "A practical student model in an intelligent tutoring system" *in Proceedings of the 11th International Conference on Tools with Artificial Intelligence*, 13-18.

[33] Fonteyne, L., Duyck, W., and de Fruyt, F. (2017) Program specific prediction of academic achievement on the basis of cognitive and non-cognitive factors. *Learning and Individual Differences* 56, 34-48. DOI: 10.1016/j.lindif.2017.05.003.

[34] Gil-Olarte, P., Palomera, R., and Brackett, M. A. (2006) Relating emotional intelligence to social competence and academic achievement in high school student. *Psicothema* 18(Suppl), 118-23.

[35] Mironova, O., Rüütmann, T., Amitan, I., Vilipõld, J., and Saar, M. (2013) "Computer science e-courses for students with different learning styles" *in Proceedings of the Federated Conference on Computer Science and Information Systems*, 735-738.

[36] Huber, SA, and Seidel, T. (2018) Comparing teacher and student perspectives on the interplay of cognitive and motivational-affective student characteristics. *PLoS One* 13(8), e0200609. DOI: 10.1371/journal.pone.0200609.

[37] Murtazina, M. S., and Avdeenko, T. V. (2021) Emotions monitoring based on EEG data in the intelligent learning systems. *Journal of Physics: Conference Series* 2032, 012030.

[38] Minichiello, A., Hood, J. R., and Harkness, D. S. (2018) Bringing user experience design to bear on stem education: a narrative literature review. *Journal for STEM Educ Res.* 1, 7–33. DOI:10.1007/s41979-018-0005-3.

[39] Kapros, E., and Koutsombogera, M. (2018) Introduction: user experience in and for learning / E. Kapros, M. Koutsombogera (eds). *Designing for the User Experience in Learning Systems. Human–Computer Interaction Series*, Springer, Cham, 1-13.

[40] Fonteyne, L., Duyck, W., de Fruyt, F. (2017) Program specific prediction of academic achievement on the basis of cognitive and non-cognitive factors. *Learning and Individual Differences* 56, 34–48.

[41] Brodmann montage areas. Available at: http://www.brainm.com/software/pubs/dg/BA_10-20_ROI_Talairach/nearesteeg.htm/ [accessed Dec. 12, 2021].

[42] Zull, J. (2002) The art of changing the brain. Sterling VA: Stylus, 280.

[43] Ishukova, E., Salmanov, V., Kalyabin, A., and Antonenko, A. (2019) "Approaches to construct a psychological portrait of users based on analysis of data in open profiles of social networks" *in Proceedings of the/ 2019 1st International Conference on Control Systems*, Mathematical Modelling, Automation and Energy Efficiency (SUMMA), 537-539.

Index

About the Editors

Dr. Archana Patel is an Assistant Professor, School of Law, Forensic Justice, & Policy Studies, (General Computer Applications/IT), National Forensic Sciences University, Gandhinagar, Gujarat, India. She has worked as a full time faculty at School of Computing and Information Technology, Eastern International University, Binh Duong Province, Vietnam. She has completed her Postdoc from the Freie Universität Berlin, Berlin, Germany. She has filed a patent entitled "Method and System for Creating Ontology of Knowledge Units In A Computing Environment" in Nov 2019. She has received Doctor of Philosophy (Ph.D.) in Computer Applications and PG degree both from the National Institute of Technology (NIT) Kurukshetra, India in 2020 and 2016 respectively. She has qualified GATE and UGC-NET/JRF exams in year 2017. Dr. Patel has also contributed in research project funded by Defence Research and Development Organization (DRDO), for the period of two year. Dr. Patel is an author or co-author of more than 40 publications in numerous referred journals and conference proceedings. She has been awarded best paper award (four times) in the international conferences. She has served as a reviewer in various reputed journal and conferences. Dr. Patel has received various awards for presentation of research work at various international conferences, teaching and research institutions. She has edited six books and served as a guest editors in many well reputed journals. Dr. Patel served as a keynote at ICOECA-2022 and ICSADL 2022. Her research interests are Ontological Engineering, Semantic Web, Big Data, Expert System and Knowledge Warehouse.

Prof. Dr. Narayan C. Debnath is the Founding Dean of the School of Computing and Information Technology at Eastern International University, Vietnam. He is also serving as the Head of the Department of Software Engineering at Eastern International University, Vietnam. Dr. Debnath has been the Director of the International Society for Computers and their Applications (ISCA), USA since 2014. Formerly, Dr. Debnath served as a Full Professor of Computer Science at Winona State University, Minnesota, USA for 28 years (1989–2017). He was elected as the Chairperson of the Computer Science Department at Winona State University for three consecutive terms and assumed the role of the Chairperson of the Computer Science Department at Winona State University for 7 years (2010–2017). Dr. Debnath earned a Doctor of Science (D.Sc.) degree in Computer Science and also a Doctor of Philosophy (Ph.D.) degree in Physics. In the past, he served as the elected President for two separate terms, Vice President, and Conference Coordinator of the International Society for Computers and their Applications, and has been a member of the ISCA Board of Directors since 2001. Before being elected as the Chairperson of the Department of Computer Science in 2010 at Winona State University, he served as the Acting Chairman of the Department. Dr. Debnath received numerous Honors and Awards while serving as a Professor of Computer Science during the period 1989–2017. During 1986–1989, Dr. Debnath served as an Assistant Professor of the Department of Mathematics and Computer Systems at the University of Wisconsin–River Falls, USA, where he was nominated for the prestigious US National Science Foundation (NSF) Presidential Young Investigator Award in 1989.